THE
EARTH IN DECAY

A History of British Geomorphology
1578–1878

THE
EARTH IN DECAY

A History of British Geomorphology
1578–1878

GORDON L. DAVIES

AMERICAN ELSEVIER PUBLISHING COMPANY, INC.
NEW YORK

First published 1969

AMERICAN ELSEVIER PUBLISHING COMPANY, INC.

52 Vanderbilt Avenue,

New York, N.Y. 10017

SBN 444-19701-x

Library of Congress Catalog Card Number 75-99798

Printed in Great Britain

Contents

Plates

Figures

Appearing in the text

Preface

IT was while reading Sir Archibald Geikie's biography of Sir Roderick Impey Murchison in the autumn of 1960 that the idea of writing this book first took shape in my mind. As it was originally conceived the book was a very different work from that now published. I then believed, in common with many another historian of the Earth-sciences, that geomorphology had been born in the late eighteenth century, and the book was to have opened in 1785 with the presentation of James Hutton's renowned theory of the Earth to the youthful Royal Society of Edinburgh. Only later did I come to appreciate the existence of the wealth of pre-Huttonian geomorphic literature which has made it possible to trace the history of the subject back to 1578. The study of this little-known early literature sheds a significant new light upon the Huttonian theory itself and places the history of nineteenth-century geomorphology in a fresh perspective.

In tracing the history of British geomorphology between 1578 and 1878 one theme predominates over all others. During these three hundred years geologists were seeking to understand the processes which have given the Earth its present configuration, and the book is very much concerned with the vicissitudes which have affected what we now know as the concept of subaerial denudation. British scholars of course had no monopoly of interest in this field, but the work of foreign scientists is considered here only if, as in the cases of Steno and Werner, their studies left a clear mark upon thought within the British Isles. Throughout the book quotations have been introduced wherever possible in order to convey something of the atmosphere of the original works, and further to this end capitalisation, punctuation and spelling have been retained as in the originals except that 'j' and 'u' have been substituted for 'i' and 'v' where necessary to conform to modern usage. The year 1878 has been selected as the terminal date for the study because by then the concept of subaerial denudation was firmly established in Britain in something approaching its modern form.

One of the chief tasks in preparing the earlier chapters of the book was the search for pre-Huttonian source material. The bibliography at the conclusion of the volume lists the titles of a number of works which are perhaps new to historians of the Earth-sciences, but the bibliography of works published before 1808 contains the titles of only twelve per cent of the works of that period which were sifted in the search for geomorphic material. All told, rather more than one thousand books, pamphlets, and papers published before 1808 have been examined, in addition to selected manuscript material in some ten British and Irish libraries.

Many friends have contributed to the present study in a variety of ways, and I am deeply grateful for all their assistance. My Dublin colleagues Professor J. P. Haughton, Mr F. H. A. Aalen, Dr J. H. Andrews, and Dr J. G. Simms all read and commented upon portions of the work as did Professor M. R. Kaatz of Central Washington State College, Ellensburg, Washington, and Professor S. I. Tomkeieff of the University of Newcastle-upon-Tyne. Professor D. E. W. Wormell of the University of Dublin cheerfully provided English translations of Latin passages from early texts. Dr C. D. Waterston of the Department of Geology in the Royal Scottish Museum, Edinburgh, was able to shed light upon the Edinburgh scene of the early nineteenth century, and his colleague Miss J. M. Sweet, M.B.E., gladly provided information arising from her own researches into the life and times of Robert Jameson. It was Miss Sweet who introduced me to Jameson's diary covering the period of Louis Agassiz's visit to Edinburgh in 1840. The diary's owner, Mrs Seton Dickson of Symington, Ayrshire, has very kindly allowed a brief passage from the diary to be quoted. The Director of the Geological Survey of Ireland gave me access to the Survey's fascinating early letter-books and has allowed me to quote from this source.

Grateful acknowledgment is also due to the staffs of many libraries, and especial mention must be made of the following: Miss M. Pollard, Rare Books Librarian, Trinity College, Dublin; the staff of the Library of the Royal Irish Academy; Mr C. P. Finlayson, Keeper of Manuscripts, Edinburgh University Library; and Mr W. H. Rutherford, Assistant Secretary of the Royal Society of Edinburgh.

Miss M. O. Hanna was reponsible for preparing the typescript of much of the book and I am very grateful for her patience and precision. I must also express my gratitude to my brother, Mr R. K. Davies, for his constant willingness to go off in search of some elusive piece of information. More than to any of these, however, I owe a debt of gratitude to my wife whose assistance has ranged from sifting incunabula to preparing the diagrams for the book.

The book was completed while I held a visiting appointment in the University of Oregon, and I must thank my many friends in Eugene for both their hospitality and their interest in the work. In particular I must thank Professor E. M. Baldwin, Professor E. T. Price, and fourteen graduate students in the Eugene Department of Geography; they were all sufficiently interested in the evolution of geomorphic thought to spend their Wednesday evenings in the Winter Quarter of 1968 in seminar discussing the subject. They, I confess, were the guinea-pigs upon whom many of the ideas embodied in this book were first tried out.

My final acknowledgment is to Dr Claudius Gilbert who was elected a Fellow of Trinity College, Dublin in 1693 and Vice-Provost of the College in 1716. He made it his life's task to assemble a large personal library consisting of works that were not already available in the library of the College. When he died in 1743 he bequeathed almost 13 000 volumes to the College, and while writing the early chapters of this book I repeatedly found myself using the volumes of the Gilbert bequest. Without his foresight and munificence the present study would have proved very much more difficult than has been the case.

G. L. Davies
Trinity College
Dublin
Ireland
December, 1968

Glossary

MANY readers of this work, as historians of science, will perhaps be more familiar with the terminology of the experimental sciences than with that of the Earth-sciences. The following glossary of geomorphic terms used in the text has been prepared in the hope that it will be of assistance to such readers.

Aggradation	The building up of a surface by the deposition of such detrital material as sand and alluvium.
Base-level	The level below which a land surface cannot be reduced by the fluvial processes. At the coast base-level and sea-level coincide.
Block Faulting	The division of a portion of the Earth's crust into blocks bounded by sets of faults mutually arranged more or less at right-angles. The various blocks have commonly moved in relation to each other.
Consequent River	A river whose direction of flow is related not to the geological structures over which it passes, but to the slope of the original surface upon which the river was initiated. Formerly known as a *Transverse River*.
Corrie	A deep, steep-walled recess in a mountain side, the work of glacial erosion. More commonly known in the United States as a *Cirque*.
Cryoturbation	Weathering and disturbance of the ground under very cold climatic conditions.
Cuesta	An asymmetrical ridge developed upon dipping strata and displaying a steep *Scarp Face* and a gentle *Dip Slope*.
Degradation	The general lowering of a surface by weathering and erosion.

Dip	The angle at which a stratum is inclined from the horizontal.
Dip Valley	A valley trending roughly parallel to the direction in which the underlying rocks are dipping.
Erratic	A glacially transported boulder.
Esker	A long narrow ridge of sand and gravel formed by a stream flowing within or beneath a glacier.
Eustatic Theory	The theory that world-wide changes of sea-level have been caused by variation in the amount of water in the world's ocean basins.
Fan	The deposit formed by a stream where it issues from a steeply graded gorge into a more gently graded valley.
Fault-line Scarp	A scarp produced along a fault as a result of differential denudation in a region where the faulting has brought weak and resistant rocks into juxtaposition.
Fault Scarp	A scarp resulting from the uplift of the region lying along one side of a fault.
Fetch	The extent of open water over which a wind blows when generating waves.
Flood Plain	An alluvial plain bordering a river.
Glacio-eustatic Theory	The theory which relates world-wide changes of sea-level to the waxing and waning of glaciers. At the beginning of a glacial period water is withdrawn from the oceans to form the glaciers, and when the glaciers melt the water returns to the oceans. See *Eustatic Theory*.
Glaciofluvial	Pertaining to streams flowing from the front of glaciers.
Growan	A sandy residue resulting from the weathering of granite.
Hanging Valley	A tributary valley which at the point of confluence is perched well above the floor of the main valley.
Honeycomb Weathering	Weathering which leaves a rock face pitted with a large number of small holes where the weaker material has been eaten out.

Inversion of Relief	In a region of newly folded rocks the anticlines (arch-like folds) form ridges while the intervening synclines (trough-like folds) form valleys. Later, after prolonged denudation, inversion of relief commonly occurs as a result of the development of valleys along the anticlines, thus leaving the synclinal structures to form the ridges.
Mesa	A plateau of small extent developed upon horizontal strata.
Moraine	Debris deposited by a glacier.
Planation Surface	A relatively smooth and approximately horizontal surface produced by denudation.
Progradation	The extension of a landmass seawards as a result of the accumulation of marine debris.
Ria	A drowned river valley.
Roche Moutonnée	A rock boss moulded by the action of a passing glacier. The feature is commonly smoothed on the side facing the oncoming ice and roughened on the opposite, downstream side.
Rock Flour	Finely powdered rock, the result of glacial abrasion.
Scarp and Vale Topography	Topography consisting of cuestas (q.v.) alternating with a series of parallel vales.
Spheroidal Weathering	Weathering in which successive shells of decayed rock peel away from a rock face to reveal a roughly spherical 'core stone' of relatively unweathered rock lying at the centre of the mass.
Stadial Moraine	A moraine (q.v.) marking a temporary stand of an ice front during its general recession.
Striations	Scratches on a rock surface resulting from the passage of a debris-laden glacier.
Strike	The bearing of the outcrop of an inclined bed upon the surface of the ground. A *Strike Valley* is a valley developed parallel to the strike, and it is therefore the valley of a particular type of subsequent river (q.v.).

Subsequent River	A river whose direction of flow is directly related to some underlying belt of geological weakness. Formerly known as a *Longitudinal River*.
Superimposition	A type of drainage evolution. The river adjusts itself to the structure of the rocks over which it first flowed, but later, when these rocks have been removed by denudation, the river persists in its original pattern of flow even though it is now flowing over rocks and structures very different from those existing at the time of the river's birth.
Talus	A sloping accumulation of debris at the foot of a steep declivity.
Till	Unsorted debris deposited directly from a glacier.
Tor	A small isolated rocky mass, the result of circumdenudation.
Trellised Drainage	A drainage pattern displaying some geometric regularity in plan as a result of many of the tributary streams being subsequents (q.v.) having their courses determined by the underlying geological structures.
Water Gap	A pass in a hill or mountain ridge through which a river flows.

Chapter One
Background Beliefs

But the corruption of philosophy by superstition and an admixture of theology is far more widely spread, and does the greatest harm, whether to entire systems or to their parts.

Francis Bacon
'Novum Organum' Bk. I, Aph. LXV.

IN the year 1485 King Richard III was slain at the battle of Bosworth Field, and his crown was placed upon the head of the victorious Henry Tudor. Historians regard the event as marking the end of the English Middle Ages, and the two following centuries were a period of transition during which Mediæval England slowly melted away under the dual impact of the Renaissance and the Reformation. A new England was in the making, and it eventually emerged – fresh, forward-looking, and recognisably modern – out of the final melting-pot of the English Civil War. Among the many changes that occurred during these two centuries there is one which transcends all others in its importance: the somewhat sterile book-learning and scholasticism of the Mediæval period was replaced by a new, virile interest in all forms of knowledge. For the first time since the demise of classical learning, secular and mundane things again began to engage serious and widespread attention, and natural phenomena, which Mediæval scholars had scarcely deigned even to notice, now became the subject of careful research. Francis Bacon, Lord Chancellor of England until his dismissal for corruption and neglect in 1621, was the vigorous apostle of the new scientific learning – the New Philosophy, as it came to be called – and William Gilbert's *De Magnete Magneticisque Corporibus* of 1600 is usually regarded as the first important fruit of the scientific renaissance in Britain. The new movement gathered momentum only slowly,

I

however, and it was not until the second half of the seventeenth century that British science fully blossomed forth. The foundation of the Royal Society in 1660 undoubtedly did much to stimulate interest in scientific pursuits, and during the closing decades of the century British science was served and graced by a remarkable number of distinguished savants – Boyle, Ray, Hooke, Newton, Flamsteed, Halley, and many others – all of whom were ebullient in their endeavours to unravel Nature's secrets. Everything was seen as demanding of explanation, and with this new spirit abroad so obvious a problem as the history of the Earth and the origin of its relief features could hardly fail to engage attention.

Those who now began to speculate about the origin of landforms were not entering upon a virgin field; many scholars in the ancient world had already displayed some interest in the subject, and even in the Mediæval world – a world where saints and angels no doubt seemed vastly more important than mountains and canyons – there were a few who paused to reflect upon the nature of the Earth's topography.[1] It was not until the Renaissance period, however, that European scholars began to take a really serious, 'scientific' interest in landforms. Among these scholars, Leonardo da Vinci, who died in 1519, perhaps deserves some priority, for his notebooks contain many astute observations on the ability of running water both to erode and to transport material. Another Renaissance writer on the subject was Georgius Agricola, the German father of mineralogy, whose *De Ortu et Causis Subterraneorum*, written in the mid-sixteenth century, contains a thoroughly sound discussion of the work of denudation and the conclusion – startling for its day – that most mountains are merely the residual masses left after Nature's destructive agents have done their work.[2] Another early writer on the subject of landforms was Bernard Palissy, the remarkable Protestant potter from Aquitaine who died in the Bastille 'un prisonnier pour la religion'. Palissy discusses the work of denudation in both his *Recepte véritable* of 1563 and his better known *Discours admirables* of 1580, and he concludes that the world would soon be in a sadly depleted state were it not for the ability of new rocks to grow in the place of those that decay.

In Britain, William Bourne, a Gravesend innkeeper, self-taught mathematician, almanac publisher, and soldier of Queen Elizabeth I, was apparently the earliest Renaissance writer to offer general observations upon the origin of landforms. In the Preface to his *Booke Called the Treasure for Traveilers*, published in London in 1578, Bourne observes that the useful traveller should note 'whether the Countrye bee a playne and champion Countrie, or Hils and Mountaynes, or lowe marsh or marishe grounde, and whether it be full of Rivers or not', and in the comparatively rare fifth part of the book he offers chapters on such subjects as the cause of marshes, the origin of sand-banks, and the formation of sea-cliffs. Bourne's book can hardly be regarded as influential, but he himself may justly be acclaimed as the precursor of a seventeenth-century English school of geomorphology. Within little more than a hundred years of Bourne's death in 1582, landforms were engaging the attention of many of Britain's leading scientists, and theories of Earth-history from the pens of English scholars were being discussed throughout Europe.

Although the roots of British geomorphology go no further back than the sixteenth and seventeenth centuries, the scientific world into which the subject was born was radically different from our own. The age that knew Galileo, Kepler, Descartes, Hooke, Newton and Boyle amply deserves to be acclaimed as 'the century of genius', but such acclaim must not blind us to the fact that, by modern standards, man was then still abysmally ignorant of the world about him. The great minds of the age represent the sturdy upward growing shoots of the rejuvenated learning, but around them there was a dense, tangled ground-layer of fallacy, superstition and nescience. Interest in astrology and alchemy, for example, was by no means dead, monarchs continued to touch their subjects for the King's Evil, and a belief in witchcraft persisted to such a degree that the hunt for English witches reached a new peak of intensity during the middle years of the seventeenth century.

Most of the myths and legends that had passed for science during the Middle Ages were brusquely dismissed long before 1700, but certain fundamental and deeply ingrained beliefs inherited from the Mediæval scholar-theologians were not so easily displaced. Some elements of the Mediæval world-picture in fact survived far into the eighteenth century to form an important part of the mental framework within which even the most enlightened devotees of the New Philosophy worked. Thus throughout the period that saw the birth of British geomorphology, scientific thought was still being influenced by ideas that had been current among the Mediæval schoolmen. Many of these ideas – ideas which now seem weird if not ridiculous – imposed

3

severe restrictions upon early geomorphic thought, and any realistic discussion of the origins of British geomorphology must be prefaced by a consideration of the seventeenth-century world-picture.

The passage of time clearly makes it just as impossible for us to see the world through the eyes of the seventeenth century as it is for the adult to recapture the outlook of his infancy. Therefore all that can be offered here is a summary of those beliefs which were widely disseminated in seventeenth-century Britain and which seem to have influenced the geomorphic thought of the day. At times in this chapter we may seem to be straying far from the realm of geomorphology, but that subject had much wider ramifications in the past than it does today. Its links with religion were particularly important, and much of our present survey is perforce concerned with some of the theological beliefs held in seventeenth-century Britain.

A Magnificent Universe in Decay

In seventeenth-century Britain, as elsewhere in Europe, religion was still a virile, potent force penetrating into every aspect of life. It was an age when people endured the restriction of their liberties, torture, and even death for their faith, when kings ruled by divine right, and when religious differences were *casus belli*. The scientists themselves were mostly possessed of a deep religious faith. Robert Hooke, for example, the inventive genius who was the Royal Society's first Curator of Experiments, never failed, we are told, to give acknowledgment to God after making a fresh discovery or solving a difficult problem. Similarly, John Ray, the father of English natural history, was also the author of *A Persuasive to a Holy Life: from the Happiness that attends it both in this World, and in the World to come*, while Newton, the towering intellectual giant of the period, felt obliged to reply to some theological criticism of the first edition of the *Principia* by later adding the famous *General Scholium* in which he discussed the nature of the deity.

To some extent the study of secondary causes, and the discovery of natural laws, conspired together to make God seem rather more remote than He had seemed to an earlier generation which had regarded Him as the prime mover behind all natural occurrences. The gradual seventeenth-century substitution of the heliocentric system for the cosy belief in a neat, finite, geocentric universe no doubt had a similar effect. These changes, however, did nothing to shake man's belief that this globe – or the sub-lunar world as it was usually termed – had been fashioned expressly for the use of the human race. Indeed, as we will see, this belief became steadily stronger as the seventeenth century progressed, and by the close of the century the world was regarded as a magnificent, lush estate which God had stocked with everything necessary for human well-being before admitting man as a freehold tenant. The entire sub-lunar world was held to reflect God's power and wisdom, His perfection in creation, and His benevolence towards mankind. Sir Matthew Hale, an eminent judge who died in 1676, liked to think of God governing the world 'with much more Accuracy and Wisdom than a Gardiner orders his Garden',[3] and Milton expressed the conviction of his age when about the middle of the seventeenth century he wrote:

> There can be no doubt that every thing in the world, by the beauty of its order, and the evidence of a determinate and beneficial purpose which pervades it, testifies that some supreme efficient Power must have pre-existed, by which the whole was ordained for a specific end.[4]

Although this belief that the divine omnipotence and beneficence was everywhere revealed in Nature was undoubtedly well developed in sixteenth- and early seventeenth-century England, it then existed, paradoxically, alongside a belief that the whole sub-lunar world – if not the entire universe – was in a state of advanced senility and decay. The idea of progress, so fundamental in modern thought, was utterly foreign to most sixteenth- and seventeenth-century minds. Instead, there was a firmly rooted belief that the world had deteriorated steadily throughout most of its history, and that the immediate future promised nothing more than further sliding into the slough of degeneracy. 'The opinion of the Worlds Decay', wrote George Hakewill early in the reign of Charles I, 'is so generally received, not onely among the Vulgar, but of the Learned both Divines and others, that the very commonnes of it, makes it currant with many, without any further examination'.[5] A few years later Henry Reynolds opened his *Mythomystes* with the words:

> I have thought upon the times wee live in, and am forced to affirme the world is decrepit, and, out of its age & doating estate, subject to all the imperfections that are inseparable from that wracke and maime of Nature.

Perhaps man has always tended to denigrate his own age and to look back fondly to the 'good old days' of some bygone era, but the sixteenth- and seventeenth-century belief in the superiority of former ages was much more than a wistful backward glance into history; it was, rather, a group inferiority complex on a gigantic scale. The history of the world was widely regarded as divisible into three distinct phases: firstly, there had been a period of generation extending from the Creation up to the Fall of Man; secondly, there was the prolonged and present period of degeneration initiated by the Fall; and thirdly, there was the eagerly awaited period of regeneration that would be ushered in by Christ's Second Coming.

To try to explain this seventeenth-century interpretation of history would take us far beyond the bounds of our present subject,[6] but there is one causal factor which must be mentioned because of its indirect influence upon the development of geomorphology. That factor is the strong Calvinistic emphasis upon the universal defiling influence of human sin. In Pre-Restoration England both the Anglican conformists, and the Puritans, regarded sin as a kind of evil, malignant germ which had polluted the world and caused a slow, insidious paralysis and decay to spread throughout Nature; a paralysis and decay which were both a manifestation of the divine wrath and a just retribution upon a wicked mankind. After the Fall, it seemed, God had renounced some of His suzerainty on Earth to the Devil, and as evidence of this it was pointed out that Hell lay within the Earth and that the Devil's minions were widespread among us in the form of witches and sorcerers. St Paul himself had observed 'that the whole Creation groaneth and travaileth in pain together' and had looked forward to deliverance 'from the bondage of corruption',[7] while even a cursory examination of the Scriptural prophecies was sufficient to reveal that the world was destined to suffer many more terrible torments before the phase of regeneration could begin. True, the decay had not yet obliterated all the wonder and former perfection of the Creation, but the sub-lunar world was believed to have deteriorated far since the fateful day when Eve ate the forbidden fruit in Eden. The Renaissance rediscovery of the wisdom and knowledge possessed by the ancients before the onset of the Dark Ages, seemed merely to confirm that the present was indeed an age of degeneration and that the human intellect had shared in the universal decay. Anything ancient seemed better than anything new, and even in the second half of the seventeenth century the Fellows of the youthful

Royal Society had to face the taunt that in their own decadent age it was foolish to try to match the peerless wisdom of those who, like Plato and Aristotle, had lived in a more perfect world closer to the Earth's golden age.

The many seventeenth-century writers who discoursed upon the decay of Nature naturally focused their attention upon the decay of man himself. Not only were men believed to be less intelligent and more evil than formerly, but they were held to be more prone to diseases, and to be stunted, short-lived dwarfs as compared with the earth's gigantic pre-diluvial inhabitants like Methuselah and Noah, who, as recounted in *Genesis*, had each lived for many centuries. Man's need of spectacles, items unknown to the ancients, was regarded as but one clear manifestation of the decay of the human faculties. The Earth itself, however, was believed to display obvious signs of having participated in the degeneration. Earthquakes, for example, were supposed to be increasingly frequent, and the world was held to be far less fertile than in the days immediately after the Creation when, as described in the Scriptures, the soil was so warm and full of nourishing juices that living creatures had actually grown from it. Similarly, the Earth's climate was held to have deteriorated as a result of the Sun and stars burning ever less brightly, while the skies were supposed to be streaked by an increasing number of comets as a portent of the fast approaching disasters which would presage the end of the present order on Earth. The notion of a universal decay even found its way into Nathanael Carpenter's *Geography Delineated forth in Two Bookes* of 1625, the foremost British geographical text of its day.[8] Because of the decay, life in the seventeenth century was felt to be far more rigorous than life had been in the pre-diluvial world where Nature had been bounteous and man had misspent his ample leisure in the lascivious living that had prompted God to cause the Flood. Shakespeare's friend Michael Drayton looked back longingly from the supposedly toil-worn, disease-ridden, and depraved seventeenth century to the wonderful pre-diluvial elysium where

> *Cover'd with grasse, more soft then any silke,*
> *The Trees dropt honey, & the Springs gusht milke:*
> *The Flower-fleec't Meadow, & the gorgeous grove,*
> *Which should smell sweetest in their bravery, strove;*
> *No little shrub, but it some Gum let fall,*
> *To make the cleere Ayre aromaticall:*

7

Whilst to the little Birds melodious straines,
The trembling Rivers tript along the Plaines.
Shades serv'd for houses, neither Heate nor Cold
Troubl'd the young, nor yet annoy'd the old:
The batning earth all plenty did afford,
and without tilling (of her owne accord).[9]

Historians of ideas are now thoroughly familiar with the seventeenth-century belief in the degeneracy of Nature, but the belief seems never to have been considered in relation to the history of the Earth-sciences. The concept is, nevertheless, fundamental to any understanding of seventeenth-century geomorphology. The implications of the belief will emerge in later chapters, and it will suffice here to note that during the sixteenth and seventeenth centuries the widespread belief in Nature's degeneration ensured a ready acceptance for the idea of topography slowly mouldering away under the attack of natural processes. The concept of denudation was, in fact, part and parcel of the concept of a general decay of Nature in all her aspects. Thus, in the 1630s, when the Bishop of Gloucester was anxious to prove the reality of Nature's degeneracy, he drew attention to the muddy, debris-laden waters of a river in flood. Here, the bishop argued triumphantly, was indisputable proof of the reality of Nature's decay and of the mouldering of the very continents themselves.[10]

Revelation through Nature and the Scriptures

The belief that God's wisdom and benevolence were everywhere evident in the Creation despite the universal decay, caused Nature to be regarded as a divine revelation scarcely less important than the Scriptures themselves. As a result, seventeenth-century science and religion became inseparably intertwined, and the study of landforms was seen as one path to God. No less an authority than John Calvin had regarded the world as a mirror in which to behold God, and he had written that 'the Lord, that he may invite us to the knowledge of himself, places the fabric of heaven and earth before our eyes, rendering himself, in a certain manner, manifest in them'.[11] In seventeenth-century England the same sentiments were echoed by Sir Thomas Browne who hailed Nature as 'the Art of God' and 'that universal and publick Manuscript, that lies expans'd unto the Eyes of all'.[12] The study of Nature was therefore deemed highly desirable as the study of God's own handiwork, and it was held that such investigations could only bring man closer to God. As Bacon wrote in a well-known passage:

> It is true that a little philosophy inclineth man's mind to atheism; but depth in philosophy bringeth men's minds about to religion. For while the mind of man looketh upon second causes scattered it may sometimes rest in them, and go no farther; but when it beholdeth the chain of them, confederate and linked together, it must needs fly to Providence and Deity.[13]

There were a few in seventeenth-century Britain who did regard the New Philosophy as a danger both to religion and to morals, but the vast majority remained convinced that since the Scriptures and Nature were both routes to God, there could be no conflict between them. 'We are not to suppose', wrote the Rev. Thomas Burnet in 1684, 'that any truth concerning the Natural World can be an Enemy to Religion; for Truth cannot be an Enemy to Truth, God is not divided against himself.'[14]

The discovery of this new approach to God through Nature initially did nothing to impair the immense status accorded to the Scriptures in the Reformed Churches. The Bible had replaced the Papacy as the infallible oracle and ultimate authority, and most Protestants believed the Scriptures to contain God's actual words dictated to unerring human scribes. Sometimes even the punctuation was held to be sacred and significant. The Bible was, of course, seen as much more than a religious manual and moral guide; it was believed to enshrine the very essence of truth on almost any subject. Even the period's keenest intellects turned to the Scriptures for inspiration and guidance, and bowed in homage before them. 'Welcome, Holy Scriptures', wrote Burnet, 'The Oracles of God, a Light shining in the darkness, a Treasury of hidden knowledge, and, where humane faculties cannot reach, a seasonable help and supply to their defects.'[15]

Among the scientists the *Book of Genesis* was especially revered as containing God's own impeccable account of the Earth's creation and early history. It, and the other books of the Pentateuch, were widely supposed to have been written under God's direction by no less a person than Moses himself, and the five books were venerated in the belief that they were by far the most ancient of all human documents. For his part in actually penning the account of the Creation and the Flood, Moses was hailed by one writer as 'the greatest Natural Philosopher that ever lived upon this Earth',[16] and throughout the seventeenth and eighteenth centuries the first few chapters of *Genesis* were the basic text for all those interested in the origin of the Earth's surface features. It had to be admitted that the Mosaic writings lack the careful attention to detail which the seventeenth-century naturalists increasingly expected, and the correct, 'scientific' interpretation of the accounts of the Creation and Flood excited much attention. Fortunately – as Milton found to his advantage when he was anxious to divorce his first wife – the careful selection of texts can make the Bible seem to support almost any proposition, and a great deal of ingenuity was displayed in trying to reconcile a growing knowledge of Nature with the Mosaic record.

No modern student of the history of the biological and Earth-sciences can fail to be surprised at the tenacity with which even the most eminent scientists of the seventeenth and eighteenth centuries clung to the belief that the first few chapters of *Genesis* afforded a reliable account of the Earth's early history. In our own day when every scholar is trained to check and re-check the validity of his source material, it seems almost inconceivable that scholars of earlier generations should have implicitly accepted the infallibility of Scripture and based a lifetime's scientific work upon this premise. But so it was. A sincere belief in the divine origin of Scripture was far too deeply rooted to be easily swept aside. Even when discrepancies between the Bible and Nature became apparent, it was generally held that they could only be the result either of man's failure to interpret the Scriptures aright, or of some human error in the perception of natural phenomena.

Few seventeenth-century scholars can ever have been seriously tempted to question the authority of Scripture, and in any case there lurked in the background certain legal restrictions which must have discouraged any tendency towards free-thought. The Press Licensing Act, for example, forbade the publication of 'any heretical, seditious, schismatical or offensive Books or Pamphlets wherein any Doctrine or Opinion shall be asserted or maintained, which is contrary to the Christian Faith, or the Doctrine or Discipline of the Church of England',[17] and under the Blasphemy Act of 1698, apostates from Christianity were incapacitated from holding any public office.[18] Admittedly there is no record of any proceedings under this latter act, but its existence certainly indicates the inflexibility of the age, and we must evidently regard the English law as comparatively lenient, for in

Presbyterian Scotland in 1697, a youth was hanged for denying the inspiration of the Scriptures.[19]

Quite apart from any such acts upon the statute-book, there were very strong social pressures which inhibited the expression of unorthodox religious opinions. It was then believed that the whole fabric of civilised society rested upon the foundation of the Christian faith, and apostasy was seen as a social danger. In a famous judgment in a blasphemy case in 1676, Sir Matthew Hale summed the position up succinctly when he informed his court that 'to say, "Religion is a Cheat", is to dissolve all those Obligations whereby Civil Societies are preserved'. Christianity, Sir Matthew observed, is 'parcel of the Laws of England; and therefore to reproach the Christian Religion, is to speak in Subversion of the Law'.[20] For two hundred years Sir Matthew's words guided the bench in cases where the truth of revealed religion had been impugned, and not until 1883 was his dictum revoked. 'For many centuries', writes Sir James Stephen, the legal historian, 'the maintenance, or even the expression of opinions, suspected or supposed to involve a denial of the truth of religion in general, was regarded in the same kind of light as high treason in the temporal order of things.'[21] Thus in the seventeenth and eighteenth centuries, to deny the veracity of any part of the Scriptures was to run the risk of being branded as an anarchist and renegade. Even if no worse fate overtook him, the avowed sceptic was assured of social ostracism.

Thomas Hobbes is perhaps the most notable seventeenth-century scholar who suffered because of the intolerance of his age. After the Plague of 1665 and the Great Fire of London in the following year – two events which were seen as divine judgments upon a depraved city – a bill against atheism and profanity was introduced into Parliament, and Hobbes's *Leviathan* was specifically mentioned as a work in need of condemnation. For a time it seemed that the heterodoxy of his masterpiece was sufficient to endanger Hobbes's freedom, if not his life, and thereafter he was unable to publish many of his works in England because they failed to obtain the necessary imprimatur. Other authors were less bold than Hobbes, but the pressures inhibiting freedom of expression are nevertheless clearly reflected in their works. In John Beaumont's *Considerations on a Book, Entituled The Theory of the Earth*, for example, published in 1693, the author, a Fellow of the Royal Society, was obviously tempted to reject the literal interpretation of *Genesis*, and he clearly harboured grave doubts about the pentateuchal chronology. In the event, however, he dared go no further than note the existence of serious objections to the chronology, and he observed that if deviations from Moses were permitted, he would choose to regard the Earth as eternal, or at least so old that its origin was quite lost to mankind. Even so eminent a figure as the astronomer Edmond Halley felt it unwise to try to reinterpret *Genesis*. Having been refused the Savilian professorship of astronomy in 1691 because of a suspicion that he held materialistic views, he in 1694 decided not to publish some papers on the Flood that he had read to the Royal Society, partly for fear lest 'he might incur the Censure of the Sacred Order'.[22]

The bibliolatry of the seventeenth- and eighteenth-century scientists can only be understood if we remember firstly, that they were subscribing to a time-honoured belief in Scriptural infallibility, and secondly, that even in the comparatively tolerant atmosphere of Restoration England, there were still very strong pressures hindering the growth of a liberal and critical outlook towards the Bible. It was difficult for an individual to attain a vantage point from which he could discern the frailty of the Scriptures, and if he did make the discovery, there were seemingly good reasons for not attempting to disseminate it widely.

Moses and the Age of the Earth

Seventeenth-century bibliolatry affected geomorphology in various ways, but it was through the Mosaic chronology that it exerted a particularly important and long-lasting influence upon the subject. The problem of the Earth's age has always had a peculiar fascination for mankind, and in the seventeenth century it was believed that the dates of all the major events in Earth-history could be learned from the Bible, provided the key of sufficient piety and scholarship was available. Chronological studies based on the Scriptures proved, to the satisfaction of most seventeenth-century scholars, that the Creation had taken place something like 130 generations ago. There could be no general agreement on the precise year of the Creation, however, because serious chronological discrepancies were found to exist among the different versions of the Scriptures. It seemed that since their bequest to man, the Scriptures had in some way become tarnished so that the chronology in the Hebrew Massoretic text now differed from that in the Samaritan Pentateuch, and both these from the chronology in the Greek Septuagint. The precise dating of the Creation thus depended upon the relative importance that a chronologer attached to the various texts. So complex did the problem of Earth-chronology appear that in 1646 Sir Thomas Browne asserted that 'without inspiration it is impossible, and beyond the Arithmetick of any but God himself.'[23] Prospective chronologers were nevertheless undeterred, and when William Hales published his three-volume manual for chronologers early in the nineteenth century, he was able to offer his readers the date of the Creation as computed by more than 120 earlier scholars, the dates ranging from a mere 3616 B.C. to 6984 B.C.[24]

Much the most famous of the many Biblical chronologers was James Ussher, the seventeenth-century Archbishop of Armagh and Primate of All Ireland. Ussher was a distinguished churchman, a Privy Councillor and confidant of Charles I, and a versatile scholar of international repute. His prestige was such that at Ussher's death in 1656, Cromwell himself ordered that the Archbishop be given a public funeral and burial in Westminster Abbey with full honours, although in the event the Protector characteristically allowed only a fraction of the cost of the funeral to come from public funds. Ussher's famous chronology was published in two parts – one in 1650 and the other in 1654.[25] According to it, the creation of Heaven and Earth had taken place 'upon the entrance of the night preceding' Sunday 23rd October in the year 4004 B.C., man and the other living creatures appearing, of course, on the following Friday. Ussher calculated that the other momentous event described in the Old Testament – the Flood – had occurred 1656 years after the Creation; he claimed that Noah and his family had joined the animals in the Ark on Sunday 7th December in the year 2349 B.C., and that as the waters subsided, the Ark finally came to rest on Mt Ararat on Wednesday 6th May in the following year.

Modern writers have tended to scoff at Ussher's chronology, and we perhaps need reminding that he was no fanatical fundamentalist or deranged scholar, but a brilliant and highly respected savant. His chronology was the result of prolonged, intensive, and careful research, and his contemporaries were no doubt grateful that he had chosen to apply his experienced mind to so seemingly important a problem. His scholarship was sound, and if he is to be censured it can only be on the somewhat unfair charge that he failed to examine the reliability of source material which he, like his contemporaries, supposed to be of divine origin.

Ussher's great reputation as a scholar ensured the chronology a widespread acceptance, and his work formed the basis of the chronology that Bishop Lloyd of Worcester inserted into the margin of a massive new folio reference edition of the Authorised Version published in London in 1701.[26] Once within the covers of the Bible, Ussher's chronology was soon regarded as possessing hardly less authority than the Scriptures themselves. Ages which knew nothing of the onetime Irish prelate, came to know his chronology as 'The Received Chronology' or even as 'The Bible Chronology'. Today, more than 300 years after its first appearance, Ussher's time-scale continues to appear both in Biblical concordances and in a few reference editions of the Bible itself.

Although it was generally accepted during the seventeenth century that the universe had existed for rather less than six thousand years, it was also believed that those few millennia had carried the Earth far beyond the median point of its life. The faithful of the seventeenth century, like those of earlier centuries, were convinced that the Parousia could not long be delayed. Milton, writing in 1641, looked forward eagerly to the day when Christ 'the Eternall and shortly-expected King shalt open the Clouds to judge the severall Kingdoms of the World',[27] while under the Commonwealth and Protectorate an extreme Puritan sect – the Fifth Monarchy Men – carried a belief in the imminence of the Second Coming far into English political life. Even at the close of the seventeenth century a scholar of repute could regard his generation as 'almost the last Posterity of the first Men, and faln into the dying Age of the World'.[28] The actual date at which the present Earthly order would be terminated was the subject of some speculation. During the first half of the seventeenth century some held that the fateful year would be 1657, partly because they believed that the Flood had occurred when the Earth was 1657 years old, and partly because 1657 is the sum of the seven Roman numerals, M, V, D, I, C, L, I, contained in the words *Mundi Conflagratio.*

At first it might seem that such cryptographic nonsense, and a belief in an impending Parousia, could hardly have influenced so mundane a subject as geomorphology, but in fact the belief did affect the subject in one very important respect; if the Second Coming was imminent, then it followed as a corollary that any consideration of the cumulative effect of prolonged future denudation was quite unnecessary. There was no need to face up to the somewhat unpleasant fact that denudation, if continued indefinitely, could result only in the eventual destruction of the continents on which man depends for his existence. Only later, in the eighteenth century, when the end of the present terrestrial order was predicted with rather less confidence, did it become necessary to explain how the ultimate destruction of the continents was to be avoided. Such destruction, it seemed, obviously could have no place in the divine plan for human happiness, and an age which no longer believed in an impending Parousia tended to stave off the date of the annihilation of the continents by minimising the effectiveness of denudation. So long as the end of the present order seemed imminent, however, no such device was necessary, and the potency of denudation could be freely admitted.

The terrestrial time-scale current during the seventeenth and eight-eenth centuries now seems absurdly brief, but we should remember that man's appreciation of time is to some extent relative, and as the oldest entity known, the universe must always have seemed highly venerable, even in an age when it was believed to have existed for rather less than six thousand years. Only when they are viewed against the time-span discovered by modern geochronology do the six millennia dwindle into trivial insignificance, and the vast majority of seventeenth-century scholars certainly felt no deficiency of time behind them. Indeed, as the end of the present terrestrial order was patiently awaited, there was some surprise that the world should have been allowed to endure for what evidently then seemed an aeon. 'The poor world is almost six-thousand years old' exclaims Rosalind in *As You Like It*,[29] and William Bourne remarked that 'the age of the worlde is of no small tyme'.[30] Be that as it may, there can be no denying that the very limited seventeenth- and eighteenth-century conception of time made an evolutionary view of Nature impossible. The notion that major changes might have been effected by natural processes working with extreme slowness was obviously inhibited so long as the period of operation of those processes was measured in thousands, rather than in millions of years.

Although the Old Testament chronology was never seriously in question during the seventeenth century, some enlightened souls did find increasing difficulty in reconciling the results of contemporary scholarship with the Mosaic writings. The discovery, for example, that the Egyptians, Chinese, and other peoples all traced their history back beyond the date at which Christian chronologers supposed the Flood to have occurred, seemed to constitute a serious challenge to the

15

Mosaic time-scale. Indeed, it now appeared that some nations claimed to trace their ancestry back to a period lying before the supposed date of the Creation itself. This discovery of the antiquity of the earlier civilisations evidently aroused renewed interest in some quarters in the Aristotelean view that the universe is eternal. Those who preferred Aristotle to Moses were of course careful not to advertise their heresy in print, but it is clear from the writings of many seventeenth-century Christian apologists that doubts about the Mosaic chronology must have been circulating in the London coffee-houses. The apologists were vigorous in their affirmation of the Mosaic infallibility, and they maintained that the annals of the ancient civilisations were not only totally unreliable, but that they had been grossly misinterpreted. At the same time the apologists threw up a barrage of arguments in support of their own brief time-scale. Was the novity of the Earth not proved, they asked, by the fact that agriculture, writing, printing, and all other important human inventions, dated from the last two or three thousand years, and was the recent origin of mankind not amply demonstrated by the continued growth of the human population and its current colonisation of hitherto empty lands?

To counter this scepticism arising out of discoveries in the Middle and Far East was not difficult, but researches very much nearer home were also tending to shed suspicion on the Biblical chronology. Scientists of the second half of the seventeenth century were beginning to perceive dimly that the story of the Earth as revealed by Nature is vastly longer than the story as told in the Bible. At this discovery they felt no elation, and unlike their successors in a much later generation, they certainly experienced no sense of science having triumphed over religious bigotry. Rather they were deeply religious men bewildered at the problem which they had unwittingly laid bare; a problem which they scarcely dared consider in their published works, but which appears time and time again in their correspondence with each other. How, they asked, does it come about that our study of God's Creation leads us to a conclusion different from that obtained by studying God's Holy Word? The devout John Ray is typical of those who were gnawed by doubts about the accuracy of the Biblical chronology. He was one of the few seventeenth-century scientists to be convinced of the organic origin of fossils, but he felt uneasy about the implications of such a view. In 1695 he wrote to his friend Edward Lhwyd, then Keeper of Oxford's Ashmolean Museum, pointing out that from such a belief 'there follows such a train of consequences, as seem to shock the Scripture–History of ye novity of the World'. He continued, as if to reassure himself: 'But whatever may be said for ye Antiquity of the Earth it self & bodies lodged in it, yet that ye race of mankind is new upon ye earth, & not older than ye Scripture makes it, may I think by many arguments be almost demonstratively proved'.[31] Although Lhwyd himself felt unable to accept Ray's views on fossils, he too harboured doubts about the Mosaic chronology, and in a letter to Ray he offered an interesting observation which seemed to support his suspicion that the Earth was of considerable antiquity. During his travels in North Wales, Lhwyd had noticed that the floors of the valleys of Llanberis and Nant Ffrancon – two valleys later famous for their display of Pleistocene landforms – are littered with many thousands of boulders. For some reason Lhwyd felt that the boulders could hardly have been transported by the Flood, and instead he adopted the explanation that they had merely rolled down from the surrounding mountains. His enquiries locally, however, revealed that only some two or three of the blocks had come down within living memory, and he intimated to Ray that 'in the ordinary Course of Nature we shall be compelled to allow the rest many thousands of Years more than the Age of the World'.[32]

These vague conjectures from Ray and Lhwyd, together with an earlier suggestion from Hooke that the study of fossils might assist in the construction of a fresh Earth-chronology,[33] led to eighteenth-century attempts to re-interpret the Creation story in a slightly less literal sense. There was of course no sudden breaking of the bonds of the Scriptural strait-jacket; rather was there an attempt to loosen some of its more troublesome fastenings so that science could wriggle into a slightly more comfortable position. It must be stressed, however, that when Hooke, Ray and Lhwyd thought of pushing back the date of the Creation beyond Ussher's 4004 B.C., they were certainly not proposing the addition of millions of years to the Scriptural age of the Earth. The astronomical figures of the modern geochronologist were in fact quite beyond the cognisance of the seventeenth-century mind, and drafts on the bank of time were then very severely restricted by the religious beliefs of the day. It was just conceivable that the pre-Adamic days mentioned in *Genesis* might not have been days in the modern sense of the word, but, as Ray freely admitted, there seemed no escaping the conclusion that all the post-Adamic years had been normal sidereal years, and that mankind was rather less than six thousand such years old. Now if this was so, and if, as was widely believed, the Parousia was imminent, then it followed that it was God's intention that mankind should exist on Earth for some six thousand years under the present order, plus a further thousand years for the Millennium. Thus mankind's total occupancy of the Earth – or, expressed another way, the Earth's total useful life – would amount to rather less than seven thousand years. This belief quite ruled out any notion that the pre-Adamic days of *Genesis* might each represent prolonged periods of time, because it seemed inconceivable that God could have needed an aeon in which to prepare a home that man was destined to occupy for only a few thousand years. Already some scholars were puzzled to know why God had needed even the six days of the hexaëmeron in which to complete the Creation; could the whole universe, they asked, not have been produced instantaneously by the divine fiat? To have suggested that the omnipotent God spent millions of years preparing the Earth for mankind would have seemed to cast such grave reflections upon the divine capability, as to have been tantamount to blasphemy. Thus even those enlightened seventeenth-century minds that were feeling their way towards a revised Earth-chronology had no conception of the true scale of Earth-history. Their notion of an extended chronology afforded no more satisfactory a back-cloth to the operation of geomorphic and other natural processes than did Ussher's time-scale. An evolutionary view of Nature was still inhibited for want of time.

The Earth-Animal and its Structure

The Scriptures were undoubtedly the paramount influence on British geomorphology during the first two hundred years of its development, but in the seventeenth century a literary influence of a very different genre can be detected in the infant subject – the influence of classical Greece and Rome. The Renaissance rediscovery of classical learning had made a profound impression upon European scholars, and many classical writers came to be regarded as almost infallible oracles whose pronouncements could be trusted implicitly. This veneration of the ancients faded somewhat during the sixteenth century, but it died a slow death, and in 1646 Sir Thomas Browne still felt it necessary to warn his contemporaries that 'the mortallest enemy unto Knowledge, and that which hath done the greatest execution upon truth, hath been a peremptory adhesion unto Authority, and more especially, the establishing of our belief upon the dictates of Antiquity.'[34] This 'adherence unto antiquity' persisted until the close of the seventeenth century, and even scientists of the calibre of Hooke and Ray were happy to support their discussions of landforms with quotations from the works of Plato, Aristotle, Strabo or Pliny. In some respects the infant geomorphology benefited from this leaning upon antiquity, but often a blind faith in the reliability of classical authors resulted in credence being given to the most weird and fanciful of stories. The seventeenth-century belief that earthquakes play a vital part in shaping the Earth's topography, for example, seems to have arisen partly from the much exaggerated stories about earthquake phenomena that were handed down from classical times.

Science in the seventeenth century was indeed a curious amalgam composed of elements drawn from the Scriptures, from ancient Greece and Rome, and from the pseudo-science of the Middle Ages. To this strange mixture was added a rapidly increasing volume of knowledge gleaned by the disciples of the New Philosophy, and the whole was surveyed through the theocentric lens of seventeenth-century religion. Even the more enlightened of seventeenth-century minds often found it difficult to separate the true elements from the false in this peculiar compound, and as a result the science of the age was festooned with many curious beliefs.

One of the strangest of the inherited beliefs, and a belief that undoubtedly influenced early geomorphic thought, was the ancient doctrine of the macrocosm and the microcosm. Man, the microcosm, was supposed to be a miniature replica of the universe, or macrocosm. 'This body', wrote the rector of St. Martin's in London's Ludgate during the reign of James I, 'is a Microcosme, & created after the rest, as an Epitome of the whole Universe and truest Mappe of the Worlde, a summarie and compendious other World.'[35] This concept, which doubtless arose out of man's desire to bring the complex universe into some simple and intelligible relationship to himself, allowed human bones to be regarded as the counterpart of the Earth's rocks, man's flesh to be likened to the soil, his hair to the grass, his pulse to the tides, his eyes to the Sun and Moon, and his bodily warmth to the terrestrial heat. Those who professed to see such parallels between the microcosm and the Earth, or geocosm, often went on to argue that since man is a miniature representation of the Earth, and alive, then the Earth too must be animated, and there were many during the seventeenth century who regarded the globe as some kind of gigantic living creature. This peculiar idea lingered on far into the eighteenth century, and in the closing decades of that century the idea evidently influenced the thought of James Hutton, who has been widely acclaimed as the founder of modern geology.

Hardly less strange than this regard of the Earth as an organism, are some of the seventeenth-century ideas about the Earth's configuration. There were still, for example, those who believed the general level of the ocean to be higher than that of the continents. In support of this view it was argued that rock, being heavier than water, must lie at a lower level; that sailors find it easier to approach land than to leave it; and that 'to such as stand on the shore, the sea seemeth to swell into the form of an hill, till it put a bound to their sight.'[36] Above all, however, Scriptural evidence was adduced to support the notion, a favourite text being taken from Psalm 104 where God is said to have set a bound to the waters so that they might not submerge the lands. What, it was asked, was the purpose of such a bound if the waters were already lower than the continents? Surely the waters were still miraculously held back. The survival of this misconception concerning the relative levels of land and sea is all the more surprising in view of the exaggerated heights often ascribed to mountains before altitudes began to be determined barometrically late in the seventeenth century. Sir Walter Raleigh was not alone in believing the Peak of Tenerife (12 192 feet) to rise 30 miles above the sea,[37] nor did John Ray see anything unreasonable in the suggestion that Cader Idris (2 927 feet) was three miles high.[38]

The modern appreciation of mountain scenery dates back no further than the early eighteenth century, and before then mountains were widely viewed as hideous blemishes upon the Earth's fair skin. Professor Marjorie Nicolson[39] has shown how this changing attitude towards mountains is reflected in poetry, literature, and painting, but, as will emerge later, the change had a profound influence upon geomorphic thought. The regard of mountains as grotesque, terrestrial deformities was probably closely related to the doctrine of the macrocosm and the microcosm; if wrinkles, boils, pimples and goitres were blemishes in the microcosm, were valleys, volcanoes, hills and mountains not equivalent blemishes in the geocosm? Gabriel Plattes writing in 1639 observed that some persons believed mountains to be 'produced by accretion in length of time, even as Warts, Tumours, Wenns, and

Excrescencies are engendered in the superficies of men's bodies',[40] and at the end of the century John Ray prefaced a teleological discussion of mountains with a vigorous denial that they are 'Warts and superfluous Excrescencies'. It followed, of course, that if mountains were ugly blemishes, then they could never have been a part of the original, immaculate Creation. This logic, coupled with the fact that mountains receive no mention in *Genesis* before the account of the Flood, encouraged the notion that the primeval globe had possessed a perfectly smooth, clear skin, and that the continents were originally topped by vast, featureless plains. Not all sixteenth- and seventeenth-century scholars subscribed to such a view; but for many, mountains were cancerous growths of diluvial and post-diluvial age; they were, further, striking evidence of Nature's degeneracy and of the defiling influence of human sin.

In their discussions of landforms, sixteenth- and seventeenth-century scholars sometimes found it difficult to distinguish between natural and man-made features. It was this difficulty that prompted Thomas Molyneux to assure the Royal Society in 1694 that there was no mortar or cement between the basaltic columns of Ireland's celebrated Giant's Causeway, and that it therefore could not be of human construction.[41] Similarly, even the most commonplace of natural events was likely to be completely misunderstood. Thus a landslide at Marcle Hill in Herefordshire in 1575 received some imaginative embellishment, and gave rise to the famous story of a mobile mountain which was christened 'The Wonder', and which is still recorded on the Ordnance Survey maps by that name.[42] William Camden recounted the story of this hill which had 'roused it selfe up, and for the space of three daies together mooving and shewing it selfe (as mighty and huge an heape as it was) with roring noise in a fearefull sort, and overturning all things that stood in the way, advanced it selfe forward to the wondrous astonishment of the beholders'.[43] Seventeenth-century British literature contains many other comparable, if less well-known stories, all reflecting a certain gullibility and ignorance of natural processes.

In view of the existence of such strange notions about the visible features of the Earth's surface, it is hardly surprising that many peculiar beliefs were current about the Earth's hidden interior. Rocks and minerals, for example, were reputed to grow within the Earth from some kind of 'petrific seed' or under the influence of 'lapidifying juices', the growth being most vigorous in the warm tropics. One

English author of the 1630s presented his readers with the somewhat alarming prospect of an Adriatic island where 'the Iron breedes continually as fast as they can worke it'.[44] Even the leading scientists of the day were not immune from such beliefs, and in one of the first questionnaires circulated by the Royal Society a foreign correspondent was asked whether he had any knowledge of the regrowth of precious stones in mines after the removal of the original gems.[45]

Fossils too were widely believed to have grown *in situ* within rocks as a result of 'some latent plastick power of the Earth'. A few men of perception, such as Hooke and Ray, were already convinced of the organic origin of fossils, but against this view there was raised an objection which seemed damning in the religious atmosphere of the age. This objection arose from the realisation that many fossils are quite unlike any modern plants or animals, and thus implicit in the supposition that fossils had an organic origin, was the belief that some species had suffered extinction. Now if this had happened, it meant either that the original Creation was overstocked and thus imperfect, or that sometime after the Creation the divine masterpiece had become disfigured through the disappearance of certain creatures. Is it likely, asked Robert Plot in 1677, 'that Providence which took so much care to secure the works of Creation in Noah's Flood, should either then, or since, have been so unmindful of some shell-fish (and of no other Animals) as to suffer any one species to be lost?'[46] With a man of Hooke's genius such a contention carried little weight, and he was prepared to allow both the extinction and generation of species long after the Creation, but lesser intellects felt the force of the argument. If they wished to retain their belief in the organic origin of fossils, they had to fall back on the argument that no extinction of species had occurred, and that modern counterparts for all the fossil forms awaited discovery in the Earth's unexplored regions. Those who adopted such a thesis would have been delighted at the modern discovery of living coelacanths, and they would, no doubt, have been among the first to lend scientific support to the story that a colony of plesiosaurs still survives in the dark depths of Scotland's Loch Ness.

Another idea that influenced seventeenth-century geomorphology was the notion that the Earth, like a maggot-eaten cheese, was riddled with 'innumerable Openings, Recesses, Fissures, Chasms, Mazes, Swallows, Water Passages, and vast Receptacles'.[47] Aristotle himself had referred to the voids and cavities within the Earth, and by the seventeenth century, belief in their existence was time-honoured. Caves such as Poole's Cavern near Buxton, and Wookey Hole, were regarded merely as the uppermost of the Earth's many vaults, and earthquakes were believed to provide further indubitable evidence of the Earth's cavernous interior. In explaining seismic phenomena, little advance had been made beyond the views expressed in Aristotle's *Meteorologica* or Seneca's *Quaestiones Naturales*, and during the seventeenth century earthquakes were invariably attributed to processes operating within the Earth's vaults. Some argued that the caverns contained water, and that earthquakes resulted from the water sapping the cavern walls and causing subterranean landslides, but most authorities believed earth-tremors to be caused by great fires burning in the caverns and either igniting natural gunpowder, or causing noxious vapours to expand and shake the crust. Volcanoes were commonly supposed to be places where such vapours escaped from their subterranean prison, and the heat and fire associated with volcanic cones was seen as proof that there were indeed fires in the vaults beneath. It had to be admitted, of course, that whereas the subterranean caverns were supposed to be universal, volcanoes and earthquakes are not. Some argued that this was because the rocks of the crust are unequally charged with gunpowder, or give off differing amounts of noxious vapour, but the ingenious Robert Plot had a rather different explanation. Obviously thinking of the Earth as an animal, he suggested that earthquakes are rare in cold regions because the Earth's pores there are closed, and in consequence the subterranean fires soon choke themselves 'with their own filthy smoak'. In warmer regions, he claimed, the pores are open, thus allowing flames to rise to the surface to cause the volcanic outbursts which, he maintained, always precede earthquakes.[48]

The caverns within the Earth were believed to play a vital part in the terrestrial economy, for, despite the now famous studies by Pierre Perrault and Edmond Halley, the idea persisted that springs are fed not by atmospheric precipitation, but by an underground circulation of water passing from the sea and through the Earth's vaults before finally reappearing at the surface. Those who subscribed to the doctrine of the macrocosm and the microcosm frequently drew a parallel between the human circulatory system and that of the Earth; the dark caverns within the globe were likened to arteries, while the surface streams returning the water to the sea were regarded as equivalent to man's superficial veins. That water was able to rise within the Earth's vaults to make its resurgence far above sea-level excited little surprise; it was believed that the fires in the caverns played the part of the human heart and vaporised the sea-water, causing it to ascend to the surface where it cooled and condensed. 'And that it is no more wonder to see springs issue out of mountains', wrote the Rev. Edward Stilling-fleet in 1663, 'then it is to see a man bleed in the veins of his forehead when hee is let blood there.'[49] According to the vaporisation theory, hot springs represented the resurgence of water that had only just been condensed, and the theory also accounted for the sea-water losing its salinity beneath the continents. It was difficult, however, to explain why, in the course of time, the Earth's vaults had not become choked with salt deposits, thus afflicting the Earth with a kind of terrestrial coronary thrombosis.

Another favourite seventeenth-century argument adduced to support the belief in subterranean cave and river systems concerned the Caspian Sea. Writers observed, as had Aristotle some two thousand years earlier, that the Caspian receives the waters of numerous rivers, yet never overflows its margins despite the lack of any surface outlet. This was regarded as clear proof of the existence of underground chan-nels draining the sea's waters away to the nearest ocean. Similarly, it was argued that the Mediterranean must have subterranean outlets because not only does it receive many inflowing rivers, but surface currents pour into its basin at both the Dardanelles and the Straits of Gibraltar. Such arguments seemed convincing in the seventeenth century, and in any case there was Scriptural evidence to support the belief in subterranean watercourses. *Exodus* refers to 'the water under the earth', and without a belief in subterranean drainage how could one interpret the verse in *Ecclesiastes* where we read: 'All the rivers run into the sea; yet the sea is not full; unto the place from whence the rivers come, thither they return again'?[50]

This review of the beliefs that influenced the early development of geomorphic thought in Britain must now be drawn to a close. Some of the concepts discussed above exerted a beneficial influence upon geo-morphic thought, but the majority of them were antiquated beliefs which tended to cramp, stultify, and warp the infant subject. Among the harmful influences the most pernicious was undoubtedly the bibliolatry of the age. Not until the eve of the French Revolution did a British scientist dare to throw *Genesis* aside in order to venture a history of the Earth based solely upon the evidence provided by Nature herself. Geology was still feeling the influence of the so-called Mosaic writings during the early decades of the last century, and even in our own day there are fundamentalist sects which reject the findings of modern geology as false, and which cling to the literal truth of the *Genesis* story. They preserve for us in a fossilised form the bibliolatry of seventeenth-century England.

REFERENCES

1. See Frank D. Adams, *The Birth and Development of the Geological Sciences*, Chap. X (Baltimore 1938).
2. Second edition, Lib. III, Cap. I (Basle 1558).
3. *The Primitive Origination of Mankind*, p. 226 (London 1677).
4. *De Doctrina Christiana*, Bk. I, Chap. II.
5. *An Apologie of the Power and Providence of God in the Government of the World*, p. 1 (Oxford 1627).
6. See John B. Bury, *The Idea of Progress* (London 1920); Victor Harris, *All Coherence Gone* (Univ. Chicago Press 1949); R. F. Jones, 'The Background of *The Battle of the Books*', pp. 10–40 in Richard F. Jones *et alii*, *The Seventeenth Century* (Stanford Univ. Press 1951); Richard F. Jones, *Ancients and Moderns*, second edition (Washington Univ. Studies, St. Louis 1961); Paul H. Kocher, *Science and Religion in Elizabethan England*, pp. 82–92 (San Marino, Calif. 1953); G. Williamson, 'Mutability, Decay, and Seventeenth-Century Melancholy', *ELH* II (1935), pp. 121–150.
7. *Romans* 8, v. 21 & 22.
8. Bk. II, pp. 7–11.
9. *Noahs Flood*, lines 59–70.
10. George Hakewill, *An Apologie or Declaration of the Power and Providence of God in the Government of the World*, third edition, Bk. V, p. 61 (Oxford 1635).
11. *Commentaries on the First Book of Moses called Genesis*, translated by J. King, I, p. 59 (Edinburgh 1847).
12. *Religio Medici*, I. 16.
13. *Of Atheism*.
14. Thomas Burnet, *The Theory of the Earth*, I, Bk. I, p. 3 (London 1684).
15. *Ibid.*, II, Bk. IV, p. 143 (London 1690).
16. Thomas Robinson, *A Vindication of the Philosophical and Theological Exposition of the Mosaick System of the Creation*, p. 54 (London 1709).
17. 13 & 14 Car. II c 33.
18. 9 & 10 Will. III c 32.
19. John M. Robertson, *A Short History of Freethought*, II, p. 181 (London 1915).
20. Sir Peyton Ventris, *The Reports of Sir Peyton Ventris*, Pt. I, p. 293 (London 1696). See also Gerald D. Nokes, *A History of the Crime of Blasphemy*, pp. 48–61 (London 1928).

21. *A History of the Criminal Law of England*, II, p. 438 (London 1883).

22. *Philos. Trans.*, XXXIII (1724–25), No. 383, pp. 118–125.

23. *Pseudodoxia Epidemica*, Bk. VI.

24. *A New Analysis of Chronology*, I, pp. 3–7 (London 1809–1812).

25. *Annales Veteris Testamenti* (London 1650), and *Annalium Pars Posterior* (London 1654). An English translation of the two works was published in London in 1658.

26. T. H. Darlow and H. F. Moule, *Historical Catalogue of the Printed Editions of Holy Scripture*, I, p. 248 (London 1903).

27. *Of Reformation in England*.

28. Burnet, *op. cit.* (1684), I, Bk. I, p. 7.

29. Act IV, Scene I.

30. *A Booke Called the Treasure for Traveilers*, Bk. V, p. 7 (London 1578).

31. Robert W. T. Gunther, *Further Correspondence of John Ray*, Ray Society Series, CXIV, p. 260 (London 1928).

32. William Derham, *Philosophical Letters Between the Late Learned Mr. Ray and Several of his Ingenious Correspondents*, p. 256 (London 1718). Reprinted in Edwin Lankester, *The Correspondence of John Ray*, Ray Society Series, p. 243 (London 1848).

33. Richard Waller, *The Posthumous Works of Robert Hooke*, pp. 335 & 411 (London 1705).

34. *Pseudodoxia Epidemica*, Bk. I.

35. Samuel Purchas, *Microcosmus, or the Historie of Man*, pp. 25 & 26 (London 1619).

36. John Swan, *Speculum Mundi or a Glasse representing the Face of the World*, pp. 194 & 195 (Cambridge 1635).

37. *The History of the World*, Bk. I, Chap. 7, p. 90 (London 1614).

38. Edwin Lankester, *Memorials of John Ray*, p. 129 (Ray Society Series, London 1846).

39. Marjorie H. Nicolson, *Mountain Gloom and Mountain Glory* (Cornell Univ. Press 1959).

40. *A Discovery of Subterraneall Treasure*, p. 5 (London 1639).

41. *Philos. Trans.*, XVIII (1694), No. 212, pp. 175–182.

42. Seventh Edition One Inch Sheet 142 at SO630364.

43. *Britain*, translated by Philemon Holland, p. 620 (London 1637). In this edition the date of the landslide is erroneously given as 1571.

44. Edward Jorden, *A Discourse of Naturall Bathes, and Minerall Waters*, second edition, p. 68 (London 1632).

45. Thomas Sprat, *The History of the Royal-Society of London*, p. 158 (London 1667).

46. *The Natural History of Oxford-Shire*, pp. 114 & 115 (Oxford 1677).

47. Bernhard Varenius, *A Compleat System of General Geography*, translated by Dugdale, I, pp. 89 & 90 (London 1733).

48. *The Natural History of Stafford-Shire*, pp. 144 & 145 (Oxford 1686).

49. *Origines Sacræ*, p. 548 (London 1663).

50. *Exodus* 20, v. 4; *Ecclesiastes* 1, v. 7.

Chapter Two

In the Beginning
1578–1705

O God! that one might read the book of fate,
And see the revolution of the times
Make mountains level, and the continent, –
Weary of solid firmness, – melt itself
Into the sea!

King Henry IV, Part Two,
Act III, Scene I.

AMONG the many Irish manuscripts preserved in Dublin there are three copies of a scientific treatise dating from about 1400 which contains a brief but sound discussion of the changes effected on the Earth's surface by denudation.[1] The text has already attracted scholarly attention, and it has been suggested that the existence of three copies of the work probably indicates that it was used for educational purposes. Surprisingly, therefore, it would seem that Irish students were being instructed in the principles of geomorphology more than five hundred years ago. Unfortunately, the treatise cannot be hailed as an early Irish contribution to geomorphic literature; it is merely a translation of some earlier Latin texts, and the bulk of the work is derived second-hand from the writings of an eighth-century Arabian scholar. A hunt through Mediæval manuscripts and incunabula in search of the earliest British or Irish scholar to make a truly original contribution to the study of landforms would be a laborious task. It might also be an unrewarding task, because the few early native British references to landforms that are known are all very trivial comments unworthy of consideration in the present context.

27

There is one point which deserves to be made at this stage: a reader will be sadly disappointed if he turns to Mediæval and Renaissance texts expecting to find topographical features being related to the nefarious activities of witches, wizards, giants, or the Devil. Stories of witches hurling hills at each other, tales of giants building islands to serve as stepping-stones, and accounts of the Devil deftly scooping out punch-bowls on mountain sides, certainly exist in folklore, but such romances were never intended to be taken very seriously. They were handed down from one generation to the next solely because of their entertainment value around the cottage fireside, and although British geomorphology was profoundly influenced by the Hebraic folklore enshrined in the Scriptures, the infant subject was quite unaffected by the native topographical legends. Indeed, no modern reader can fail to be impressed at the sound understanding often displayed by the earliest British writers on the subject of landforms.

Early Writers on Landforms

William Bourne's *Booke Called the Treasure for Traveilers* of 1578 contains what is evidently the earliest noteworthy British discussion of landforms, and the work makes a convenient starting point for our study. Bourne was a leading citizen of the Thames-side borough of Gravesend; there he was born about 1535, and there he died in 1582. By trade he was an innkeeper, but he also served as a train-band gunner in the local forts which had been built to guard the river-approach to London. Ballistics perhaps presented Bourne with some intriguing problems, for in spite of a somewhat rudimentary education, he had a keen interest in applied mathematics, and he was the author of several successful technical treatises. These works, as Bourne himself explained, were never intended for the scholar; they were rather manuals of practical science for the use of the shipwrights, gunners, and navigators who must have been his daily companions. The *Treasure for Traveilers* was one such work; it was a general handbook for all voyagers and wayfarers. The volume is now comparatively rare, and not all the surviving copies contain the fifth and final book wherein Bourne offers what, with characteristic modesty, he terms his 'simple opinions' on such topics as 'the naturall causes of Sands in the Sea and rivers, and the cause of marish ground, and Cliffes by the sea Coasts, and rockes in the Sea'. Although his understanding of natural processes was somewhat elementary, his full acceptance of the reality of

28

denudation, and of its importance in shaping topography, must entitle Bourne to an honourable place in the history of British geomorphology.

During the decades following the appearance of Bourne's book, many British writers offered observations upon the origins of topography, but especially noteworthy is the work of Nathanael Carpenter. Carpenter was born in 1589 at Northleigh in Devonshire, and he studied at Oxford where he was befriended by James Ussher, the chronologer and future Archbishop of Armagh. Ussher persuaded Carpenter to accept the office of schoolmaster to the King's wards in Dublin, and it was there that Carpenter died about 1628. During his comparatively short life, Carpenter wrote various philosophical and mathematical works, but such reputation as he has with posterity rests upon his *Geography Delineated Forth in Two Bookes* which was first published in Oxford in 1625. In its own day this work was never very influential, but today it is acclaimed by historians of geography as a pioneer British treatise on the theoretical aspects of the subject. The book admittedly leans heavily upon the works of earlier continental authors, but in the present context this does little to diminish the interest of a volume which devotes four of its twenty-eight chapters to discussions of such topics as rivers, lakes, mountains, valleys, plains, continents, and islands. Carpenter's *Geography* in fact contains the first adequate British discussion of the principles of geomorphology, and it forms a landmark in our story. Carpenter himself saw no particular merit in the book, and on his death-bed he bitterly regretted that his interest in such mundane subjects as geography had drawn him away from divinity, that queen of the sciences. He lamented, as he put it, that he had 'formerly so much courted the Maid instead of the Mistress'.[2]

Carpenter's book was the outstanding British contribution to geomorphology during the first half of the seventeenth century, but later, as interest in landforms quickened, the literature of the subject grew apace. At Amsterdam in 1650, Bernhard Varenius, a young German geographer, published his famous *Geographia Generalis* which contains a consideration of 'changes on the terraqueous globe', and a discussion of such landforms as lakes, bays, straits, rivers, and mountains. The work was soon circulating widely in England where it became a standard text for the young gentlemen attending Newton's lectures at Cambridge. Two Latin editions of the work were published there under Newton's own editorship before the first English translation appeared in 1693. Another famous text, and a classic of geological literature, appeared in Florence in 1669 when Nicolaus Steno, a Danish anatomist, published his *Prodromus*, which includes an interpretation of the geological history of the Tuscan landscapes.[3] Henry Oldenburg, the Royal Society's first Secretary, immediately recognised the importance of the work, and in 1671 he published a now rare English translation.[4]

Other Fellows of the Royal Society were also curious about the history of the Earth's surface, and many papers on the subject are scattered through the Society's *Philosophical Transactions* which began to appear in 1665. There were two Fellows at this period whose geomorphic writings are of outstanding interest: Robert Hooke and John Ray. To the amazingly versatile Hooke must perhaps go the credit for delivering the earliest geomorphic lectures presented in Britain, for in 1664 and 1665 he discoursed at London's Gresham College upon the relationship between landforms and seismic activity,[5] and during the following thirty years he entertained the Royal Society with an interminable series of lectures upon the same theme. The polymathic Hooke was one of the most brilliant men of a glittering age of British science, but he had the misfortune to be overshadowed by Newton, and after Hooke's death his many accomplishments were forgotten. Much happier is the story of John Ray, the blacksmith's son from Black Notley in Essex who became the father of English natural history, and who has received international acclaim for his achievements in the biological sciences. Ray probably devoted even more attention to landforms than did Hooke, and although he perhaps lacked Hooke's brilliant insight into the true nature of Earth-history, his writings, and especially his *Three Physico-Theological Discourses* of 1692, are a most valuable source of material on late seventeenth-century geomorphic thought.

The interest evinced in Earth-history during the second half of the seventeenth century reached its climax during the century's last two decades. It was then that three remarkable and quite fanciful theories of the Earth made their appearance, each of them purporting to be a detailed and rational account of the momentous events described in the opening chapters of *Genesis*. Thomas Burnet's *Telluris Theoria Sacra* published in two parts, one in 1681 and the other in 1689, was the first of the theories to come off the presses. It was followed in 1695 by John Woodward's *An Essay Toward a Natural History of the Earth*, and in 1696 by William Whiston's *A New Theory of the Earth*. These three theories will be examined fully in the next chapter, and they excited a great deal of contemporary interest and controversy. Strangely, this controversy failed to stimulate further studies in Earth-history, and British interest in landforms waned rapidly during the opening decades of the eighteenth century. The reasons for this eclipse of geomorphology will emerge later, and for the moment it suffices to note that the death of Ray in 1705 serves as a convenient terminus to that first phase in the history of British geomorphology which forms the subject of this chapter.

Although interest in geomorphology was widespread during the seventeenth century, one class of author – the topographers – paid scant attention to the subject. In vain do we search through the many sixteenth- and seventeenth-century county surveys for the earliest examples of British regional geomorphology. Indeed, many of the topographers were so obsessed with archaeology, inscriptions on tombstones, genealogy, and architecture, that they scarcely troubled even to describe the relief of their chosen region. A book such as Dr Gerard Boate's *Irelands Natural History* of 1652, for example, offers nothing more enlightening than the following enigmatic account of Ireland's physique:

> For some parts are goodly plain Champain, others are Hilly, some Mountainous, and others are composed of two of these sorts, or of all-three together, and that with great variety, the which also is very great, in those three un-compounded sorts.[6]

Even in the stimulating atmosphere of the late seventeenth century, Robert Plot – professor of chemistry at Oxford, keeper of the Ashmolean Museum, and the best known of the county natural historians – failed to provide his readers with anything but the most sketchy of topographical descriptions. His large folio volumes on Oxfordshire and Staffordshire contain ample discourses upon the plants, fossils, notable echoes, strange diseases, and physiological abnormalities occurring within the two counties, but he evidently felt topography to be unworthy of attention. The same neglect of topography is reflected, of course, in the maps of the period which, if they attempt to show relief at all, do so by means of crude 'mole-hill' symbols scattered over the map in an almost haphazard fashion.

There is, nevertheless, ample published material available from which to reconstruct a picture of sixteenth- and seventeenth-century British

31

geomorphology. In addition to the authorities already mentioned, there were many other authors who made at least passing reference to the history of the Earth's surface, and in this chapter some of this material has been welded together into a synthetic account of early British geomorphology. In this way a wearisome and repetitious chronological account of the contributions of individual authors is avoided, but at the same time it is necessary to introduce a word of caution: to collect statements from a medley of contexts and authors – from the work of a Biblical scholar or Christian apologist in one place, and from that of a pioneer geologist or botanist in another – and then to twine all the threads together to represent what we now know as geomorphology, may give a completely false impression of the state of the subject during the sixteenth and seventeenth centuries. It must be remembered that the science of geomorphology is a modern creation, and even the late seventeenth century, which scarcely knew geomorphology's geological parent, certainly knew nothing of a science of landforms. Interest in the subject there certainly was, but if the present chapter conveys the impression that there was then a coherent and organised body of opinion on the subject of landforms, that impression is quite false. It must be stressed that the picture of seventeenth-century British geomorphology here presented is entirely the synthesis of the present writer, and no single scientist of that age saw the subject as it is here displayed. What follows is a twentieth-century edifice built out of sixteenth- and seventeenth-century materials.

In modern geomorphology it is axiomatic that landforms are the result of the interaction of two sets of forces. Firstly, there are the forces of earth-movement – the so-called endogenetic forces – which operate within the earth and which are responsible both for the folding and faulting of rocks, and for the slow raising or lowering of the Earth's surface to create a new topography. Secondly, there are the external forces of denudation – the so-called exogenetic forces – which slowly, but relentlessly, sculpt and wear down the relief produced by earth-movements. A reasonably clear understanding of the relative importance of these two sets of processes was not arrived at until comparatively recently, and during the sixteenth and seventeenth centuries thought on the subject was certainly not sufficiently far advanced for landforms to be viewed clearly in terms of sets of constructive and destructive forces. The notion that there are two sets of such forces is nevertheless implicit in much of the early writing on landforms, and for this reason it is here convenient to make a dichotomous approach to sixteenth- and seventeenth-century geomorphology. Attention will first be given to early notions about the forces responsible for forming an initial topography, and then we will examine the sixteenth- and seventeenth-century appreciation of the denudational, or exogenetic, processes.

Early British writers on the subject of landforms attributed the raising of the Earth's topography to one or other of three processes. Firstly, there were those who held that the Earth's present relief was only two days younger than the globe itself, and who believed that the Earth's mountains, valleys, and plains had all been moulded – perhaps even literally – by the divine hand on the third day of the Mosaic hexaëmeron. Secondly, there were those who maintained that the surface of the pre-diluvial globe had been ploughed up by the surging, tearing torrents of the Flood, and who were convinced that the Earth's topography had been largely, if not totally, reformed beneath the Flood waters. Finally, there were those who believed that during its history the face of the Earth had been shattered by numerous gigantic earthquakes, and that the Earth's present topography is in consequence essentially the result of seismic convulsions. These three theories were never mutually exclusive, and most of those who speculated about the origin of landforms were happy to regard some features as coeval with the Earth, while others were seen as the work of the Flood or ot earthquakes. The emphasis placed upon the three processes varied from author to author, and during the course of the seventeenth century there was also a general shift of opinion; in the early decades of the century it seems to have been widely accepted that the Earth's present topography was essentially of primeval origin, but gradually this view became less popular, and writers of the second half of the century usually emphasised the morphological significance of earthquakes, or proclaimed the debt that the Earth's topography owed to the Flood. Each of the three theories will now be examined in turn.

Topography as Old as the Creation

The notion that the Earth's topography is a permanent, immutable feature of the globe, and that it dates back to the Creation, is no doubt a very natural and ancient one. The idea was certainly widespread in sixteenth- and seventeenth-century England, and Nathanael Carpenter is one who subscribed to this view. 'Mountaines, Valleyes, and Plaines', Carpenter wrote, 'were created in the Earth from the beginning, and

33

few made by the violence of the Deluge.'[7] He defined a mountain as 'a quantity of Earth heaped above the ordinary height of the Land',[8] and he perhaps thought – very reverently of course – of the Creator making mountains just as the modern child makes sand-pies upon a beach. A few years later, in 1636, Henry Hexham's edition of Mercator's *Atlas* was a little more explicit about the divine creative process, for it claimed that the ocean basins, the continents and all the continental topography had been formed by great winds on the third day of Creation.[9] The same notion was in the mind of Sir John Pettus, deputy-governor of the royal mines, when in 1670 he wrote of the Creation and observed:

> . . . that when God breathed upon the Face of the Waters, that was a Petrefying [unfortunately the word was printed as 'Putrefying'] Breath; and that such Waters as were quiet and calm turned into Plains or Levelled Earth, and the Boisterous Waters into Hills and Mountains, according to the proportion of the Billows, and their Spaces into Vallies, which have ever since continued in those wonderful and pleasant Dimensions.[10]

Somewhat later the Rev. Erasmus Warren came forward with another suggestion; with the doctrine of the macrocosm and the microcosm in mind, no doubt, and perhaps with his own painful experience of sun-burn, he wondered whether hills and mountains might not be huge blisters raised upon the moist surface of the youthful continents when they were first exposed to the intense heat of the newly created Sun.[11] Newton had a rather different idea; in a letter to Thomas Burnet he suggested that the Earth's pristine relief might have been formed by unequal precipitation in a chaotic, primordial fluid just 'as saltpetre dissolved in water, though ye solution be uniform crystallises not all over ye vessel alike, but here & there in long barrs of salt'. Newton went on to offer Burnet a rather strange analogy: 'Milk', he observed, 'is as uniform a liquor as the chaos was. If beer be poured into it, and ye mixture let stand till it be dry, the surface of ye curdled substance will appear as rugged and mountainous as the earth in any place.'[12]

Few of Newton's contemporaries would have dared to liken the divine creative act to the pouring of beer into milk, and many probably felt it sacrilegious to inquire too closely into the methods that God had used to shape the Earth's original relief. It was not for man to pry into God's secrets, and in a catechism on the Book of *Genesis* published in

1620, one Alexander Rosse, a royal chaplain, answered the question 'How made God the drie land to appeare?', with the simple, non-committal response: 'By causing the earth, which before was plaine, to swell with mountaines.'[13] Similarly in Book VII of *Paradise Lost* Milton wrote:

> . . . *when God said*
> *Be gather'd now ye Waters under Heav'n*
> *Into one place, and let dry Land appeer.*
> *Immediately the Mountains huge appeer*
> *Emergent, and thir broad bare backs upheave*
> *Into the Clouds, thir tops ascend the Skie.*

Many must have felt this to be about as far as one could safely go towards understanding the divine *modus operandi*.

Although this belief that some if not all of the Earth's relief is coeval with the globe gradually lost support as the seventeenth century progressed, it long persisted in the very influential text by Varenius. Even in the eighteenth century, a discussion of the origin of mountains in English editions of *Geographia Generalis* continued to be prefaced by the statement: 'This is a great Question with some Philosophers, but others think it superfluous, and not fit to be enquired into; because they suppose Mountains to have had a Being since the Creation.'[14] Indeed, the much-lauded Varenius himself, while offering somewhat contradictory statements on the subject, clearly leaned towards the view that the Earth's major mountains, together with many straits and bays, all date from the Creation.[15]

During the seventeenth century there were three objections which could be raised against the theory that some, if not all, of the Earth's topography dates back to the Creation. The first objection arose from the sixteenth- and seventeenth-century acceptance of the reality of denudation. Could topography which mouldered under the attack of natural processes really have had a divine origin? Was it not inconceivable that God should allow any of His own exquisite handiwork to be defaced by the action of such puny, mundane forces as rain and rivers? Varenius was one who evidently felt the force of this argument, for despite his own assertion that many topographical features are indeed coeval with the Earth, he elsewhere observed of some mountains that 'since we can perceive a natural Decay and Corruption of them, we may judge they do not proceed from a supernatural Origin'.[16] The

35

present writer knows of no British author before the 1690s who reasoned in this way, however, and in fact such an argument can have carried
little weight in a Britain where it was generally admitted that the
entire creation was corrupt and decadent as a result of man's depravity.
That mountains shared in the decay probably seemed neither to prove
or disprove their divine origin; it merely showed that God had a
punitive plan for mankind, and further demonstrated the extent to
which the once perfect universe had been contaminated by human sin.

The second objection to the belief arose from the notion that
mountains are grotesque blemishes on the Earth's skin. Mountains,
observed John Dennis in 1693 after crossing the Alps, 'are not only
vast, but horrid, hideous, ghastly Ruins'.[17] Clearly such imperfections
could never have existed in the original, immaculate Creation. Indeed,
as one writer observed, to suppose that mountains had a divine origin
'appeareth to bee an Opinion, whereby great dishonour may reflect
upon the Creator'.[18] As we have already seen, this logic encouraged
the belief that the primeval continents had been topped by vast featureless plains, and this view found its fullest expression in Burnet's
Telluris Theoria Sacra of 1681. Burnet argued forcefully that the ugly,
contorted, and broken structure of mountains proved conclusively
that they had originated in a world already deeply contaminated by
human sin, and he looked longingly back to a time when the primeval
globe had displayed 'the beauty of Youth and blooming Nature, fresh
and fruitful, and not a wrinkle, scar or fracture in all its body; no
Rocks nor Mountains, no hollow Caves, nor gaping Chanels, but even
and uniform all over'.[19]

Both the above objections to the belief in the existence of primeval
topography have a Scriptural or theological basis, but the third objection rests on rather more scientific grounds, and arose out of the seventeenth-century interest in fossils. The first chapter of *Genesis* makes it
clear that the Earth's primeval continents appeared on the third day of
the Creation, whereas the oceans were not stocked with creatures
until the fifth day. Now if, as some enlightened souls claimed, fossils
represented the remains of former marine organisms, then it followed
that any rocks containing fossils could not be a part of the original
Creation. By the close of the seventeenth century it was generally
acknowledged that fossiliferous strata were to be found high in the
Alps, Apennines, and Pyrenees, and thus implicit in the acceptance of
the organic origin of fossils was the admission that Europe, at least,

was not a part of the primeval world. There were many, of course, who stubbornly adhered to the belief that fossils were nothing more than sports of Nature, but before the close of the seventeenth century men of the stature of Hooke, Steno, and Ray had become convinced of the organic origin of fossils, and were fully prepared to concede that large-scale changes in the distribution of land and sea must have taken place subsequent to the Creation. Slowly, as the true nature of fossils began to be generally understood, and as the full extent of fossiliferous strata came to be appreciated, the notion that some of the Earth's pristine topography still survived was gradually abandoned. John Beaumont wrote in 1697:

> I am of opinion there is no Mountain on the Earth now, that is an original Mountain, or that existed when the World first rose, and concluded with Aristotle, that the Sea and Land have chang'd places, and continue so to do.[20]

It of course had to be admitted that Scripture makes no reference to the formation of a second generation of continents, but here the story of the Flood – the sole universal catastrophe mentioned in Scripture – was seized upon as a very convenient solution to the problem. As a belief in the survival of pristine topography waned, there arose in its place a belief that during the Flood the Earth's topography had been completely remodelled at the divine behest.

A Diluvial Metamorphosis

There has probably always been a tendency for Christian peoples to ascribe some of the topographical features about them to the ravages of the Flood, but during the sixteenth and seventeenth centuries there were wide differences of opinion as to the nature of the Flood and its work. On the one hand there were those who freely admitted the catastrophic nature of the Flood, but who denied it any significance in the shaping of topography, and on the other hand there were those who regarded the Earth's present topography as essentially of diluvial origin.

The writers who denied the Flood any topographical importance limited its work to the destruction of all life forms save those gathered in the Ark, and they emphasised a passage such as that in *Genesis* 6, verse 17, where Noah learns of the forthcoming catastrophe which is 'to destroy all flesh, wherein is the breath of life, from under heaven'.

37

Nathanael Carpenter belonged to this school of thought, and in 1625 he presented the case for believing the Flood waters to have been placid, and therefore quite incapable of remodelling the Earth's surface.[21] Carpenter observed, for example, that Noah had accepted the dove returning to the Ark with an olive leaf as a sign that the Flood waters were abating, and he must therefore have known that the leaf came from a tree that was still firmly rooted in the ground. Now if the Flood was incapable of uprooting a slender olive tree, Carpenter asked, how could it possibly have thrown up huge mountain ranges, or converted former sea-beds into dry land? In any case, he observed, a pillar reputed to have been erected by Seth some 1426 years before the Flood was still standing in the first century A.D. when its existence was recorded by Flavius Josephus. Again, Carpenter reasoned, the use in *Genesis* of present-day place and river names when referring to pre-diluvial sites seems further, indubitable proof that the modern world is essentially the same as that known to Noah's ancestors. For more than two centuries these arguments adduced by Carpenter continued to be bandied about in the often heated controversy over the role of the Flood in the shaping of topography, and the olive leaf argument appeared as late as the 1830s in so reputable a work as Lyell's great *Principles of Geology*.

Undoubtedly Carpenter's thesis that the Flood waters were calm and impotent, accorded well with a literal interpretation of *Genesis*, but at the opposite end of the thought-spectrum there were those who regarded the Flood as a divine instrument which in a single climacteric act had destroyed not only all pre-diluvial life, but also the pre-diluvial world itself. Those of this persuasion were not without Scriptural evidence in support of their belief, for in *Genesis* 6, verse 13, God refers to the Earth's evil inhabitants and promises: 'I will destroy them with the earth'.

In his description of the county of Pembroke completed in 1603, the Welsh topographer George Owen wrote about 'the violence of the generall flood, which at the departinge thereof breake southward and tare the erthe in peeces and seperated the Ilands from the Contynent, and made the hilles and valleies as we now finde them',[22] and such views are typical of those held by many of his contemporaries. The most vehement believers in a diluvial metamorphosis of the world were those who regarded mountains as grotesque blemishes, and who imagined the pre-diluvial continents to have been topped only by expansive plains. They pictured the Flood as a vicious, swirling body of waters sweeping over the globe, tearing up the smooth surface of the exquisite pristine continents, twisting and shattering the Earth's rocks, and finally leaving the debris to form the world's present topography. Post-diluvial mankind, it seemed, was condemned to eke out its miserable existence surrounded by the diluvial wreckage of the luxurious ante-diluvial paradise. This view found expression in the ninth song of Michael Drayton's *Poly-Olbion* written in 1613:

> Tell us, ye haughtie Hills, why vainly thus you threat,
> Esteeming us so meane, compar'd to you so great,
> To make you know your selves, you this must understand,
> That our great Maker layd the surface of the Land,
> As levell as the Lake untill the generall Flood,
> When over all so long the troubled waters stood:
> Which, hurried with the blasts from angry heaven that blew,
> Upon huge massy heapes the loosened gravell threw:
> From hence we would yee knew, your first beginning came.

This belief in a diluvial metamorphosis gained widespread support as the seventeenth century progressed, and as the growing appreciation of the true nature and extent of fossiliferous strata made it increasingly unlikely that any of the Earth's pristine relief could have survived down to modern times. Fossils were now regarded as relics of pre-diluvial creatures which the rising Flood waters had swept on to the continents, and which there became entombed in newly formed rocks as the world underwent its transformation. The following passage from the pen of a Yorkshire correspondent of the Royal Society in 1700 typifies the late seventeenth-century view on this aspect of the Flood's work:

> It appears also that in that Deluge the Earth suffered wonderful great violence and force, that Seas were raised into Mountains, and Mountains sunk into Seas, that Beds of Shells had sometimes such Weights of Earth and pieces of Mountains, and Rocks flung upon them, that they were crushed in pieces, and others squeez'd flat.[23]

Robert Hooke believed the changes to have been of an even more drastic nature. He argued that the Flood was of far too short a duration to account for the formation of all the world's fossiliferous strata, and in a lecture to the Royal Society in 1688 he maintained that the catastrophe had witnessed a wholesale interchange of land and sea. As the pre-diluvial land masses subsided, he claimed, the beds of the pristine

39

oceans, together with their accumulated banks or shells, had been raised to form our present continents.[24] One final, and rather more moderate, late seventeenth-century estimate of the work of the Flood is worth quoting from the pen of John Ray. He held that efforts to discover the true nature of the Flood could only be conjecture, but he ventured to suggest that it had performed such work as:

> In some places adding to the Sea; in some taking from it; making Islands of Peninsulæ, and joining others to the Continent; altering the Beds of Rivers, throwing up lesser Hills, and washing away others.[25]

Thus towards the close of the seventeenth century, the Flood began to assume an extremely important place in explanations of topography. It retained this importance for more than a century, and the supposed work of the Flood will repeatedly engage our attention in later chapters. Those who claimed the Flood as an important topography-forming agent had to concede that Moses is silent on the subject of a diluvial metamorphosis, but they no doubt felt that here was a perfect illustration of the way in which science could illuminate the Scriptures. Earlier scholars had regarded the Flood merely as a universal inundation, but now the realisation of the true nature of fossiliferous strata had revealed the enormity of the diluvial catastrophe. In the age of Noah, it seemed, mankind had become so heinous that God had been forced not only to drown the Earth's inhabitants, but also to destroy the contaminated primeval continents, and to remould the entire surface of the globe. It is no wonder that science was hailed as a valuable adjunct to revealed religion!

Topography from Earthquakes

Seismic convulsions are the third and final topography-forming process recognised during the sixteenth and seventeenth centuries. Among scholars there was then a widespread and deeply ingrained belief that gigantic earthquakes periodically shatter the Earth's crust, raising some earth-blocks and depressing others. Even those who believed in a diluvial metamorphosis of the globe, readily admitted that many modern topographical features owed their formation to post-diluvial earthquakes. In his *Treasure for Travellers* William Bourne offered a few comments on the work of earthquakes, and he observed that after a shock 'there may bee some hilles or cliffes standyng up much higher then the grounde was before. And so by this meanes the places that have bene dry land, may become sea and water, and in lyke manner that place that hath beene water, may become dry lande.'[26]

The earliest British writer to discuss the effect of earthquakes at any length was apparently John Swan, a Cambridge author of whom little is now known. In his *Speculum Mundi*, published in Cambridge in 1635, Swan explained that earthquakes are used to strike a fear of God into mankind, to cause noxious, lethal vapours to escape from the Earth's interior as punishment for the wicked, and to effect many changes in the Earth's configuration. These changes, he noted, include the raising and destruction of mountains, the removal of mountains from one place to another, the conversion of plains into plateaux, the creation and submergence of islands, the formation of new rivers and volcanoes, and the opening of straits across former isthmuses. Seismic shocks of this latter type, Swan suggested, must have been responsible for creating the Straits of Dover, Gibraltar, Messina, and many others.[27]

The foremost British exponent of the view that earthquakes have played a major part in the shaping of topography was Robert Hooke. Hooke believed that most of the Earth's relief features, including the very continents themselves, had all been produced by seismic and volcanic activity, and he presented his views in a rambling dissertation entitled *A Discourse of Earthquakes, and Subterraneous Eruptions. Explicating the Causes of the Rugged and Uneven Face of the Earth* which he completed in 1668. The discourse was published posthumously in 1705 along with some of Hooke's many lectures on the same subject,[28] and one passage from the work is worth quoting both as a summary of Hooke's personal opinions and as a specimen of the kind of beliefs widely current among his contemporaries. Hooke wrote:

> [Earthquakes] have turn'd Plains into Mountains, and Mountains into Plains; Seas into Land, and Land into Seas; made Rivers where there were none before, and swallowed up others that formerly were; made and destroy'd Lakes, made Peninsuls Islands, and Islands Peninsulas; vomited up Islands in some places, and swallowed them down in others; overturn'd, tumbl'd and thrown from place to place Cities, Woods, Hills, etc. cover'd, burnt, wasted and chang'd the superficial Parts in others; and many the like strange Effects, which, since the Creation of the World, have wrought many very great changes on the superficial Parts of the Earth, and have been the great Instruments or Causes of placing Shells, Bones, Plants, Fishes, and the like, in those places, where, with much astonishment, we find them.[29]

41

In his *Micrographia* of 1665 Hooke describes an experiment which he conducted in an effort to reproduce topography in miniature by means of what he regarded as a tiny simulated earthquake.[30] At the time of making his experiment Hooke was seeking the origin of the lunar topography, and in an earlier experiment he had produced craters similar to those on the Moon by allowing heavy bodies to fall into a mixture of pipe-clay and water. He refused to accept that the lunar craters could be the result of impact, however, for he could conceive of no objects that could possibly have bombarded the Moon, and in any case, as he sagaciously observed, we have no reason to believe that the Moon is made of pliable pipe-clay. In his second experiment, therefore, he boiled what he describes as alabaster, and noted that after removing it from the fire, the last bubbles to rise to the surface and burst left pits similar to the Moon's craters. This, he believed, indicated the true origin of the scarred lunar surface; it was the result of innumerable seismic and volcanic explosions caused by hot vapours escaping from within the Moon. He regarded this experiment as equally revealing of the origin of terrestrial landforms, and it deepened his conviction that noxious gases within the Earth are the root-cause of all topography. Hooke, as we have seen, was probably responsible for the earliest geomorphic lectures delivered in Britain, and his experiments in selenomorphology must – paradoxically – entitle this remarkable man to be hailed as our first experimental geomorphologist.

Hooke, and the many who shared his views about the seismic origin of topography, had to admit that modern earthquakes are trivial as compared with the gigantic convulsions necessary to raise mountain ranges such as the Alps or Pyrenees. John Ray perceived the presence of a major problem here, for in a work published in 1673 he wrote:

> In general since the most antient times recorded in History, the face of the Earth hath suffered little change, the same Mountains, Islands, Promontories, Lakes, Rivers still remaining, and very few added, lost or removed. Whence it will follow, that if the Mountains were not from the beginning, either the World is a great deal older than is imagined or believed, there being an incredible space of time required to work such changes as raising all the Mountains, according to the leisurely proceedings of Nature in mutations of that kind since the first Records of History: or that in the primative times and soon after the Creation the earth suffered far more concussions and mutations in its superficial part than afterward.[31]

Hooke himself would certainly have agreed with Ray's latter suggestion, for he was convinced that during the Earth's life, seismic outbursts had become both less powerful and less frequent. This he attributed partly to the subterranean fires dying down, partly to there being fewer 'spirituous, unctuous and combustible or inflammable juices' in the ageing Earth, and partly to the senile Earth having grown a tough, callous skin that was stiff and less liable to fracture.

There are three factors to be borne in mind when seeking to understand and account for the wide sixteenth- and seventeenth-century acceptance of the theory that earthquakes play a major part in shaping topography. Firstly, since the Earth was believed to be riddled with innumerable fissures and caverns, it seemed quite reasonable to suppose that portions of the crust would collapse periodically into the vaults beneath, leaving the jagged, broken edges of the tilted earth-blocks to form mountains. Secondly, even seventeenth-century scientists still leaned heavily upon the writings of the ancients, and these writings, emanating from the Mediterranean earthquake-zone, abound in somewhat exaggerated stories of the effects of seismic and volcanic activity. In his *Timaeus*, for instance, the revered Plato had related how Atlantis had been swallowed into the ocean during one terrible day and night of earthquakes, and the Elder Pliny, who was himself killed during the eruption of Vesuvius in A.D. 79, had told of an earthquake which caused two mountains to rush together and fall upon each other. Strabo had claimed that all off-shore islands had been torn from the neighbouring mainland by seismic shocks, and in the final scientific utterance of the classical world Seneca wrote:

> The earthquake produces a thousand strange sights, changing the aspect of the ground, levelling mountains, elevating plains, exalting valleys, raising new islands in the deep.[32]

Statements such as these, coming from highly esteemed sources, must have exerted a very strong influence on sixteenth- and seventeenth-century thought, and the writings of the ancients were frequently adduced to support a belief in the topographical importance of earthquakes. Had there been less reverence for the ancients, northern Europeans possessing little first-hand experience of earthquakes and volcanoes, might have ascribed rather less importance to such phenomena. This was certainly the opinion of John Woodward, in many ways the outstanding British geologist of his age, for in 1695 he chided his

43

contemporaries for having blindly accepted the classical authors as reliable and authoritative on such matters.[33]

British scholars did of course receive reports of contemporary earthquakes and volcanic outbursts, but these often gave somewhat exaggerated accounts of the topographical effects of the phenomena. This was a third and final reason for the belief that seismism is of vital importance in the development of topography. Typical of the over-drawn descriptions that circulated after contemporary earthquakes are those current after the admittedly disastrous tremor that destroyed Port Royal in Jamaica in 1692. The Royal Society printed accounts of the earthquake[34] in which we find it stated that 'a great Mountain split, and fell into the Level Land, and covered several Settlements', that the mountains were 'strangely torn and rent; insomuch that they seem to be of quite different Shapes now from what they were'; and that the earthquake threatened 'to cast this Island into it's first Chaos, or at least into a new Model or Shape, different from that which Nature first gave it; breaking one Mountain, and thereof making two or three; and joining two Mountains, and making thereof one, closing up the unhappy Valley betwixt'. Any lingering doubts about the potency of earthquakes must have been dispelled by accounts such as these. If modern earthquakes could wreak such havoc, what might not have been accomplished by the mighty convulsions that shook the Earth in its youth?

Catastrophism in Geomorphology

By the close of the seventeenth century there were probably few who still believed in a primeval origin for the Earth's topography, and it was generally accepted that landforms were essentially the result of diluvial, seismic, or volcanic action. Thus violent, cataclysmic disturbances came to occupy an important place in explanations of topography, and there emerged what later historians of science termed the doctrine of catastrophism. The doctrine reached its fullest develop-ment long after the seventeenth century, and its cramping influence was still being felt by the earth and biological sciences far into the nineteenth century.

The most important factors which led to the seventeenth-century interpretation of landforms in terms of catastrophes have been dis-cussed above, but there remains one fundamental factor which must be re-emphasised: catastrophism arose directly out of the very limited

44

conception of time that prevailed in an age of bibliolatry. Any notion that slow-acting processes might have far-reaching, cumulative effects was obviously inhibited by the belief that all the events of Earth-history had to be crammed into rather less than six thousand years. Floods, earthquakes, and volcanic outbursts seemed the only means whereby great changes could have been wrought on the Earth's surface in the time available, and catastrophism was a result of the attempt to reconcile a shadowy inkling of the complexity of Earth-history with the Scriptural Earth-chronology. In this respect it is interesting to compare the gigantic earthquakes which Hooke invoked to explain the presence of marine fossils high in mountain ranges, with the kind of earth-movement which the modern geologist might offer by way of explanation of the same phenomenon. Hooke and the modern geologist differ not so much in the type of earth-movement they envisage, as in the intensity they ascribe to its operation; what Hooke had to regard as uplift effected within a few days, if not within a few hours, his modern counterpart is free to claim as the cumulative effect of processes operating through perhaps tens of millions of years.

Today, the term 'catastrophist', even in the hands of historians of science, has become something of a term of derision, and the branding of the early naturalists as catastrophists is to some degree both unfair and misleading. They certainly did attach great importance to catas-trophes in their attempts to explain topography, but most of them never for a moment doubted the reality or efficiency of such slow-acting processes as fluvial and marine erosion. Indeed, during the sixteenth and seventeenth centuries those who are now stigmatized as catastrophists had, almost without exception, a very sound grasp of the principles of subaerial denudation. They were fully aware of what denudation might accomplish if given sufficient time, but it seemed that in the six millennia available, the effects of processes such as fluvial and marine erosion must have been insignificant as compared with the havoc supposed to have been wrought by catastrophes.

The Basis of a Belief in Denudation

The idea of topography slowly mouldering away under the attack of natural agents was thoroughly acceptable even to the most con-servative of sixteenth- and seventeenth-century minds, and an examina-tion of several hundred multifarious British works of the period has revealed no author before the 1690s who seriously doubted the potency of Nature's destructive forces. There is today a widely current belief that James Hutton, John Playfair, and other late eighteenth-century geologists were the first to appreciate fully the effectiveness of Nature's denudational processes, but this belief is quite false. An understanding of the work of denudation was almost universal in sixteenth- and seventeenth-century Britain, and writers such as Hooke and Ray, who because of their acceptance of denudation have sometimes been hailed as the precursors of Hutton and Playfair, were in fact thoroughly typical of their age.

Apart from the evidence of their own eyes, there were a number of factors which must have made it well-nigh impossible for sixteenth- and seventeenth-century scholars to overlook the existence of what we now know as the exogenetic forces. There was, for instance, the influence of the ancients. The works of Herodotus, Aristotle, Strabo, Ovid, Philo, and others – works then thoroughly familiar to all British scholars – contain many astute observations on the decay of rocks, and the erosive activities of running water and the sea. There seem no grounds, however, for supposing the existence of a causal connection between the Renaissance rediscovery of classical learning, and the widespread sixteenth- and seventeenth-century admission of the reality of denudation. The knowledge that the ancients had believed in denudation no doubt made it easier for Renaissance scholars to subscribe to the same idea, but this was not just another case of 'adherence unto antiquity'. The Renaissance scholars in fact inherited a belief in denudation direct from their Mediæval predecessors. When Leonardo da Vinci, Georgius Agricola, Bernard Palissy, William Bourne, and other Renaissance scholars discussed denudation, they were merely developing ideas which had been current in Europe for many centuries.

Professor Frank D. Adams has already offered some excellent examples of Mediæval works which reveal a sound understanding of the principles of denudation,[35] and brief quotations from two Mediæval texts which circulated in the British Isles will suffice here to demonstrate that the reality of denudation was by no means a Renaissance rediscovery. The first quotation comes from one of the most influential of the Mediæval encyclopædias – from *De Proprietatibus Rerum*, which was written about 1250 by Bartholomaeus Anglicus, an English Franciscan. Down to about 1600 this work remained one of the principal natural-history texts, and it is therefore interesting to find that it contains a reference to what would now be termed mountains of circumdenudation. Some authorities, Bartholomaeus notes, believe the 'cause of mountaines & of valleyes be nought els but moving of waters that dig and weare the soft parts of the earth, & the hard parts that maye not be digged, be made Mountaines, and places that wer digged deepe, wer made for the sea and for rivers'.[36] The second quotation comes from the early Irish text to which reference was made earlier. There we read – as did Irish students more than five hundred years ago – that the rain 'flows constantly about the earth rooting up the soil, and everything dissoluble and non-resistant that it finds in the earth it carries from place to place in the rivers, and the force of the rivers carries off the same things to the sea, and the bottom of the sea is filled with them'.[37] With such a heritage behind them, it would have been surprising had scholars of the sixteenth and seventeenth centuries renounced the concept of denudation.

In Britain, at least, there was really no danger of such a renunciation, because the concept of denudation was entirely in accord with some of the most deeply-rooted tenets of the age. Since the Earth was not eternal, for example, it seemed logical that it should decay, for, as Aristotle had observed in *De Caelo*, 'generated things are seen always to be destroyed'. Again, a belief in denudation was encouraged by the doctrine of the macrocosm and the microcosm; if in senility the human body loses its youthful perfection, was it not only natural to expect the body of the senile Earth to become decayed, worn, and scurvy? Plants and animals, observed Hooke in a lecture in 1699, 'decay and grow old, as we call them, to grow stiff, and dry, and rough, and shrivelled; all which Marks or Symptoms may plainly be discovered also in the Body of the Earth'.[38]

Above all, a belief in denudation was in harmony with the widespread notion that Nature was everywhere in decay as a result of the corrupting influence of human sin. A generation which believed that the Sun and stars were fading, that the Earth's fertility and climate were steadily deteriorating, and that man himself was becoming progressively more depraved, was fully prepared to admit that the Earth's present topography is merely the decayed wreckage of some bygone world. The concept of denudation was an integral part of the belief in a decadent universe, and the decay of topography is frequently mentioned in sixteenth- and seventeenth-century discussions of Nature's debasement. In 1580, for example, Francis Shakelton wrote:

47

... doe we not see the yearth to be changed and corrupted: Some-tymes by the inundation of waters: Sometymes by fiers: And by the heate of the Sunne: And doe we not see that some partes of the same doe ware old, and weare awaie even for verie age.[39]

John Davies of Hereford, in similar vein, wrote the following lines in his *Muse's Sacrifice* of 1612 as part of a general discussion of the universal decay:

> The Worlds Parts are decaid (as doth appeare)
> Etna, Parnassus, and Olympus too
> Are not so eminent as erst they were;
> and all that's done, seems quite now to undoe.

Then, as observed earlier, when the Bishop of Gloucester defended the belief in a decadent Nature against George Hakewill's critique in the 1630s, the Bishop adduced the turbid waters of a river in flood in support of his case. Such waters, the Bishop urged, are indubitable proof of a decay so complete that even the continents are not immune from its influence.[40]

Although most scholars doubtless saw denudation as a manifestation of God's punitive plan for mankind, some did manage to regard it as a process which in the long run would benefit the Earth. If hills and mountains were ugly diluvial blemishes, was the slow, relentless activity of denudation not working to restore the continents to their pristine regularity? Robert Hooke may have inclined to this belief, for he observed 'that the Earth itself doth, as it were, wash and smooth its own Face, and by degrees to remove all the Warts, Furrows, Wrinckles and Holes of her Skin, which Age and Distempers have produced'.[41]

With arguments such as these to hand, the case for accepting the reality of denudation must have seemed quite unanswerable, and scrip-tural support for the concept was probably deemed hardly necessary. Any who did search the Scriptures for texts bearing upon the subject must have encountered a reference in *Habakkuk* to 'everlasting moun-tains' and 'perpetual hills', but on the other hand, they prob-ably rested assured of the fact of denudation when they found the statement in *Job* that 'The waters wear the stones'; the assertion in *Isaiah* that 'every mountain and hill shall be made low'; and the promise in the *Psalms* that the Earth and Heavens 'shall wax old like a garment'.[42]

Now we can turn to some examples of references to denudation drawn from the literature of the period. The range of denudational processes then recognised was of course somewhat restricted – nothing was known, for example, of the erosive work of ice – and it is here convenient to consider the subject under two heads: firstly, the work of rain and rivers, and secondly the work of the sea.

The Work of Rain and Rivers

Here pride of place must go to William Bourne; in his *Treasure for Traveilers* he noted that in rivers swollen with flood water 'the swiftnes of the running of the water dooth fret away the bankes', and he suggested that marshes and meadows alongside rivers, and sand-banks in the sea, had all been formed of material washed down from the neighbouring highlands.[43] It was Nathanael Carpenter in 1625, however, who was apparently the first English writer to offer a general discussion of the work of denudation, although his views on the subject admittedly owe much to the slightly earlier work of a German Jesuit named Josephus Blancanus. Carpenter, like Bourne, well understood the erosive power of running water; he pointed out that rainwater visibly washes soil down from the mountains, while rivers are seen 'by litle and litle continually to fret and eat out the feet of mountains'.[44] The height of a mountain, he claimed, depends largely upon its ability to withstand 'this violence of the water', although he admitted that if time allowed, even the mightiest peaks would be humbled and that 'the whole Earth should in the end be over-whelmed with waters, as in the beginning'.[45] He was aware that some of the debris washed down from the mountains comes to rest upon the adjacent lowlands before ultimately being swept into the sea, and he suggested that this deposition upon the lowlands was the explanation for the burial of ancient buildings and monuments. When it eventually reaches the sea, Carpenter observed, some of the debris may cling to the coast and cause estuarine silting, and as an example of this process he turned to his native Devon where, he noted, the harbour of Totnes on the River Dart, once accessible to the largest vessels, could no longer receive even a small boat except at full tide.

Bernhard Varenius, in his *Geographia Generalis*, had rather less to say on the subject of denudation than did Carpenter, and his discussion of landforms is in many respects inferior to Carpenter's. Indeed, some passages in Varenius strike the modern reader as somewhat inept, and they certainly reveal a failure to understand the true nature of a river valley. One of his propositions, for example, reads: 'The Chanels of Rivers the nearer they are to their Fountains, are generally so much the higher; and most of them are depressed gradually towards their Mouths', and elsewhere he argued that rivers go underground 'because they meet with elevated Ground which they cannot overflow, and therefore are forced to glide into the next Grotto they meet with'.[46] In spite of such fatuous statements, Varenius did understand that vast quantities of debris are removed from the continents by rivers, and he claimed that in the Hwang Ho, sand and gravel amount to as much as one third-part of the waters. Lakes, straits and bays, Varenius observed, are slowly filled in with this river-borne sediment, becoming firstly shallows, then a fen, and finally dry ground. He followed Herodotus and Aristotle in believing that 'Egypt is stretched further and further every Year' by deposition on the Nile delta, and he claimed that similar progradation was taking place around the mouths of such rivers as the Indus, Ganges, and Rio de la Plata.

The same idea was in Nicolaus Steno's mind, for Henry Oldenburg's translation of the *Prodromus*, published in 1671, contains the following passage:

> Yet this is certain that a great parcel of the Earth is every year carried into the Sea (as is obvious to him, that shall consider the largeness of Rivers, and the long passages through the Midland Countries, and the innumerable number of Torrents; in a word, all the declivities of the Earth) and consequently that the Earth carried away by the Rivers and joyned to the Sea-shores does every day leave new Lands fit for new Inhabitants.[47]

Two years later, John Ray provided his readers with some examples of the process described by Varenius and Steno. In the vicinity of Venice, and in the Camargue, he claimed, the sea has clearly been pushed back by the deposition of riverain debris. In the latter region, he recorded 'we were told [in 1665] that the Watch-Tower had in the memory of some men been removed forward three times, so much had been there gained from the Sea'.[48] Similarly, Ray believed that the wide plains on either side of such rivers as the Thames and the Trent had been produced by the silting up of former arms of the sea, and that the English Fens, Egypt, China and the Low Countries had all been won from the sea by the deposition of river-borne debris.

Ray lived in an age when the notion of a general, punitive decay of Nature was fast disappearing from the scene, and as the author of *The Wisdom of God Manifested in the Works of the Creation*, he saw the masterful workings of a benevolent God even in so mundane a thing as denudation. He wrote:

> . . . the Rain brings down from the Mountains and higher Grounds a great quantity of Earth, and in times of Floods spreads it upon the Meadows and Levels, rendring them thereby so fruitful as to stand in need of no culture or manuring.[49]

The Low Countries nevertheless presented Ray with something of a problem: Varenius had recorded the existence of a bed of sand and cockle-shells 100 feet below the surface at Amsterdam, and Ray believed that all the material above the shells was riverain silt which had been deposited on a former sea-bed, but he was puzzled by the amount of silt present. 'Which yet, is a strange thing' he observed, 'considering the novity of the World, the Age whereof according to the usual Account is not yet 5,600 years.'[50]

Another persuasive writer on the subject of denudation was Thomas Burnet. His theory of the Earth may have been fanciful, but his thorough grasp of the role of denudation is revealed in the following passage from the English version of his theory published in 1684.

> [Mountains are consumed insensibly by] the Winds, Rains, and Storms, and heat of the Sun without; and within, the soaking of Water and Springs, with streams and Currents in their veins and crannies. These two sorts of causes would certainly reduce all the Mountains of the Earth, in tract of time, to equality; or rather lay them all under Water: for whatsoever moulders or is washt away from them, is carried down into the lower grounds, and into the Sea, and nothing is ever brought back again by any circulation: Their losses are not repair'd, nor any proportionable recruits made from any other parts of Nature. So as the higher parts of the Earth being continually spending, and the lower continually gaining, they must of necessity at length come to an equality.[51]

A few years later very similar views were expressed by Richard Bentley in his Boyle Lectures. In his fourth lecture Bentley was anxious, for reasons which need not concern us, to prove that the world was becoming progressively damper. He argued as follows:

... the tops of Mountains and Hills will be continually washed down by the Rains, and the Chanels of Rivers corroded by the Streams; and the Mud that is thereby conveyed into the Sea will raise its bottom the higher; and consequently the declivity of Rivers will be so much the less; and therefore the Continents will be the less drain'd.[52]

It should be stressed that Bentley was a cleric, a classical scholar, and a literary critic, but certainly not a scientist. That he should have used an argument such as this, clearly demonstrates that the role of denudation was understood far beyond the still thin ranks of the naturalists. Certainly the two last quotations reveal that the concept of base-levelling – usually regarded as a nineteenth-century innovation – was in fact accepted almost three hundred years ago.

This is a convenient point at which to note that the concept of denudation found a prominent place in the armoury of weapons used by the late seventeenth-century Christian apologists in combating the resurgent, and seemingly dangerous Aristotelean heresy of the Earth's eternity. If the Earth really were eternal, the apologists argued, then denudation would long since have destroyed all the Earth's mountains and reduced the continents to featureless plains. This argument is an ancient one, but it was very widely quoted during the seventeenth century. One writer to use it was Sir Matthew Hale, who has the doubtful distinction of being one of the last English judges to send women to the scaffold for sorcery. In a work published in 1677 Sir Matthew observed:

That if the World were eternal, by the continual fall and wearing of Waters all the protuberances of the Earth would infinite Ages since have been levelled, and the Superficies of the Earth rendred plain, no Mountains, no Vallies, no inequalities would be therein, but the Superficies thereof would have been as level as the Superficies of the Water.[53]

A few years later Thomas Burnet offered the same argument, and he was so impressed with the potency of denudation that he claimed a period of some ten thousand years would be ample for the destruction of all the Earth's present relief.[54] It had to be admitted, of course, as Sir Matthew Hale was honest enough to point out, that this use of mountains as a proof of the Earth's novity was invalidated if Nature possesses some device whereby the globe's topography is periodically revived. What is so interesting in the present context is that the concept of denudation was so generally accepted and understood during the seventeenth century that it was widely employed to bolster the Mosaic chronology.

Although the potency of subaerial denudation was generally admitted in sixteenth- and seventeenth-century Britain, no effort was made to understand the precise *modus operandi* of the denudational agents. Similarly, there are very few examples of attempts to explain specific features of the landscape in terms of denudation. Careful, detailed observations had scarcely begun to be made, and writers were much more concerned with landforms in general than with particular local examples. A few instances of early endeavours to relate actual topographical features to denudation may be mentioned as specimens of their type. In 1661, for example, Joshua Childrey related what he regarded as the barrenness of the English chalk Downs to the fact that the loose earth 'is continually washed away by great rains'. Similarly, he observed that in Gloucestershire 'the hills, and sides of hills are the most wet and clayie. The cause doubtless is the same with this, to wit, That the rains that fall, wash by degrees the uppermost mould down into the Valleys, because it is more loose and light; but leaves the under-clay behind, because more stiff and fast, and so very hardly to be tempted away'.[55] Across the Irish Sea, Dr Thomas Molyneux, a noted physician, invoked denudation when seeking to explain how the antlers of the so-called Irish Elk are always found deeply embedded in the ground. Their burial, he claimed, is the result of debris being swept down from the adjacent highlands for, he remarked, 'in a very long Course of Time, the higher Lands being by degrees dissolved by repeated Rains, and washt and brought down by Floods, covered those Places that were scituated lower with many Layers of Earth: For all high Grounds and Hills, unless they consist of Rock, by this means naturally lose a little every year of their Height'.[56]

John Ray was so convinced of the rapidity of denudation that he offered his readers the following apocryphal story gleaned during his travels:

> I have been credibly informed that whereas the steeple of Craich in the Peak of Derbyshire in the memory of some old men yet living could not have been seen from a certain Hill lying between Hopton and Wirksworth, now not only the Steeple, but a great part of the Body of the Church may from thence be seen: Which without doubt comes to pass by the sinking of a Hill between the Church and place of View.[57]

Ray's friend, Edward Lhwyd, is one other seventeenth-century writer who deserves mention for the careful attention that he paid to the landforms of a particular region. During his travels in Wales, Lhwyd reached the conclusion that the higher a mountain is, the steeper are its slopes. 'This', he wrote, 'I can ascribe to nothing else but the Rains and Snow which fall on those great Mountains, I think, in ten times the Quantity they do on the lower Hills and Valleys.' Lhwyd accepted the great sheets of scree which mantle so many cliff faces in Snowdonia as clear evidence of the potency of denudation, and he observed that in the North Wales valleys of Llanberis and Nant Ffrancon 'the People find it necessary to rid their Grounds often of the Stones which the Mountain Floods bring down'. 'I affirm', he continued, 'That by this means not only such Mountains as consist of much Earth and small Stones, or of softer Rocks, and such as are more easily dissoluble, are thus wasted, but also the hardest Rocks in Wales.'[58]

The Work of the Sea

An understanding of the erosive potential of the sea is perhaps only to be expected among an island-people, and British literature of the sixteenth and seventeenth centuries certainly abounds with references to the subject. Again priority must be given to William Bourne who discussed the origin of sea cliffs in the following passage:

> My opinion is thys, as the age of the worlde is of no small tyme, so in process of tyme the often sufferynges of the bellowes of the Seas have beaten away the feete of those hilles, that are by the sea coastes. And so undermyning it, although it were of harde stone, yet the wayght of that which was undermined hanging over, in rayny wether, or after great frost, must needes fall downe into the Sea. And then that soyle or substaunce that fell downe, in process of time was beaten or washed away agayne, by the often soussing of the bellowes of the sea, in the time of great wyndes and stormes. And then the stuffe so fallen down, being washed and consumed away, the sea doth begin to undermine it agayne, by litle and lytle.[59]

Bourne went on to explain that although the blocks and stones which fall into the sea from a cliff face may initially be angular, they are soon rounded as a result of the sea's ceaseless activity. They are, he observed, 'tossed to & fro by the waves and bellowes of the sea, a great number of them together, the one doth so fret and rubbe or grinde against the other, that it must needs rubbe or fret away al the sharpe edges of those stones, how hard soever the stone is'.[60] His understanding of marine erosion and attrition was thus just as sound as his appreciation of the work of rain and rivers.

The failure of the English topographers to interest themselves in landforms was mentioned earlier, but the ravages of the sea around the British coast did elicit some comment from them. John Norden, for example, believed that the rocks of Cornwall could never have withstood the onslaught of the Atlantic had they been less resistant.[61] His Welsh contemporary, George Owen, referred to the sea off Pembroke 'dealeinge soe unkindely with this poore Countrey as that it doth not in anye where seeme to yeld to the lande in anye parte, but in everye corner thereof eateth upp parte of the mayne'.[62] John Speed, alluding to the Welsh coast further north, wrote of the Irish Sea 'whose rage with such vehemency beateth against her bankes, that it is thought and said, some quantity of the Land hath been swallowed up by those Seas',[63] and William Camden affirmed that the same sea had similarly 'slashed, mangled, and devoured' part of the Lancashire coast.[64] Still further north, Martin Martin claimed that in his native Hebrides the islands of Coll and Tiree had been severed from each other by the fury of the Atlantic, and he suggested that in the still more exposed Outer Isles, Harris and North Uist had been torn asunder by the same agent.[65]

Some of the most interesting seventeenth-century writing on the subject of coastlines is contained in a little historical work published in Antwerp in 1605 by Richard Verstegan (alias Richard Rowlands) a Roman Catholic fugitive from Elizabethan England.[66] To Verstegan it seemed inconceivable that God could have created the continents with their present ragged coastlines, because 'Almightie God the cause and conductor of Nature, in creating the world did leave no parte of his woork imperfect or broken'.[67] He admitted that the indented coastlines of the modern continents were to some extent the result of both the Flood and seismic shocks, but he believed that the chief responsibility lay with marine erosion. Whenever coastal cliffs occur, he observed, 'it is a plain sign that the violence of the sea hath so worne and eaten out the sydes of them beaneath at the bottom; that the upper-part for want of under-propping, hath falne down'.[68] Islands too, he claimed, detract from the beauty of the Earth, and they could therefore have formed no part of the original Creation. He regarded most of them as of

55

post-diluvial origin, and he was convinced that the insulation of Britain was certainly a post-diluvial event. He claimed that the primeval Anglo-French isthmus had been breached either by an earthquake, or by the human excavation of a canal for defensive or commercial purposes, and that marine erosion had then speedily widened the original gap to form the present Straits of Dover. Before the rupture of the isthmus, he argued, the waters of the North Sea stood rather higher than those of the English Channel, so that the opening of the Straits of Dover caused a fall in the North Sea's level, and the emergence of parts of its bed. The Low Countries, and the lowlands on either side of the Thames Estuary, he suggested, are examples of the emergent floor of the old sea.

Nathanael Carpenter incorporated Verstegan's views on coastlines into his *Geography Delineated*,[69] and he affirmed his own belief that most islands have been severed from the neighbouring continents by post-diluvial marine erosion. Had the severance taken place earlier, he asked, how could the descendants of the animals in the Ark have reached their present insular habitats? Although Carpenter never doubted the erosive potential of the sea, there is in this context one noteworthy idea which creeps into his work: he believed that the sea preserves a nice equilibrium, so that its erosional achievements in one place are exactly balanced by its depositional effects in another. In this way he supposed the amount of land and sea on the globe to remain constant.[70] Others, including John Ray, cherished the same idea, and later this concept of a state of equilibrium among the natural processes assumed some significance in geomorphology. During the seventeenth century, however, the notion was of little importance, and twenty-five years after the publication of Carpenter's work, Varenius certainly had no illusions about a balance between marine erosion and deposition; indeed he was fully prepared to admit the possibility of marine planation on a global scale. 'If the Ocean continually wash away the Shores and lay them in deep Places', he wrote, 'at last all the high Parts will come down and be washed away, and the Sea come in on the whole Earth.'[71]

Although the ability of the sea to wear away the land was generally recognised during the seventeenth century, little attempt was made to explain how the sea performs its erosive work. The damage inflicted by breaking waves armed with boulders must have been obvious, but it was generally believed that tidal scour is at least as important as wave-action. Joshua Childrey, for example, wrote of the Cornish coast in 1661:

The cause of the devouring of the Land by the sea, I conceive to be its being a Promontory lying open to the merciless stormes and weather, and withall, lying in a place where two currents meet and part; I mean the Tide as it comes in, and returnes out of the Sleeve, or narrow Seas, and the Irish Seas, and Seavern.[72]

A more extreme view found expression in a letter written by Dr. Wallis, the Professor of Geometry at Oxford, and published in the *Philosophical Transactions* in 1701.[73] Wallis followed Verstegan in accepting the onetime existence of a land-bridge between England and France, but he maintained that the isthmus had been breached solely by the action of tides rushing northward up the English Channel and southward down the North Sea. The Low Countries, Romney Marsh, and the lowlands around the Thames Estuary, Wallis claimed, had all been constructed out of the debris worn from the neighbouring isthmus by the tidal scour.

It was implicitly accepted in sixteenth- and seventeenth-century Britain that with the exception of the Flood, the waves had always worked at their present level. Scripture makes no mention of any post-diluvial flooding of the continents – indeed such a possibility seemed precluded by God's promise to Noah after the Flood that 'the waters shall no more become a flood to destroy all flesh' – and the notion that sea-level might once have been higher – or lower – than at present was then quite foreign to most minds. Gabriel Plattes was an exception. In two works published in 1639[74] this now forgotten writer on agricultural subjects urged that the entire landscape of the British Isles had been moulded by the sea during some submergence other than the Flood. The hills and dales, he claimed, 'doe shew plainely the worke of the water, even as the Claw of a Bear, or a Lion, doth shew by his print that a Bear or a Lion hath been in such a place'. He continued:

... all England hath bin Sea; by the hills, and dales, and unevennesse of the ground; being evidently graven by the water, whose propertie is to weare the ground deepest, in such places where the earth is most loose, as it is in all vallies; and to spare it most, in all rockie and firme grounds; of which sort the Mountains are.[75]

In his own day, Plattes was evidently unique, but two hundred years later the notion that the topography of the British Isles had been sculpted chiefly by marine action achieved wide popularity with some of the most eminent of nineteenth-century geologists.

Our survey of sixteenth- and seventeenth-century British geomorphology is now complete. Many of the ideas then current about landforms now seem bizarre, but alongside stories of mobile mountains and a belief in a cataclysmic Deluge, there is to be found a considerable volume of sound understanding of the true nature of landforms. Nowhere is the perception of the early writers more evident than in their discussions of the work of denudation. Here, however, we encounter a tantalising problem. Because of the embryonic state of the Earth-sciences, much sixteenth- and seventeenth-century writing on landforms lacks precision, and in consequence we are repeatedly left wondering exactly what relative importance scholars then attached to denudation, as opposed to catastrophes, in the shaping of topography. They clearly had no doubts as to the ability of denudation to remodel the Earth's surface if the Parousia were sufficiently long delayed, but much less clear is the role which they ascribed to denudation in the development of present landscapes. Were they prepared to accept a fluvialistic interpretation of our landscapes, or did they regard them as essentially catastrophic in origin? No satisfactory answer can be given to this question. Had they paused to consider the matter, most early writers would perhaps have preferred a catastrophic explanation of topography, but on the other hand, they showed no tendency to belittle the achievements of denudation. It was generally admitted that all the alluvium and other debris lying upon the world's lowlands was the product of denudation, and implicit in such a belief was the notion of prodigious denudation in the world's mountain regions. Thus many seventeenth-century scholars would doubtless have been ready to admit the fluvial origin of mountain valleys, and perhaps even the circumdenudational origin of most mountain topography.

At first sight it might seem that the cramping influence of the Mosaic chronology must have inhibited any attempt at a fluvialistic interpretation of landscapes, but this was not the case. If, as was then believed, great flood-plains and deltas had all been formed from the spoils of denudation within only six millennia, then clearly, it seemed, denudation must be an extremely potent force – so potent a force that in the 1680s Burnet believed a further ten thousand years of denudation would be ample for the elimination of all the Earth's present mountain topography. Thus during the sixteenth and seventeenth centuries, the Old Testament time-scale, far from inhibiting a belief in denudation, actually encouraged the notion that denudation is a much more vigorous force than is in fact the case. The question of whether a period of six millennia – a period, let it be remembered, which seemed vastly longer in the seventeenth century than it does today – was really sufficiently long for denudation to have accomplished all that was being ascribed to it scarcely arose, for the date of the Creation and the reality of denudation were both regarded as indubitable facts, and it appeared unnecessary to examine one in the light of the other. To have conducted such an examination would to most minds have appeared just as stupid and futile as to inquire whether a week was sufficiently long to contain seven days.

Edward Lhwyd and John Ray, however, did ponder upon the question of secular rates of denudation in relation to the length of the Mosaic time-scale. The two discoveries which raised this problem in their minds have already been mentioned. Firstly, there was Lhwyd's realisation that if the boulders in Llanberis and Nant Ffrancon had rolled down from the adjacent mountains as infrequently in the past as they do today, then their accumulation on the valley floors in their present number must have occupied far longer than the period allowed by the Old Testament chronology. Secondly, there was Ray's difficulty arising from the presence at Amsterdam of 100 feet of sediments which he interpreted as riverain mud. If this sediment, Ray asked, and if all the other debris lying upon the world's lowlands and in the sea, has been swept down from the mountains within a period of less than six millennia, then why have the changes wrought by denudation during the last few centuries been so insignificant? Ray accepted the notion of the ultimate base-levelling of the continents, but the world, he noted, 'doth not in any degree proceed so fast towards this period as the force and agency of all these Causes together seem to require'.[76] True, there was the case of the hill near Craich, but elsewhere the effect of denudation seemed quite imperceptible. For reasons already explained, Ray could not solve this dilemma by proposing a vast extension of the Biblical time-scale. Instead, he advanced the idea that denudation is indeed a very rapid process, but that its ravages are largely hidden from us by some unknown renovating agent which immediately repairs the damage inflicted by rain and rivers.[77] This peculiar notion made its appearance early in the seventeenth century, but Ray was apparently the earliest scientist of repute to embrace the idea. The belief in a mysterious, restorative process is one of a number of new ideas which began to influence British geomorphology at the close of the seventeenth century, and the emergence of the belief offers a convenient moment at which to terminate this chapter.

REFERENCES

1. M. Power, *An Irish Astronomical Tract*, Irish Texts Society, XIV (London 1912).
2. Anthony Wood, *Athenæ Oxonienses*, I, p. 441 (London 1691).
3. *Nicolai Stenonis De Solido Intra Solidum Naturaliter Contento Dissertationis Prodromus.*
4. Nicolaus Steno, *The Prodromus to a Dissertation Concerning Solids Naturally Contained within Solids* (London 1671).
5. Robert Hooke, *Lectures de Potentia Restitutiva, or of Spring*, p. 48 (London 1678).
6. *Irelands Natural History*, p. 78 (London 1652).
7. Nathanael Carpenter, *Geography Delineated Forth in Two Bookes*, Bk. II, p. 166 (Oxford 1625).
8. *Ibid.*, Bk. II, pp. 165 & 166.
9. Gerardus Mercator, *Atlas*, p. 19 (Amsterdam 1636).
10. *Fodinæ Regales*, Introduction (London 1670).
11. *Geologia: or, a Discourse Concerning the Earth before the Deluge*, pp. 209 & 210 (London 1690).
12. Herbert W. Turnbull, *The Correspondence of Isaac Newton*, II, pp. 329–334 (Cambridge 1960).
13. *The First Booke of Questions and Answers upon Genesis*, p. 5 (London 1620).
14. Bernhard Varenius, *A Compleat System of General Geography*, translated by Mr. Dugdale, I, p. 129 (London 1733). All the quotations from Varenius used in this chapter are taken from this edition.
15. Bernhard Varenius, *Geographia Generalis*, Lib. I, Cap. IX, Prop. 8; Cap. XIII, Prop. 4 (Cambridge 1672).
16. *Ibid.*, Lib. I, Cap. IX, Prop. 8.
17. *Miscellanies in Verse and Prose*, p. 139 (London 1693).
18. Gabriel Plattes, *A Discovery of Subterraneall Treasure*, pp. 4 & 5 (London 1639).
19. Thomas Burnet, *The Theory of the Earth*, I, Bk. I, p. 67 (London 1684).
20. *Considerations on a Book Entituled The Theory of the Earth*, p. 30 (London 1697).
21. Carpenter, *op. cit.* (1625), Bk. II, pp. 167–169.
22. George Owen, 'The First Booke of the Description of Penbrookshire in Generall 1603', *Cymmrodorian Record Series*, I, p. 82 (London 1892).
23. *Philos. Trans.*, XXII (1700), No. 266, pp. 677–687.
24. Richard Waller, *The Posthumous Works of Robert Hooke*, pp. 413–416 (London 1705).
25. John Ray, *Three Physico-Theological Discourses*, p. 125 (London 1693).
26. William Bourne, *A Booke Called the Treasure for Traveilers*, Bk. V, p. 17 (London 1578).
27. *Speculum Mundi or a Glasse representing the Face of the World*, pp. 229–238 (Cambridge 1635).
28. Waller, *op. cit.* (1705), pp. 279–328.
29. *Ibid.*, p. 312.
30. *Micrographia*, pp. 242–246 (London 1665).
31. John Ray, *Observations made in a Journey Through part of the Low-Countries, Germany, Italy and France*, pp. 126 & 127 (London 1673).

32. Seneca, *Quaestiones Naturales*, Bk. VI, 4. See John Clarke, *Physical Science in the Time of Nero* (London 1910).

33. *An Essay Toward a Natural History of the Earth*, pp. 62 & 63 (London 1695).

34. *Philos. Trans.*, XVIII (1694), No. 209, pp. 78–100.

35. Frank D. Adams, *The Birth and Development of the Geological Sciences*, Chap. X (Baltimore 1938).

36. *De Proprietatibus Rerum*, Lib. XIV, Cap. II. The quotation is from an English translation of the early sixteenth century.

37. Power, *op. cit.* (1912), Chap. VIII.

38. Waller, *op. cit.* (1705), p. 427.

39. *A Blazyng Starre or Burnyng Beacon*, p. Aiiif (London 1580).

40. George Hakewill, *An Apologie or Declaration of the Power and Providence of God in the Government of the World*, Bk. V, p. 61 (Oxford 1635).

41. Waller, *op. cit.* (1705), p. 348.

42. *Habakkuk* 3, v. 6; *Job* 14, v. 9; *Isaiah* 40, v. 4; *Psalm* 102, v. 26.

43. Bourne, *op. cit.* (1578), Bk. V, Chap. I.

44. Carpenter, *op. cit.* (1625), Bk. II, p. 175.

45. *Ibid.*, Bk. II, p. 178.

46. Varenius, *op. cit.* (1672), Lib. I, Cap. XVI, Prop. 6 & 10.

47. Steno, *op. cit.* (1671), pp. 106 & 107.

48. Ray, *op. cit.* (1673), p. 8.

49. *The Wisdom of God Manifested in the Works of the Creation*, p. 91 (London 1701).

50. Ray, *op. cit.* (1673), p. 8.

51. Burnet, *op. cit.* (1684), I, Bk. I, pp. 37 & 38.

52. A Sermon [being the fourth of the first Series of Boyle Lectures, entitled *A Confutation of Atheism*], p. 24 (London 1693).

53. *The Primitive Origination of Mankind*, p. 95 (London 1677).

54. Burnet, *op. cit.* (1684), I, Bk. I, p. 38.

55. Joshua Childrey, *Britannia Baconica*, p. 58 (London 1661).

56. *Philos. Trans.* XIX (1695–97), No. 227, pp. 489–512.

57. Ray, *op. cit.* (1673), p. 8.

58. William Derham, *Philosophical Letters Between the Late Learned Mr. Ray and Several of his Ingenious Correspondents*, p. 255 (London 1718). Reprinted in Edwin Lankester, *The Correspondence of John Ray*, Ray Society Series, p. 242 (London 1848).

59. Bourne, *op. cit.* (1578), Bk. V, Chap. III.

60. *Ibid.*, Bk. V, Chap. IV.

61. *A Topographical and Historical Description of Cornwall*, pp. 3–5 (London 1728).

62. Owen, *op. cit.* (1892), p. 2.

63. *The Theatre of the Empire of Great Britaine*, Bk. II, p. 117 (London 1611).

64. *Britain*, p. 754 (London 1637).

65. *A Description of the Western Islands of Scotland*, pp. 51 and 271 (London 1703).

66. *A Restitution of Decayed Intelligence*, pp. 95–112 (Antwerp 1605).

67. *Ibid.*, p. 98.

68. *Ibid.*, p. 99.

69. Carpenter, *op. cit.* (1625), Bk. II, pp. 119 & 120; 184–190.

70. *Ibid.*, Bk. II, p. 184.
71. Varenius, *op. cit.* (1672), Lib. I, Cap. XVIII, Prop. 12 and 18.
72. Childrey, *op. cit.* (1661), p. 27.
73. *Philos. Trans.*, XXII (1701), No. 275, pp. 967–979.
74. Plattes, *op. cit.* (1639); and *A Discovery of Infinite Treasure, Hidden since the Worlds Beginning* (London 1639).
75. Plattes, *op. cit.*, *A Discovery of Infinite Treasure*, pp. 36 & 44.
76. Ray, *op. cit.*, p. 229.
77. *Ibid.*, p. 229.

Chapter Three

A Generation of Theorists
1668-1696

... ay, sir, the world is in its dotage; and yet the cosmogony or creation of the world has puzzled philosophers of all ages. What a medley of opinions have they not broached upon the creation of the world!

Ephraim Jenkinson
in 'The Vicar of Wakefield'.

EAGER for 'natural knowledge', and increasingly confident of their intellectual capacity, late seventeenth-century scholars found that the somewhat bald *Genesis* account of the Creation and the Flood merely whetted their appetite for a fuller understanding of these momentous events. Moses, it seemed, had pruned his narrative to the bare minimum, so that the simplest of peoples might understand the story without being overwhelmed with incomprehensible detail, but the scholars of a sophisticated age now began to demand amplification of what Moses had written. Quite oblivious to their ignorance of Nature, numerous writers plunged headlong into the task of devising comprehensive 'scientific' histories of the Earth's surface. *Genesis* of course continued to be accepted implicitly, but within the Mosaic framework scholars felt free to use their powers of reasoning to augment and rationalise the *Genesis* story. They believed themselves to be recapturing some of the detail which had doubtless been revealed to Moses, but which, at the divine command, he had later omitted from the Scriptural record. Those responsible for these early essays in Earth-history – theories of the Earth as they came to be called – were the precursors of the historical geologists of our own day, but in the seventeenth century the study of

Earth-history already seemed possessed of a venerable pedigree; the theorists saw themselves as treading in the footsteps of Moses himself, 'the greatest Natural Philosopher that ever lived upon this Earth'.

Continental authors turned to Earth-history rather earlier than did their British counterparts, and notable foreign contributions to the subject are contained in such volumes as Descartes's *Principia Philosophiæ* of 1644, and Kircher's *Mundus Subterraneus* of 1664. Surprisingly, these early European works aroused little or no interest in Britain, and it was not until the publication of Nicolaus Steno's *Prodromus* in Florence in 1669, that British scholars at last began to concern themselves with the subject.

Nicolaus Steno and the Tuscan Landscapes

Nicolaus Steno (alias Niels Steensen) was born in Copenhagen in 1638. As a student he devoted himself to medicine and anatomy, and he studied at the universities of Copenhagen, Amsterdam, and Leiden. He soon acquired a considerable scientific reputation, and about 1665 he was appointed physician to the court of Grand Duke Ferdinand II at Florence. There, in 1667, he was received into the Roman Catholic Church, and his conversion was a turning point in his life; henceforth his interest in science waned, as theology and church affairs increasingly engaged his attention. The renowned *Prodromus* of 1669[1] was his last important scientific work, and he never completed the geological treatise which the *Prodromus* was intended to introduce. In 1675 he took holy orders, and soon after, the Pope made him titular Bishop of Titiopolis and Vicar Apostolic of Hanover. After living an ascetic priestly life, and filling various ecclesiastical offices, he died at Schwerin in Mecklenburg in 1686, his passing regretted by Catholics and Protestants alike. His body, described as a package of books, was secretly returned to Florence for burial, and after lying in the crypt of the church of Saint Lorenzo for more than 250 years, his body was exhumed in 1953, carried in solemn procession through the streets of the city, and then re-interred in the same church in a side-chapel renamed the Steno Chapel in his honour.

Steno was drawn to geology through an interest in fossils. He was one of those who were convinced that these were the remains of former plants and animals, and his theory of the Earth was largely an attempt to explain how such bodies had become entombed within solid rock. He nevertheless regarded two other phenomena which he had observed in the Tuscan rocks as equally demanding of explanation: firstly, why are the fossiliferous rocks arranged in superposed strata, and secondly, why are those strata commonly bent and broken?

He claimed that all strata had been formed as a result of precipitation in water, and he suggested that rocks could be divided into two categories. On the one hand, there are those which are fine-grained, homogeneous, and unfossiliferous, and he believed that these had been precipitated from a universal fluid which covered the Earth during the early stages of the Creation. On the other hand, there are the coarse, variegated, fossiliferous rocks. These he regarded as clastic sediments of much younger age, formed on the sea-bed from debris swept off the decaying continents by rivers, and incorporating the remains of plants and animals. He claimed that among both the ancient and the newer rocks the differences between neighbouring strata were the result either of periodic variations in the type of material imported by marine currents, or of the sorting of materials according to their density, so that the heaviest debris settled first, to be followed later by successively lighter grades in ordered sequence. One aspect of the sedimentary problem, however, defeated him: like many a later geologist, he failed to explain what caused the strata to become consolidated on the sea-bed.

Steno recognised that the upper beds in any series are normally the youngest (the modern law of the order of superposition), and that at the time of their formation, strata must have been horizontal. He attributed the present disarray of the rocks either to the bending and fracturing of the crust by subterranean explosions, or to portions of the crust collapsing into the caverns beneath. All the Earth's topography had been produced, he claimed, by these two violent processes, and, like Hooke, he regarded all mountains, except volcanoes, as merely the fractured edges of tilted earth-blocks.

In the present context, much the most important part of the *Prodromus* is its conclusion, where Steno offers an illustrated account of the physiographic history of Tuscany. This pioneer essay in historical geology is particularly interesting because Steno regarded the Tuscan landscapes as a key to the whole of Earth-history. What is true of Tuscany, he claimed, is true for all our continents, and Grand Duke Ferdinand, to whom the *Prodromus* is dedicated, was no doubt flattered to learn that his tiny kingdom was a microcosm of the entire world. The six diagrams which Steno drew to illustrate his theory are reproduced in Figure 1 and they represent the following stages:

65

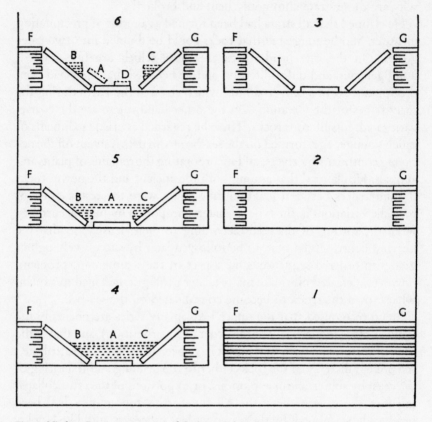

Fig. 1 *Nicolaus Steno's interpretation of the history of the Tuscan landscapes. Redrawn from Steno's* Prodromus *of 1669*

1. This is a portion of the Earth during the Creation, and before the formation of the continents. The globe is entirely submerged beneath a universal fluid from which precipitation is taking place to form superimposed, horizontal beds of sedimentary rock (F-G). In the absence of continents, plants, and animals, the individual strata are of uniform composition, and free from fossils and all other heterogeneous material.

2. The newly formed strata have emerged from the fluid to form the Earth's smooth, pristine continents. Now the chief activity is taking place beneath the continents where water is sapping the continental foundations, eating away the strata, and opening up great caverns.

3. Large areas of the continents have now been completely undermined, and the roofs of the great caverns collapse (I). The resulting valleys are drowned by the sea. This is the Flood, which occurred some four thousand years ago.

4. In the Flood waters, new sedimentary strata are formed (B-A-C). These younger rocks incorporate plant and animal remains from the ante-diluvial world as well as debris washed off the neighbouring land-masses.

5. These new rocks emerge from the waters at the close of the Flood, and are themselves undermined as a second generation of caverns develops.

6. A period of renewed crustal collapse. Admittedly this period is not specifically mentioned in Scripture, but it was this collapse which reduced the youngest of the Earth's strata to their present disordered state (A-D), and which formed the Earth's modern topography.

Steno has been widely acclaimed as one of the founders of modern geology, and certainly many novel ideas were brought together in the *Prodromus*. His recognition of the organic origin of fossils, his understanding of the true nature of sedimentary strata, his appreciation of the work of a marine transgression, his formulation of the law of order of superposition, and his illustration of what we now know as a geological unconformity, certainly makes the *Prodromus* a remarkable work for its day. Steno's English contemporaries were quick to notice the significance of the work. Martin Lister, for example, referred to the *Prodromus* in a letter to the Royal Society in August 1671,[2] and Henry Oldenburg's English translation of the work was published in the same year.[3] John Ray, who met Steno at Montpellier in the 1660s, makes frequent reference to the *Prodromus* in his own writings, and late in the century, John Woodward, one of the English theorists, was actually accused of plagiarising Steno's masterpiece.

Thomas Burnet and the Sacred Theory

From the wisdom of Steno, we must now turn to the fanciful, but ingenious and extremely popular theory of the Earth devised by Thomas Burnet, the earliest of a trio of important English seventeenth-century theorists. Burnet was a divine who, thanks to the influence of the Duke of Ormonde, became master of the Charterhouse in 1685, and his theory first appeared in two volumes published in 1681 and 1689 under the title *Telluris Theoria Sacra*. The work was much admired, not least for its eloquent Latin style, and King Charles II was so impressed that he encouraged Burnet to prepare an English edition of the theory. This appeared in two handsome folio volumes; the first, in 1684, was dedicated to the King himself, and the second, in 1690, was dedicated to Queen Mary, the consort of William III.[4]

Although Burnet had travelled widely on the continent in the company of various of the English nobility, he was a scholar of the old school. He evidently knew nothing of the natural world which his contemporaries were revealing by their experiments and field-studies, and he rested content in the belief that a full knowledge of Earth-history could be obtained by applying a leavening of reason to a scholarly understanding of the Scriptures and the classics. 'For while I went on Philosophising', Burnet wrote, 'new Light still broke in, both from the Holy Scriptures, and the Monuments of the Ancients',[5] and he quoted very liberally from both sources in support of his beliefs. In formulating his theory, Burnet worked upon the largest canvas conceivable; he not only undertook to unravel the Earth's past history, but he also foretold the Earth's future down to the establishment of the Millennium. Something of the ambitious scope of the work is revealed by the title of the English edition: *The Theory of the Earth: Containing an Account of the Original of the Earth, and of all the General Changes which it hath already undergone, or is to undergo, till the Consummation of all things*. In one chapter he offered his readers the excitement of learning 'How the Sea will be diminish'd and consum'd. How the Rocks and Mountains will be thrown down and melted, and the whole exterior Frame of the Earth dissolv'd into a Deluge of Fire', but such fancies need not concern us; we must confine our attention to that part of the theory which is concerned with the Earth's past history.

Burnet's theory arose from his attempt to answer one simple question: where is there sufficient water to have caused a Flood which submerged even the highest mountains to a depth of fifteen cubits? This problem had long taxed scholars, and Burnet estimated that the drowning of the present globe would require at least eight times as much water as is contained in the modern oceans. 'I can as soon believe, that a Man could be drown'd in his own Spittle,' wrote one of Burnet's contemporaries, 'as that the World should be deluged by the Water in it.'[6] Like most of those who had examined the problem, Burnet rejected as unscriptural the notion that the Flood had covered only the inhabited parts of the globe, and he also refused to invoke a special creation of waters followed by their miraculous post-diluvial annihilation. Led on by this will-o'-the-wisp problem, Burnet approached the question from a fresh angle: if water cannot be found in sufficient quantities to drown the present, mountainous continents, he reasoned, then it follows that at the time of the Flood the earth must have possessed a configuration completely different from that of the modern world. Here Burnet resorted to the old idea that the pristine world had been devoid of any mountain topography, and no other English writer urged this view with more vigour. The drowning of such a featureless world, Burnet claimed, needed no more water than is contained in the modern oceans.

Burnet's theory opens with an account of the formation of this mountainless, ante-diluvial globe. According to the theory, it originated some six thousand years ago from a globular mass of chaotic liquid containing 'all sorts of little parts and particles of matter, mixt together, and floating in confusion, one with another'. Slowly, the heavier material in the liquid fell towards the centre of the mass, where it consolidated to form a spherical nucleus. The remaining portions of the chaotic fluid then separated out into the heavier substances, which became the terrestrial fluids, and the lighter substances, which became the Earth's atmosphere. Further separation followed among the terrestrial fluids as what he termed the 'fat', 'oily' and 'light fluids' rose to the surface to float upon the 'lean and more Earthy' fluids which were principally water. At this stage the Earth's atmosphere 'was as yet thick, gross, and dark; there being an abundance of little Terrestrial particles swimming in it still', but gradually these particles fell to Earth, where they mixed with the fat and oily fluids and congealed to form a thick hard skin on the surface of the terrestrial fluids. This was the original crust of the Earth. It sealed almost all the terrestrial fluids securely within the globe, and thus created the watery abyss, or great deep, referred to in *Genesis*. How else, Burnet asked, can we explain the Psalmist's words that God founded the world 'upon the seas, and established it upon the floods', or that He 'stretched out the earth above the waters'?[7] (Plate II.)

Burnet reasoned that since the precipitation of earthy particles was everywhere equal, then the crust formed above the abyss must have been universal and devoid of any relief. Mountains, valleys, lakes, ocean basins, islands, and all other topographical features, he claimed, were quite unknown in the ante-diluvial world. The Earth was then just a perfectly smooth, oceanless ball. He discoursed at some length upon the delights of the ante-diluvial world, and he reveals himself as a thorough-going disciple of the doctrine that the universe is today in a senile and decadent state.

Surprisingly, Burnet believed that the ante-diluvial Elysium contained the seeds of its own destruction. The light rainfall, he claimed, was insufficient to prevent the Earth's skin from drying out and cracking beneath the powerful rays of the youthful Sun. Similarly, the Sun's heat vaporised some of the fluids in the abyss, and the vapours caused further crustal cracking as they sought to escape from their prison – a process which he likened to the bursting of the shell of a boiled egg by the air and moisture trapped within. It was at this stage, he claimed, about 1600 years after the Creation, that God became angry with man for his evil ways and resolved to engineer a flood which would destroy the whole world save for the Ark and its contents.

He argued that the rain of forty days and nights mentioned in *Genesis* could have contributed very little water to the Flood, and he urged that the real cause of the catastrophe was the opening of the 'fountains of the great deep' mentioned by Moses. He interpreted the passage as meaning that God now allowed the great crustal cracks to extend down into the abyss so that the immense volumes of water trapped there were suddenly released. The rain of forty days and nights nevertheless had its part to play, for it facilitated the final extension of the cracks 'not only by softning and weakning the Arch of the Earth in the bottom of those cracks and Chasms which were made by the Sun, and which the Rain would first run into, but especially by stopping on a sudden all the pores of the Earth, and all evaporation, which would make the Vapours within struggle more violently'.

He presented his readers with a graphic and awesome account of the scene as the turbulent waters escaped from the gloomy abyss via the huge crustal fractures, and he gave full vent to his considerable literary powers as he described the Flood surging over the globe in tidal waves of such magnitude that only divine providence saved the Ark from annihilation. So great was the shock, he claimed, that the whole Earth was shaken on its axis, which now, for the first time, became inclined to the plane of the ecliptic. The consequential seasonal variation of climate, it seemed, was but one detrimental effect of the diluvial catastrophe.

Burnet belonged to the school of thought which maintained that the Earth had undergone a complete diluvial metamorphosis, and he regarded the Earth's present topography merely as wreckage revealed as the Flood waters abated. An orator, Burnet observed, might describe the Earth as 'a beautiful and regular Globe', and as 'the darling and favourite of Heaven', but as a philosopher he believed that he saw the world for what it really is – a gigantic and hideous ruin. Locally, he admitted, the Earth may possess some beauty, but viewed in general ' 'tis a broken and confus'd heap of bodies, plac'd in no order to one another'. According to the theory, the ruin of the Earth had taken place when the exit of the waters from the abyss left the already deeply fissured crust largely unsupported, so that the individual crustal blocks were free to subside into the void beneath. Such foundering could not everywhere be of the same magnitude, because of the shortening of the Earth's radius involved, and it was to this differential collapse that Burnet attributed all the Earth's present topography. He regarded mountains as regions of minimum collapse, and ocean basins, at the other extreme, as areas where the foundering had taken place on such a vast scale that the original crust had completely disappeared into the abyss, leaving the abyssal waters exposed to view. The ragged coastlines of the modern continents, he claimed, are the result of the subsiding oceanic-blocks being torn away from their slightly more stable continental neighbours, and islands are merely splinters of the latter blocks torn off during the catastrophe. This theory, Burnet boldly asserted, offers the only reasonable explanation of the origin of the Earth's topography.

Burnet likened the fate of the primeval world to the collapse of a building, and he observed that just as there are vacuities amidst fallen masonry, so too there are great caverns in the terrestrial ruins. These caverns played an important part in his theory; they are, he claimed, great reservoirs into which the water descended as the Flood abated, thus allowing dry land to re-appear. This retreat of the waters, he urged, continued long after the main body of the Flood had disappeared, and all lakes, bogs, and fens are a legacy of the Flood testifying to its very recent departure. He even suggested that the recession of the sea from some coasts in modern times was a result of further diminution of the oceans caused by water still finding its way into empty caverns very deep in the crust.

In devising his theory, Burnet probably owed more to Descartes than he cared to admit; the Frenchman had already invoked differential crustal collapse into a watery abyss as an explanation of the Earth's topography.[8] To most of its English readers, however, the theory seemed both novel and highly ingenious. It was lavishly presented with much literary and dialectical skill, and even some of the most eminent scholars were quite blinded to the theory's absurdities. The theory, as we have seen, had its patrons in the Royal Family, and praise was showered upon it from many quarters. Samuel Pepys was one of the earliest to read the English edition, and in June 1684 he wrote to John Evelyn thanking him for having drawn his attention to the work. Pepys felt the whole hypothesis to be so ingenious and rational 'that I both admire and believe it at once'.[9] The theory was the subject of a laudatory article in an early number of *The Spectator*,[10] and even Newton was impressed; he discussed the theory with Burnet before its publication, and in January 1680–81 he wrote to the theorist to say that of the origin of the sea, rocks, and topography 'I think you have given the most plausible account'.[11] Something of the theory's popularity can be gauged from the fact that it ran through ten editions before 1759. The eleventh edition appeared in 1826, and Burnet's stately prose earned the work a reprint as recently as 1965.

In spite of this acclaim, the theory was not without its critics. It needed no great scholarship to discern that the theory is in conflict with Scripture at a number of important points. Burnet's conception of an ante-diluvial world devoid of oceans and mountains, for example, was difficult to reconcile with a specific reference to the creation of seas and marine creatures in the Mosaic hexaëmeron, or with Moses's assurance that the Flood waters rose to stand fifteen cubits above the tops of the mountains. The theory's incompatibility with *Genesis* was pointed out by Herbert Croft, the aged Bishop of Hereford, in 1685. The Bishop branded the theory as a 'Philosophick Romance' full of 'extravagant fancies', 'vain Fopperies', and 'fabulous Inventions', and of Burnet himself he remarked: 'either his Brain is crakt with over-love of his own Invention, or his Heart is rotten with some evil design'.[12] This evil design, the Bishop claimed, could only be the subtle subversion of both the Scriptures and the Church. Erasmus Warren, another of Burnet's clerical critics, advised the theorist that no human mind could hope to fathom the mysteries of Earth-history. 'Though we have Moses', Warren wrote, 'yet I believe we must stay for Elias, to make out to us, the true Philosophical modus of the Creation, and Deluge'.[13]

The theory's most able critic was John Keill, the Oxford mathematician, and it was he who first fully demonstrated the theory's absurdity.[14] Keill carefully dissected the theory with all the scientific tools at his disposal, and he convincingly demonstrated the fanciful nature of the various processes which Burnet had invoked. The theorist himself was unabashed, but his reply to Keill only evoked another, and more vigorous critique from the Oxford mathematician.[15] This controversy between Burnet and Keill is of some interest as revealing two conflicting seventeenth-century attitudes towards Earth-history. On the one hand, there was Burnet, scientifically ignorant, yet anxious to offer a rational 'scientific' account of the incidents narrated in the early chapters of *Genesis*. On the other hand, there was Keill, a deeply religious Newtonian, who was convinced that many Scriptural events are not amenable to scientific explanation. Indeed, Keill thought it impious to import science into religion, and to employ mundane natural processes to account for events which the Bible implies were miraculous. In particular, Keill warned Burnet that to show how the Flood could have been caused by ordinary mechanical processes, without any divine intervention, could only encourage the atheists in their belief that God was a superfluous concept. Thus, paradoxically, Keill, the scientist, was eager to invoke supernatural forces in explanation of events in Earth-history, while Burnet, the cleric, was equally eager to demonstrate the entire sufficiency of natural processes. Keill was certainly not alone in his adherence to the old view that Earth-history has been controlled by periodic divine intervention, and his dispute with Burnet serves as a reminder that even at the end of the seventeenth century there were still those who rejected any theory of the Earth which banished the miraculous and employed only secondary causes. For Keill

73

and his kin *Genesis* remained an entirely satisfactory account of Earth-history; it needed neither explanation nor amplification.

Bishop Croft, in his critique of Burnet's theory, urged the ecclesiastical authorities to bring the theorist to book for his heterodoxy, but Burnet's peculiar interpretation of the Scriptures initially did nothing to harm his reputation. Indeed, shortly after the appearance of the second volume of the English edition of the theory, Archbishop Tillotson secured Burnet preferment to the office of Clerk of the Closet at the court of William III, and one historian asserts that Burnet was once considered a likely successor to Tillotson at Canterbury.[16] In 1692, however, Burnet, perhaps emboldened by the favourable reception widely accorded to his somewhat unorthodox theory, published a new work in which he openly rejected the literal interpretation of *Genesis*, argued that the story of the Fall was mere allegory, and suggested that the days of the hexaëmeron were not days in the normal sense of the word.[17] Burnet had now completely misjudged the intellectual climate of his day. He was deemed to have far overstepped the permitted limits of free-thought, and his hopes for ecclesiastical preferment were dashed. He apologised for any offence he might have given, but feeling against him was so strong that the king was obliged to remove him from his office at court, and Burnet spent the closing years of his life quietly in the Charterhouse where he died in September 1715.

John Woodward and a Liquefying Flood

Burnet's theory encouraged others to speculate about the Earth's origin and subsequent history, and it was under the stimulus of the *Sacred Theory* that John Woodward, in 1695, published *An Essay Toward a Natural History of the Earth*. Woodward was a layman without a university education. He was born in Derbyshire in 1665, and is said to have come of a good family, although some have doubted this because when he left school, Woodward aspired to no higher station than an apprenticeship to a London linendraper.[18] The London apprentices of the day, however, were a respectable and privileged group, and Woodward certainly seems to have moved in high circles, for his scholarly habits attracted the attention of Dr. Peter Barwick, the personal physician to King Charles II. Barwick took Woodward into his own household and gave him tuition in medicine and other subjects. By 1692 Woodward had acquired sufficient reputation – or influence – to ensure his election to the Chair of Physic at Gresham College. This post he held until his death in 1728. Woodward was elected to Fellowship of the Royal Society in 1693, and a few years later both the Archbishop of Canterbury and the University of Cambridge conferred upon him the degree of M.D. Thus from a somewhat unlikely beginning, Woodward enjoyed a meteoric rise to fame and honour in the world of science, and his advancement is all the more remarkable in view of the fact that he seems to have been a rather repulsive character. He was renowned for his eccentricities, pomposity, irritability, and bad manners, and his personal vanity was such that he had mirrors fixed in all his rooms so that he might lose no opportunity of gazing at himself. To one of his contemporaries he was 'that ill-natured piece of formality',[19] and to another that 'affected and pedantic mountebank'.[20] He was frequently the butt of the wits of his day, and during his later years he seems to have been widely disliked. When, as early as 1700, a scurrilous anonymous pamphlet appeared ridiculing the Royal Society, Woodward was immediately suspected of being the author, although he warmly – and it seems honestly – denied the charge. Ten years later, he was again in trouble with the Society when he was expelled from the Council after insulting the Secretary and refusing to apologise. Woodward claimed that he had been provoked by the Secretary pulling 'grimaces very strange and surprising' at him across the council table, but the action he brought against the Council for re-instatement was unsuccessful.[21]

Woodward was a man of many parts; in addition to his professional interest in medicine, he dabbled in archæology and ethnology, and his botanical studies give him some claim to be regarded as one of the founders of experimental plant-physiology. It is Woodward the geologist, however, who is best remembered by historians of science. Woodward's interest in Earth-history was first aroused by the chance discovery of some fossils in the rocks (the Lias and the Oolite) around Sherbourne, in Gloucestershire, during a visit to the seat of Sir Ralph Dutton, Dr. Barwick's son-in-law.[22] Woodward immediately resolved to travel throughout England 'to get as compleat and satisfactory information of the whole Mineral Kingdom as I could possibly obtain'.[23] Unlike Burnet, Woodward was a naturalist of the new school, eager to come to grips with Nature in the field. 'Observations', he wrote, 'are the only sure Grounds whereon to build a lasting and substantial Philosophy',[24] and he was evidently the first fully to appreciate the vital importance of field-work in geological studies. His enlightened

75

approach to the subject is apparent from those of his writings where he discusses the techniques necessary for the field-geologist. He considers, for example, the observational methods to be used in mines, and the best localities in which to search for exposures, while he stresses the importance of using a field-notebook, of measuring the dip of strata, and of carefully labelling all specimens collected.[25] In his *Essay* of 1695 this pioneer English field-geologist has left the following account of the method he used as he worked his way systematically through the English countryside:

> I made strict enquiry wherever I came, and laid out for intelligence of all Places where the Entrails of the Earth were laid open, either by Nature (if I may so say) or by Art, and human Industry. And wheresoever I had notice of any considerable natural Spelunca or Grotto; any digging for Wells of Water, or for Earths, Clays, Marle, Sand, Gravel, Chalk, Cole, Stone, Marble, Ores of Metals, or the like; I forthwith had recourse thereunto: and taking a just account of every observable Circumstance of the Earth, Stone, Metal, or other Matter, from the Surface quite down to the bottom of the Pit, I entered it carefully into a Journal, which I carry'd along with me for that purpose. And so passing on from Place to Place, I *noted* whatever I found memorable in each particular Pit, Quarry, or Mine: and 'tis out of these *Notes* that my Observations are compiled.[26]

Woodward built up a large museum of minerals, rocks, and fossils by his own collecting and by purchase, and at the end of his life he claimed to have in his cabinets more than 2,800 different geological specimens from England alone.[27] The published catalogue of his collection certainly shows him to have possessed a remarkable knowledge of English geology,[28] but among his contemporaries the collection was a cause for some ridicule. When Woodward was lampooned at the Theatre Royal in 1717 in John Gay's vulgar comedy *Three Hours after Marriage*, he was portrayed in the guise of Dr Fossile, 'the Man that has the Raree-Show of Oyster-shells and Pebble stones', and who is delighted at the prospect of adding to his museum a monstrous antediluvian knife, fork, and spoon recently dug out of a Babylonian mine.

On completing his geological travels in England, it was Woodward's intention to make a geological excursion on the continent, but his plans were upset by the outbreak of the War of the Grand Alliance in 1689.

Thwarted, but undismayed, he devised a geological questionnaire which he circulated among experienced travellers, and the replies convinced him that the conclusions he had already drawn from his own field-studies were of universal validity. Content in this belief, he in 1695 published his theory of the Earth, in order, he explained, to satisfy the curiosity and demands of his friends. The book was intended merely as an introduction to a larger work which Woodward promised would follow, and in many places in his theory he asks his readers to accept dogmatic assertions pending the appearance of full, reasoned arguments in the later volume. Like Steno – and like many another author who has promised great things to come – Woodward failed to produce his *magnum opus*. In 1697 he was reported to be hard at work on it,[29] and a few papers said to have been detached from the projected volume were published during his lifetime,[30] but the book was still far from complete when Woodward was buried in Westminster Abbey in 1728.

Woodward assured his readers that in formulating his theory he had been 'guided wholly by Matter of Fact', and he claimed that he used no material 'but what hath due warrant from Observations; and those both carefully made, and faithfully related'. He no doubt sincerely believed that he had achieved this dispassionate approach to Earth-history, but in fact his mind was so conditioned by the bibliolatry of the age that he achieved nothing more than the examination of Nature through a pair of Mosaic spectacles. His field observations did, nevertheless, lead him to two important, if not very original, conclusions. Firstly, he discovered that much of the Earth's surface is underlain by strata of various types, the strata being separated from one another by what he termed 'horizontal or parallel Fissures'. Secondly, his experience of fossils convinced him that they are, as he put it, 'the very Exuviæ of Animals' which had somehow become entombed within rocks. His knowledge of fossils was sufficient to raise a novel problem in his mind: how does it come about that the English fossils are a heterogeny of life-forms drawn from all parts of the globe, so that the remains of native organisms are found in juxtaposition to those of elephants and tropical fish? Such a mixture of fossils, Woodward believed, could not possibly be accounted for by merely invoking changes in the distribution of land and sea. He felt that a much more drastic process must have been involved, and his whole theory stemmed from an attempt to explain how various parts of the Earth had acquired a coat of strata containing such a varied assortment of native and exotic fossils.

Woodward's theory, like Burnet's, had the Flood as its centrepiece, and it is upon Woodward's version of the diluvial metamorphosis that our attention must focus. He agreed with Burnet that the Flood waters had emerged from an abyss deep within the globe, but the two theorists differed radically in their views about the way in which the Flood had destroyed one world and created another. Woodward's interpretation of events during the Flood is best told in his own words, although it must be borne in mind that the word 'fossil' was then used not in its modern, restricted sense, but with reference to any organic remains, rocks, or minerals dug from the earth. He explained:

> That during the time of the Deluge, whilst the Water was out upon, and covered the Terrestrial Globe, All the Stone and Marble of the Antediluvian Earth: all the Metalls of it: all Mineral Concretions: and, in a word, all Fossils whatever that had before obtained any Solidity, were totaly dissolved, and their constituent Corpuscles all disjoyned, their Cohæsion perfectly ceasing. That the said Corpuscles of these solid Fossils, together with the Corpuscles of those which were not before solid, such as Sand, Earth, and the like: as also all Animal Bodies, and parts of Animals, Bones, Teeth, Shells: Vegetables, and parts of Vegetables, Trees, Shrubs, Herbs: and, to be short all Bodies whatsoever that were either upon the Earth, or that constituted the Mass of it, if not quite down to the Abyss, yet at least to the greatest depth we ever dig: I say all these were assumed up promiscuously into the Water, and sustained in it, in such manner that the Water, and Bodies in it, together made up one common confused Mass.[31]

When the Flood had completed this work of destruction, and had become a thick, turbid liquid holding the surface of the ante-diluvial world in solution and suspension, the second, creative phase of the Flood began. Slowly, day by day, Woodward maintained, the debris was released from the liquid and sank down until it reached the globe's insoluble core. There the debris immediately consolidated to form the rocks of a new world. According to Woodward, this precipitation took place in a carefully ordered sequence, the order being 'that of the different specifick Gravity of the several Bodies in this confused Mass, those which had the greatest degree of Gravity sinking down first, and so setling lowest; then those Bodies which had a lesser degree of Gravity fell next, and settled so as to make a Stratum upon the former; and so on, in their several turns, to the lightest of all, which subsiding last, settled at the Surface, and covered all the rest'.[32] In this way he explained the existence of variegated strata superimposed upon each other, and he invoked the same sorting by density to account for the distribution of fossils within the rocks. He regarded fossils as the remains of ante-diluvial plants and animals which had for some reason proved immune to the Flood's solvent action, but which had been taken up into suspension from various parts of the world and then mixed together in the chaos of the turbid liquid. Finally, they too were sorted and precipitated according to their densities along with the rock-forming debris. Thus, Woodward claimed, fossils of the same density, and therefore of the same type, were precipitated contemporaneously, and in this manner he sought to explain his important field-observation that certain types of fossil characterise particular strata.

Woodward believed that all the rocks formed during the Deluge had been laid down in expansive, horizontal sheets, and he attributed the present ruptured and disordered state of the strata to late-diluvial earth-movements, although he failed to explain the cause of these earth-movements, just as he failed to explain so much else connected with his theory. These earth-movements are nevertheless important in the present context because Woodward believed them to have been responsible for forming all the Earth's present topography. Like Steno and Burnet, he regarded mountains merely as tilted earth-blocks, and he wrote that 'the more eminent Parts of the Earth, Mountains and Rocks, are only the Elevations of the Strata; these, wherever they were solid rearing against and supporting each other in the posture whereinto they were put by the bursting or breaking up of the Sphere of the Earth'. Only regions underlain by strong rocks, he noted, have the strength to stand up in this way to form mountains. 'Countries which abound with Stone, Marble, or other solid Matter', he observed, 'are uneaven and mountainous: and that those which afford none of these, but consist of Clay, Gravel, and the like, without any Stone etc. interposed, are more champaign, plain, and level.'[33]

Although Burnet and Woodward both believed that the Earth had undergone an immense diluvial transformation, the two theorists had very different conceptions of the Deity, and this caused their views on the Flood to differ in one extremely important respect. Burnet adhered to the older, Puritan view of God, and regarded Him as mankind's angry, vengeful judge who had ordained that there should be a progressive, punitive decay of the entire universe. He saw the Deluge as merely one terrible episode in this universal degeneration of Nature,

79

and he believed that man was condemned to eke out his miserable present existence amidst the grotesque ruins of the magnificent ante-diluvial world which his ancestors had so grossly misused. Woodward, on the other hand, writing more than a decade later, lived at the dawn of a new era. After the Restoration, the old Puritan conception of the Deity rapidly faded, and with it there passed the belief in a decadent universe. In place of the wrathful God of the Puritans, there emerged a new conception of the Deity as a benign, compassionate Being who was the architect of a magnificent Creation. This was the view to which Woodward subscribed. Like many of his contemporaries he leaned strongly towards a teleological interpretation of Nature, and because of this he felt impelled to seek the divine benevolence even in the tumult of the Flood. He felt it inconceivable that a merciful God could have used the Flood solely as a vicious, destructive vehicle for the divine vengeance, and he firmly rejected both Burnet's belief that the Flood had been an unmitigated catastrophe, and his claim that the present Earth is a chaotic ruin utterly devoid of plan.

Woodward urged that the primeval world had been fashioned as a home for man in his state of pristine innocence, but that after the Fall this world had proved increasingly unsuited to human needs. Burnet had regarded fallen man as undeserving of the ante-diluvial Elysium, but in Woodward's mind there was no question of deserts; he felt merely that the primeval paradise and an evil mankind had in the long run proved incompatible, and he saw the Flood as the instrument of re-creation through which God had given man a world more in keeping with his enfeebled state. Even amidst the ferment of the Deluge, he claimed, God, 'a most wise and intelligent Architect', was working with 'exquisite Contrivance and Wisdom' to shape a new world containing no 'Blemishes, no Defects: nothing that might have been altered for the better: nothing superfluous: nothing useless'.[34] To Woodward it must have seemed almost as though God Himself had erred in creating a primeval world which had eventually proved an unsuitable human habitat. Be that as it may, he certainly believed that when considered in relation to man's fallen state, the post-diluvial world was in many respects an improvement upon its predecessor. So obsessed was he with the Flood as an agent of re-creation, that he almost completely overlooked those punitive aspects of the event which had been so strongly emphasised by Burnet and other writers. Woodward's interpretation of the Flood is epitomised in the following passage:

For the Destruction of the Earth was not only an Act of the profoundest Wisdom and Forecast, but the most monumental Proof, that could ever possibly have been, of Goodness, Compassion, and Tenderness, in the Author of our Being; and this so liberal too and extensive, as to reach all the succeeding Ages of Mankind: all the Posterity of Noah: all that should dwell upon the thus renewed Earth to the End of the World; by this means removing the old Charm: the Bait that had so long bewildered and deluded unhappy Man: setting him once more upon his Legs: reducing him from the most abject and stupid Ferity, to his Senses, and to sober Reason: from the most deplorable Misery and Slavery, to a Capacity of being happy.[35]

Woodward's adoption of the new conception of the Deity, his rejection of the idea of a progressive decay of Nature, and his regard of the Flood as an agent of re-creation, forced him to a fresh assessment of the post-diluvial changes on the Earth's surface. If the new world which emerged from the Flood waters had been carefully designed to meet human needs, and if it was man's gift from a beneficent God, then it followed that any post-diluvial changes – whether caused by earthquakes, volcanoes, or denudation – could only have blurred the fabric of the divine masterpiece and made it less capable of serving its intended purpose. He was convinced that no such changes could have been allowed to occur. In particular he protested vigorously at the belief that post-diluvial earthquakes and volcanic outbursts have played a major part in shaping the Earth's present topography. 'I must needs freely own', he wrote, 'that when I first directed my Thoughts this way, 'twas a matter of real Admiration to me, to find that a Belief of so many, and such great Alterations in the Earth, had gained so large footing, and made good its ground so many Ages, in the World; there being not the least signs nor footsteps of any such thing upon the face of the whole Earth: no tolerable Foundation for such a belief either in Nature, or History.'[36] Woodward deserves much credit for this bold rejection of a popular belief, and as we have seen already, he scolded his contemporaries sharply for their blind adherence to the ancients on the subject of earthquakes and volcanoes.

Much less commendable were his views on denudation, or deterration, as he termed it. Like his contemporaries, he was thoroughly familiar with the concept of denudation, but it presented him with a grave problem. Denudation was clearly a destructive, and therefore, he supposed, evil process, and he found difficulty in understanding how such a process could possibly exist within the order of a beneficent Deity. He was anxious to admit to his theory only those processes which seemed beneficial to mankind, and he observed that the changes which earlier writers had attributed to agents such as denudation 'are without use, and have no end at all, or, which is worse than none, a bad one: and tend to the damage and detriment of the Earth and its Productions'.[37] If, as his teleological view of Nature led him to believe, all mountains, valleys, and plains had been carefully moulded during the Deluge as integral parts of a wonderful Creation, then surely they would never be allowed to moulder away. For this reason he was very strongly tempted to reject the whole idea of denudation, but a consideration of his attempt to grapple with this problem is best reserved for the next chapter where it can be considered in its full historical perspective.

Woodward's theory, like Burnet's, received a mixed reception.[38] Some of its critics claimed that it ran counter to *Genesis*, and they were quick to draw attention to Woodward's many inconsistencies and dialectical failings. Others approached the theory experimentally, and after examining rock specimens from different mine-levels, they were able to cast doubt upon Woodward's claim that the heavier rocks always lie deepest in the crust.[39] Woodward was even accused of plagiarising Steno who, it was pointed out, had already suggested that different strata might have originated through the sorting of debris according to density. In his correspondence John Ray deplored the arrogant and presumptuous manner in which the theory had been presented,[40] and Edward Lhwyd regarded the theory as so full of absurdities as to be scarcely worthy of attention. He wished that Woodward's friends, instead of offering encouragement, had persuaded him 'to forbear troubling ye world with such whimsies' and expressed the hope that Woodward's work 'will make men preferre Natural History, to these romantic theories, which serve to no other use, but to give us some shew of ingenuity in ye inventors'.[41] Another of Woodward's detractors likened the theory in satirical verse to some culinary operations he had witnessed:

> *Thus I've observed*, pro re natâ
> *A Kitchen-Wench of Bread lay Strata,*
> *Eggs, Suet and Plums in plenteous store;*
> *But in a Moment of an Hour,*
> *Milk in a Deluge vast comes flowing,*
> *And dissipates all she'd been doing.*[42]

Woodward evidently corresponded with Leibnitz on the subject of the theory, but otherwise he made little attempt to defend his conception of Earth-history. He was perhaps sufficiently vain to feel the critics beneath his notice, and he certainly had the satisfaction of knowing that his work was being widely read both in Britain and on the continent; three English editions of the theory and one Latin edition appeared during his lifetime, and soon after his death it was translated into French and Italian.

Woodward undoubtedly deserves more credit than he has been accorded by historians of science. Because of the fanciful nature of his *Theory*, he has been consigned to the same compartment of scientific history as Burnet, but this is hardly a fate which Woodward deserves. He should be remembered, rather, as the first geologist to work his way systematically through the English countryside, seeking out exposures, collecting specimens, and building up from his notebooks a picture of the geology of England which must have been unique in its day. As we have seen, he was well aware that some strata contain characteristic fossils, and had he been able to shed his Mosaic shackles, he might have perceived some of the implications of his discovery. Woodward might in fact have become the founder of modern stratigraphy more than half a century before the birth of William Smith, but the grip of Moses was too strong; instead of becoming the father of English geology, Woodward merely became the first of a long line of English field-geologists who made valiant attempts to harmonise their observations of Nature with the Mosaic record. It is nevertheless fitting that Woodward's name should be perpetuated in the title of Britain's oldest academic post in the Earth-sciences. When he died, his will directed that most of his estate should be sold, and part of the proceeds used to buy sufficient land to yield an annual income of £150. This land was to be conveyed to the University of Cambridge which was instructed to use most of the income to pay the salary of a lecturer who, in addition to serving as the curator of the collection of geological specimens which Woodward bequeathed to the University, should also read at least four lectures annually upon any subject treated of in Woodward's works. Thus was born the Woodwardian Chair of Geology.

William Whiston and the Comets

The third of the seventeenth-century English theorists was William Whiston, a young, but distinguished Cambridge mathematician who in

1696 published a theory of the Earth, price six shillings bound, under the following title: *A New Theory of the Earth, from its Original, to the Consummation of all Things. Wherein the Creation of the World in Six Days, the Universal Deluge, and the General Conflagration, as laid down in the Holy Scriptures, are shewn to be perfectly agreeable to Reason and Philosophy.* Whiston too had been profoundly impressed by Burnet's theory, and he offered a vindication of it as part of the exercises for his degree at Cambridge.[43] Further study and reflection, however, convinced him that the theory was not entirely satisfactory, and he reached the conclusion that its chief deficiency arose through Burnet having completely overlooked the important role of comets in Earth-history. Whiston believed that the globe had been fashioned from a comet, that the close approach of a comet had caused the Flood, and that at the end of time the proximity of yet another comet would cause the Earth to be consumed in a vast conflagration. In devising his own theory, Whiston merely sought to improve upon Burnet's work by allowing comets to assume what he believed to be their rightful place in the Earth's story.

The theories of Burnet and Whiston differ only slightly so far as the general structure of the Earth is concerned. Whiston accepted, for example, that the continents are suspended over a watery abyss which came into existence at the Creation as the heavy chaotic liquids sank towards the centre of the sphere, leaving the lighter solids to form the Earth's crust. His explanation of topography, on the other hand, was novel; he attributed all relief features to density differences which became apparent as the crust solidified. In a manner strikingly reminiscent of the modern theory of isostasy, he pictured the continents as being composed of a number of blocks, or columns, floating in the abyss so that the heavy columns sink deeply into the fluid to form valleys and plains, while the lighter columns are barely immersed and stand up as mountains. The higher a mountain is, he observed, 'the more rare, porous, and light its Column; and the lower any Valley, the more fix'd close, dense, and solid its Column must needs be suppos'd'.[44] He adduced two proofs of his contention that mountains are more hollow, and therefore less dense, than lowlands. Firstly, he pointed out that volcanoes and earthquakes occur chiefly in mountainous regions, thus proving that subterranean vaults are more numerous there than beneath plains. Secondly, he claimed that all rivers are nourished by water ascending from the abyss, and the fact that rivers rise amidst mountains, he urged, proves that 'the Vapours appear to have a more free and open vent or current up the Mountainous Columns, than the neighbouring ones'.[45] Whiston believed that these density differences had caused the Earth to be diversified with topography ever since the Creation, and he rejected Burnet's notion of a perfectly smooth ante-diluvial globe. He did nevertheless have doubts about the existence of ante-diluvial ocean basins, because their formation would have required the accumulation of high-density material over vast tracts of the Earth's crust, and he saw no reason why differential accumulation should have occurred on such a scale.

For Whiston, as for the other theorists, the Flood loomed large as by far the greatest cataclysm in Earth-history, but here he introduced his other touch of novelty; whereas Burnet and Woodward had conjured the Flood waters up from within the Earth, Whiston sought them in the heavens. He claimed that on 28th November in the year 2349 B.C. a comet passed so close to the Earth that some of the vapours in the comet's tail condensed and fell to Earth as a torrential downpour. This he believed to have been the chief cause of the Flood, and he later decided that the comet responsible was the very one which had flashed across the English skies in 1680 and earned itself a discussion in the *Principia*.[46] The remainder of the Flood waters came, as described in *Genesis*, from the opening of the fountains of the deep, and in this event, too, the comet played a part. Its approach, he claimed, set up stresses which caused the Earth's crust to crack, whereupon the weight of Flood waters already present depressed some of the fracture-bounded earth-blocks and squeezed the water up out of the abyss. He clearly invoked this process in an effort to preserve some conformity between his theory and *Genesis*, but after criticism, he conceded its impossibility, and in the later editions of the *Theory* he relied more heavily upon the comet's tail as a source of the Flood waters. He explained the eventual disappearance of the Flood as being partly the result of evaporation, and partly the result of the fluid sinking back down the crustal cracks as the diminishing weight of the Flood waters allowed the abyss to assume something like its former proportions. The abyss was unable to accommodate all the waters, however, and he regarded the modern oceans as a diluvial residue.

In some respects Whiston borrowed from Woodward. He held, for example, that during the Flood the ante-diluvial world had been completely buried beneath debris precipitated by the turbid flood waters, and that this debris had consolidated to form horizontal strata arranged in a density sequence. Similarly, he explained the present disarray of the strata as being the result of the crustal blocks being raised and lowered during the withdrawal of the Flood. On these subjects, however, Whiston's writing is somewhat vague, and in the present context his *Theory* is less important than the treatises of Burnet and Woodward.

Whiston lacked Burnet's persuasive eloquence, and his theory is tediously and ineffectively presented. Ray thought it 'pretty odde & extravagant'[47] and, like Burnet, Whiston attracted the incisive attention of John Keill, who demonstrated that the theory could be reconciled neither with Moses nor with physical science.[48] Whiston nevertheless had his enthusiastic supporters; John Locke praised the *Theory*, and observed that 'I have not heard any one of my Acquaintance speak of it, but with great Commendation as I think it deserves',[49] and even Newton signified his approval.[50] The *Theory* ran through many editions, and it certainly did nothing to damage Whiston's reputation as a scientist, because in 1701 he became Newton's deputy at Cambridge and two years later he succeeded Newton as Lucasian Professor. His tenure of the chair was nevertheless short; he became an Arian, and the University expelled him for his heterodoxy in 1710. He devoted the remaining forty-two years of his long life to the advancement of primitive Christianity. A stream of religious pamphlets and books came from his pen, and he must have cut a rather pathetic figure as he travelled through England with the models of various Jewish temples which served to illustrate the lectures he eagerly delivered to such as would listen.[51] His interest in science survived his changed circumstances, and he continued to give public lectures on comets, meteors, and earthquakes, but when he died in 1752 he was an outcast from scientific society, excluded from the Royal Society, he claimed, by the opposition of his erstwhile friend, Newton. Perhaps Whiston's most permanent memorial is in the *Vicar of Wakefield*, because it seems likely that Goldsmith had Whiston in mind as he moulded the character of Dr Primrose.

Robert Hooke and a Cyclic Earth-History

The theories of Burnet, Woodward, and Whiston have always had a peculiar fascination for historians as bizarre freaks of pseudo-science, and they have never been allowed to lapse into obscurity. It is from relative oblivion, however, that we must rescue Robert Hooke, our final seventeenth-century theorist. Hooke never formulated a theory of the Earth in the normally accepted sense, but his writings show him to have possessed a remarkably sound grasp of the fundamentals of Earth-history, and his ideas are deserving of careful attention.

Hooke's name has already cropped up many times in the present work, more especially in connection with his advocacy of the seismic origin of topography, but this is a convenient point at which to examine this remarkable man a little more closely. Hooke was born at Freshwater, in the Isle of Wight, in 1635,[52] and although a sickly child, a genius for mechanics soon became apparent in him. His early handiwork included a working clock made entirely in wood, and a fully-detailed ship-model containing a device which caused it to fire off its guns as it sailed across the local harbour. His health improved sufficiently to allow him to attend Westminster School and Christ Church, Oxford, and it was there, through an association with Robert Boyle, that Hooke began his long scientific career. His inventions soon earned him a reputation, and in 1662 he was appointed Curator to the infant Royal Society. As such, it was his responsibility to provide the Fellows with three or four experiments at each of their meetings, and it is regrettable that for the remainder of his life he had to pander to the dilettantish interests of some of the Society's more influential aristocratic members. Another onerous duty devolved upon Hooke after the Great Fire when he became one of the City Surveyors charged with the task of rebuilding London. Partly because of these many calls upon his time, and partly because of the fertility of a mind which was ever urging him into new fields, he rarely brought any of his work to perfection. No portrait of Hooke is known, but his surviving diary for the years 1672 to 1680 reveals him as a sociable and warm-hearted individual. His earliest biographer[53] nevertheless assures us that late in life Hooke was 'melancholy, mistrustful, and jealous', and at the time of his death in 1703 he was almost totally blind and in a state of advanced, but premature, senility.

Hooke had a prolonged and bitter quarrel with Newton, and he had the misfortune to live out his life in the shadow of Newton's even greater genius. After his death, Hooke's former associates in the Royal Society were happy to let all recollection of his contentious character and bent, unkempt form, slip from their memory as they basked in the reflected glory of the long Newtonian presidency which began in the year of Hooke's death. Hooke's many remarkable achievements were soon forgotten, and it is only since the tercentenary of his birth that attempts have been made to rescue him from obscurity. Today it is increasingly recognised that Hooke was one of the most prolific inventive geniuses that the world has yet seen, but his geological writings are still relatively unknown.

Those of Hooke's views of interest in the present context are found on a few pages of the rambling *Discourse of Earthquakes* which was published among his posthumous works in 1705.[54] The *Discourse* evidently formed the substance of one of Hooke's lecture courses, and according to Richard Waller, his posthumous editor, the manuscript of the *Discourse* was completed on 15th September 1668. Ample internal evidence supports Waller's assertion, and there is no evidence of any subsequent revision of the manuscript apart from the insertion of one reference to a paper published in 1701.[55] Indeed, the idea of Hooke returning to revise a manuscript hardly seems in keeping with his mercurial character, and the *Discourse* published in 1705 probably differs in no important respect from the manuscript completed in 1668. The *Discourse* therefore pre-dates Steno's *Prodromus* by one year, and this greatly enhances the interest of Hooke's work.

Hooke's theory of Earth-history rested upon four fundamental tenets. Firstly, there was his conviction that fossils are the remains of former marine organisms, and his regard of fossiliferous strata in the midst of the continents as clear evidence that major changes in the distribution of land and sea must have taken place long after the Creation. 'Parts which have been Sea are now Land', he wrote, 'and divers other Parts are now Sea which were once firm Land.' Even Great Britain, he observed, 'may have been heretofore all cover'd with the Sea, and have had Fishes swimming over it'. These interchanges of land and sea he of course attributed to the seismic convulsions upon which he laid such emphasis. The second tenet was his very firm belief in the efficiency of denudation, and the third was his conviction that material worn from the continents eventually comes to rest in the ocean basins where it is soon consolidated to form new strata. His understanding of this latter process was remarkably enlightened for its day. Steno had avoided the issue altogether, but Hooke suggested that the consolidation was effected in one of four ways. Firstly, new rocks might be formed by fusion resulting from 'some kind of fiery Exhalation arising from subterranean Eruptions or Earthquakes'. Secondly, they might be formed by saline substances, 'working by Dissolution and

88

Congelation, or Crystallization, or else by Precipitation and Coagulation'. Thirdly, there was the influence of 'some glutinous or bituminous Matter, which upon growing dry or setling grows hard, and unites sandy Bodies together into a pretty hard Stone'. Lastly, Hooke believed that clastic material might become cemented by 'a very long continuation of these Bodies under a great degree of Cold and Compression'[56]

The final tenet of Hooke's geological creed is the most interesting and important. He maintained that all Nature is in a state of flux and yet is held in balance.[57] 'All things', he observed, 'almost circulate and have their Vicissitudes.' Planets, he noted, tend to fly away from the Sun, but are pulled back by 'a magnetick or attractive Power that keeps them from receding'; generation creates, but death destroys; winter reduces the growth of summer; night refreshes what day has scorched; and water circulates continually from the sea, through the atmosphere as vapour and rain, and back to the sea in rivers. He applied this notion of a balance in Nature, and of natural cycles, to the surface of the Earth itself. He believed, for example, that as earthquakes raise one portion of the Earth's surface, another sinks in compensation, and he suggested that the British Isles had been elevated in this way as Atlantis sank. Similarly, he argued that fluvial erosion can dissect elevated blocks of the crust so that they become a series of residual hills which, in their turn, are reduced 'back again to their pristine Regularity, by washing down the tops of Hills, and filling up the bottoms of Pits'. Hooke in 1668 thus had an inkling of the so-called Normal Cycle of Erosion which W. M. Davis evolved more than two hundred years later.

In his *Discourse*, Hooke attempts to weld these four fundamental tenets into a coherent, cyclic theory of Earth-history, but unfortunately his writing on the subject is vague and diffuse, and it is difficult to find passages which may be quoted to illustrate his views. It is nevertheless clear that he had a sound if shadowy understanding of the cycle which is known to modern geologists as the geostrophic cycle. The following passage does afford some insight into the remarkably advanced nature of Hooke's thought. Having referred to a low-lying part of the Earth's surface being buried beneath debris washed down from higher ground, he continued:

> This Part being thus covered with other Earth, perhaps in the bottom of the Sea, may by some subsequent Earthquakes, have since been thrown up to the top of a Hill, where those parts which it was by the former

means covered, may in tract of time by the fall and washing of Waters, be again uncovered and laid open to the Air, and all those Substances which had been buried for so many Ages before, and which the devouring Teeth of Time had not consumed, may be then exposed to the Light of the day.[58]

This passage reveals Hooke's sound grasp of the importance of successive phases of sedimentation and denudation, and his understanding that some of the Earth's topography has been exhumed from beneath a cover of younger sediments. He went on to suggest that some parts of the Earth might have been subjected to a whole series of denudational episodes, interspersed by periods of sedimentation; ''tis probable', he observed, 'there may have been several vicissitudes of changes wrought upon the same part of the Earth'. Few, if any, of Hooke's contemporaries can have possessed so enlightened an understanding of the cyclic nature of events at the Earth's surface.

Eyles has suggested that Steno may have been made aware of Hooke's views through his correspondence with some third party (perhaps Henry Oldenburg) who had attended Hooke's lectures, and that Steno may have been influenced by these views while preparing his *Prodromus*.[59] It is certainly tempting to compare the beliefs embodied in Hooke's *Discourse* with those in the *Prodromus*, but such comparison is hardly fair to Hooke; the *Prodromus* is a polished and carefully reasoned monograph, whereas the *Discourse* is in the raw state in which Hooke left it at his death. Had Hooke himself groomed the *Discourse* for publication he would no doubt have developed some of its more pregnant suggestions, and thus secured a place of honour for himself among the fathers of geology. Even from the evidence available in the *Discourse*, it is clear that the neglected Hooke had an understanding of Earth-history which is in some respects superior to that of the much-lauded Steno. The Dane, for example, was at pains to demonstrate a complete accordance between the Scriptures and his interpretation of the Tuscan landscapes, whereas Hooke, although deeply religious, was convinced that the Earth had experienced many more vicissitudes than are recorded in the Pentateuch. Similarly, Steno evidently adhered to the Old Testament chronology, while Hooke was sufficiently enlightened to suggest that a study of fossils might yield a longer and more realistic time-scale. Again, Steno regarded the Flood as the sole marine transgression in Earth-history, and he evidently held that most, if not all, fossiliferous strata had been deposited in its waters.

Hooke, on the other hand, asserted that the Deluge had been of far too short a duration to have been responsible for the formation of fossiliferous rocks, and although he believed in a wholesale diluvial interchange of land and sea, he also believed that land and sea had changed places on many other occasions. He was convinced, for instance, that the emergence of the Alps, Apennines, and British Isles were all postdiluvial events.

In the *Discourse* Hooke wrote: 'There are some other Conjectures of mine yet unmention'd, which are more strange than these; which I shall defer the mentioning of till some other time', but that other time evidently never came, and his later geological writings display little of the perspicacity that is so evident in his earlier work. One can only regret that the ideas embodied in the *Discourse* were never developed, for it is clear that by 1668 Hooke had formulated the shadowy outline of a theory of the Earth that is almost identical with the renowned theory which James Hutton presented to the Royal Society of Edinburgh in 1785. Indeed, Hooke's theory is superior to Hutton's in that Hooke had a far sounder understanding of the processes responsible for cementing sedimentary strata than did Hutton more than a century later. Hooke's notion of old continents slowly being worn down by denudation, and of their being replaced by new lands raised from the ocean bed, represents the acme of seventeenth-century thought on Earth-history, but Hooke was too advanced for his age. His ideas made no impact on his contemporaries, and they lay forgotten for almost two centuries.

Although the theories of Burnet, Woodward, Whiston, and, to a lesser extent, Steno, were the most widely read accounts of Earth-history in seventeenth- and eighteenth-century England, there were many other writers who ventured into the same field. Indeed, after the appearance of Burnet's theory in 1681, the devising of new theories and the criticising of those already extant, became a popular intellectual pastime in England. Even John Ray indulged himself; his *Three Physico-Theological Discourses*, first published in 1692, was inspired by Burnet's theory, and the book contains long discussions of the Creation, the Flood, and the coming dissolution of the world. Ray wisely refrained from attempting too detailed an account of Earth-history, however, and his theory, in common with most of its contemporaries, lacks the flourish of detail which made the works of Burnet, Woodward, and Whiston so attractive and plausible. Reference to the works of some of the lesser theorists will be found in the bibliography of the present volume, but the theories themselves need not detain us; they were pale, imitative works, and their authors had nothing original to contribute to the understanding of landforms.

Fanciful though they now seem, the theories of Steno, Burnet, Woodward, and Whiston remained current for more than a century, and even as late as 1831 Dr. John Macculloch, the pioneer of Scottish geology, still regarded them as meriting serious attention.[60] The theories owed their undeservedly long life to the stagnation which afflicted the infant geological science in Britain throughout most of the eighteenth century. When, after 1778, a new generation of theorists arose, they could look back to the scholars of Burnet's day as their own immediate predecessors.

REFERENCES

1. *Nicolai Stenonis De Solido Intra Solidum Naturaliter Contento Dissertationis Prodromus* (Florence 1669).
2. *Philos. Trans.*, VI (1671), No. 76, pp. 2281–2284.
3. *The Prodromus to a Dissertation Concerning Solids Naturally Contained within Solids* (London 1671). Another translation, by J. G. Winter, is in *University of Michigan Studies, Humanistic Series*, XI, Pt. II (1916).
4. Thomas Burnet, *The Theory of the Earth* (London 1684 & 1690).
5. *Doctrina Antiqua de Rerum Originibus*, pp. 1 & 2 (London 1736).
6. William Nicholls, *A Conference with a Theist*, Pt. II, p. 184 (London 1703).
7. *Psalm* 24, v. 2; 136, v. 6.
8. *Principia Philosophiæ*, Pars Quarta (Amsterdam 1644).
9. *Private Correspondence and Miscellaneous Papers of Samuel Pepys*, edited by J. R. Tanner, I, p. 23 (London 1926).
10. No. 146 (17 August 1711).
11. Herbert W. Turnbull, *The Correspondence of Isaac Newton*, II, p. 329 (Cambridge 1960).
12. *Some Animadversions upon a Book intituled The Theory of the Earth*, Preface (London 1685).
13. *Geologia: or, a Discourse Concerning the Earth before the Deluge*, Preface (London 1690).
14. John Keill, *An Examination of Dr. Burnet's Theory of the Earth, Together with some Remarks on Mr. Whiston's New Theory of the Earth* (Oxford 1698).
15. Thomas Burnet, *Reflections upon the Theory of the Earth, occasion'd by a Late Examination of it* (London 1699). John Keill, *An Examination of the Reflections on The Theory of the Earth. Together with a Defence of the Remarks on Mr. Whiston's New Theory* (Oxford 1699).

16. John Oldmixon, *The History of England, during the Reigns of King William and Queen Mary*, pp. 95 & 96 (London 1735).

17. Thomas Burnet, *Archæologiæ Philosophicæ*, Lib. II, Cap. VII & VIII (London 1692).

18. John W. Clark, and Thomas McK. Hughes, *The Life and Letters of the Reverend Adam Sedgwick*, I, pp. 166–187 (Cambridge 1890); V. A. Eyles, 'John Woodward, F.R.S. (1665–1728)', *Nature, Lond.*, CCVI (1965), pp. 868–870; John Ward, *The Lives of the Professors of Gresham College*, pp. 283–301 (London 1740).

19. John Nichols, *Illustrations of the Literary History of the Eighteenth Century*, I, p. 806 (London 1817).

20. Zacharias C. von Uffenbach, *London in 1710*, p. 176 (London 1934).

21. Charles R. Weld, *A History of the Royal Society*, I, pp. 337 & 352–355 (London 1848). David Brewster, *Memoirs of the Life, Writings, and Discoveries of Sir Isaac Newton*, II, pp. 243–247 (Edinburgh 1855).

22. John Woodward, *An Attempt Towards a Natural History of the Fossils of England*, I, Pt. II, p. 1 (London 1729).

23. John Woodward, *An Essay Toward a Natural History of the Earth*, p. 4 (London 1695).

24. *Ibid.*, p. 1.

25. *Brief Instructions for Making Observations in all Parts of the World* (London 1696); *Fossils of all Kinds, Digested into a Method, suitable to their mutual Relation and Affinity*, Number XI (London 1728).

26. Woodward, *op. cit.* (1695), pp. 4 & 5.

27. Woodward, *op. cit.* (1728), p. x.

28. Woodward, *op. cit.* (1729).

29. John Harris, *Remarks on some late Papers, Relating to the Universal Deluge: and to the Natural History of the Earth*, Preface (London 1697).

30. John Woodward, *The Natural History of the Earth*, translated by B. Holloway (London 1726).

31. Woodward, *op. cit.* (1695), pp. 74 & 75.

32. *Ibid.*, pp. 29 & 30.

33. *Ibid.*, pp. 80 & 81.

34. *Ibid.*, pp. 151, 83, 150.

35. *Ibid.*, p. 94.

36. *Ibid.*, pp. 54 & 55.

37. *Ibid.*, p. 227.

38. See Lester M. Beattie, *John Arbuthnot*, Chap. III (Harvard Univ. Press, Cambridge 1935); Charles E. Raven, *John Ray*, pp. 449–451 (Cambridge 1950); Ward *op. cit.* (1740), pp. 285–290.

39. *Philos. Trans.*, XXVII, 1710–1712, No. 336, pp. 541–544; Charles Leigh, *The Natural History of Lancashire, Cheshire, and the Peak, in Derbyshire*, pp. 66 (Oxford 1700); William Derham, *Physico-Theology*, pp. 67 & 68 f (London 1713).

40. Robert W. T. Gunther, *Further Correspondence of John Ray*, Ray Society Series, CXIV, p. 256 (London 1928).

41. Robert W. T. Gunther, *Early Science in Oxford*, XIV, pp. 268 & 269 (Oxford 1945).

42. Anon., *Tauronomachia*, p. 4 (London 1719).

43. William Whiston, *A Vindication of the New Theory of the Earth from the Exceptions of Mr. Keill and Others*, Preface (London 1698).
44. William Whiston, *A New Theory of the Earth*, Bk. I, p. 62 (London 1696).
45. *Ibid.*, Bk. II, p. 78.
46. William Whiston, *The Cause of the Deluge Demonstrated* (London 1714).
47. Gunther, *op. cit.* (1928), p. 277.
48. Keill, *op. cit.* (1698 and 1699).
49. *The Works of John Locke Esq.*, III, p. 556 (London 1727).
50. Whiston, *op. cit.* (1698), Preface.
51. *Memoirs of the Life and Writings of Mr. William Whiston* (London 1749).
52. Margaret L'Espinasse, *Robert Hooke* (London 1956).
53. Richard Waller, *The Posthumous Works of Robert Hooke*, pp. i–xxvii (London 1705).
54. *Ibid.*, pp. 279–328.
55. G. L. Davies, 'Robert Hooke and his Conception of Earth-History', *Proc. Geol. Ass., Lond.*, LXXV (1964), pp. 493–498.
56. Waller, *op. cit.* (1705), pp. 290; 293 & 294.
57. *Ibid.*, pp. 312 & 313.
58. *Ibid.*, p. 314.
59. 'The Influence of Nicolaus Steno on the Development of Geological Science in Britain', pp. 167–188 in Gustav Scherz, *Nicolaus Steno and his Indice* (Copenhagen 1958). Also in *Acta hist. Sci. nat. med.*, XV (1958).
60. *A System of Geology*, pp. 389–392 (London 1831).

Chapter Four

Relapse and Recovery
1705-1807

It may be well to forewarn our readers, that in tracing the history of geology from the close of the seventeenth to the end of the eighteenth century, they must expect to be occupied with accounts of the retardation, as well as of the advance of science. It will be our irksome task to point out the frequent revival of exploded errors, and the relapse from sound to the most absurd opinions.

*Sir Charles Lyell
writing in 1830 in
the first volume of
his 'Principles of
Geology'.*

THE history of British geomorphology during the hundred years after Ray's death in 1705, falls readily into two unequal periods. Firstly, from 1705 to 1778, there was a period of relapse when the Earth-sciences lay stagnant and forgotten, and secondly, from 1778 to 1807, there was a period of recovery during which interest in Earth-history rapidly regained all its former vitality. Hitherto the eighteenth-century hiatus in the history of geomorphology has commonly been mistaken for the subject's primordial state, and the late eighteenth-century revival has similarly been mistaken for the subject's birth.

The long period of relapse was just one manifestation of the general lethargy which overtook European science soon after 1700. The great advances of the previous century seem to have left science exhausted, and as the giants of Newton's generation passed from the scene, the whole pace of scientific progress suddenly slackened. There were, nevertheless, three more specific factors which caused British scholars to shun the Earth-sciences in favour of studies in other fields. Firstly, there was now a strong desire to base all science upon an experimental basis, and neither geology nor geomorphology seemed to offer much that might be tested in the laboratory. Secondly, those who did incline towards studies in natural history were strongly imbued with the Linnaean desire for classification, but here again, and with the notable exception of mineralogy, the Earth-sciences provided little scope for such taxonomic work. Finally, after the controversy over the theories of Burnet, Woodward, and Whiston had abated, there was a reaction against such speculative studies. In criticising Woodward's theory, Edward Lhwyd had expressed the hope that such whimsies 'will make men preferre Natural History, to these romantic theories', and his wish was all too amply granted. New editions of the chief seventeenth-century theories of the Earth continued to appear with some regularity throughout the period of the relapse, but it was very late in the century before a British author assumed the mantle of Burnet, Woodward and Whiston.

For fifty years after 1705 not one British scholar made a contribution of any significance to the literature of geomorphology, and the period produced little that can stand comparison with, say, the systematic discussion of landforms contained in Nathanael Carpenter's *Geography Delineated* of 1625. In tracing the history of seventeenth-century geomorphology, it was possible to lean heavily upon the writings of some of the leading scholars of the day, but the student of British geomorphology during the period of the relapse must have recourse to works of a very different genre. From the writings of men of the standing of Hooke, Newton, and Ray, he must pass to scraps of material culled from the publications of untutored travellers, second-rate topographers, and long-forgotten clerics. Even such a hack-work as Oliver Goldsmith's *History of the Earth, and Animated Nature*, of 1774, must be considered an important source for the study of eighteenth-century British geomorphology. Goldsmith toiled over this work for five years after dissipating the 800 guineas paid him in advance by an indulgent publisher, but the first volume of the work contains one of the most readable and complete discussions of landforms to be published between 1705 and 1778.

The revival of interest in the Earth-sciences began first on the continent of Europe as a result of the work of such geologists as Bergman, Desmarest, Guettard, Lehmann, Pallas, de Saussure, and Werner. In

Britain the beginning of the revival was marked by the publication in 1778 of *An Inquiry into the Original State and Formation of the Earth* by John Whitehurst, a Derbyshire clock-maker and a pioneer speleologist. Others soon followed Whitehurst into the study of Earth-history. The chief among this new generation of British geologists were Jean André de Luc, a Genevese expatriate living at Windsor; James Hutton, a gentleman-scientist from Edinburgh; John Williams, a none-too-successful Welsh mineral-prospector; Richard Kirwan, an Irish chemist and mineralogist; and Robert Jameson, who held the chair of Natural History in Edinburgh from 1804 until 1854. The letters, papers, and books which poured from the pens of these scholars come to the historian of the Earth-sciences like welcome rain after a prolonged drought.

By the last decade of the eighteenth century, British interest in Earth-history was fully restored, and geology at last began to take rapid strides forward. The speculative approach to the subject was finally abandoned, and geology became a true science firmly grounded upon painstaking field-observation. This transformation of its geological parent into a field-science had important repercussions upon geomorphology, but the present chapter is concerned with the history of geomorphology before the transformation took place. The year 1807 has been adopted as the chapter's terminal date because it was then that the Geological Society of London was founded. This important event was both an outward sign of the new scientific interest in geology, and a stimulus to the further development of the subject into a precise field-science.

In tracing the history of geomorphology in Britain during the period between 1705 and 1807, it is convenient to show first how a slight loosening of the Mosaic shackles gave the late eighteenth-century geologists a somewhat less cramped terrestrial time-scale against which to set their studies. Secondly, attention will be focussed upon eighteenth-century ideas about the forces responsible for raising new features upon the Earth's surface, and finally we will examine the eighteenth-century understanding of the work of denudation. So far as the raising of new topography was concerned, eighteenth-century scholars had little fresh to offer. They still saw topography as the result of diluvial, seismic, and volcanic catastrophes, or as the product of some now extinct process which had operated during the Creation. On the other hand, eighteenth-century scholars took a very different view of denudation from that adopted by their seventeenth-century predecessors. This was the result of the emergence about 1700 of a religio-scientific problem which necessitated a reassessment of the earlier belief in the potency of Nature's destructive forces. The efforts made to resolve the problem seriously impaired belief in the reality of denudation, and this unfortunate occurrence was one of the two most important eighteenth-century events in the history of British geomorphology. The other important event was the appearance after 1778 of a fresh crop of theories of the Earth, but a full discussion of these theories will be reserved for the next two chapters.

The Hexaëmeron Reinterpreted

The bibliolatry of seventeenth-century Britain survived far into the eighteenth century, and an implicit faith in the literal interpretation of the Mosaic chronology remained widespread until at least 1760. There still seemed no reason to doubt Ussher's conclusion that the Earth was rather less than six thousand years old, and those who sought to remind their contemporaries of the high antiquity claimed by the ancient civilisations, received short shrift. When a work on Chinese history was reviewed in the *Philosophical Transactions* in 1730, the Royal Society's critic roundly condemned all chronologies attributing high antiquity to the Chinese nation as absurd and false. 'Such Chimeras', he wrote, 'deserve not the Pains of refuting. They are equally repugnant to good Sense, the Rules of Criticism, and to Religion.'[1]

We saw earlier that such late seventeenth-century figures as Hooke, Lhwyd, and Ray, had all sensed the existence of a contradiction between the Mosaic chronology and the age of the Earth as revealed by Nature herself. Among the scholars of the first half of the eighteenth century, however, only Edmond Halley is known to have entertained similar suspicions. Significantly, he was the last survivor of that group of brilliant savants which had included Hooke and Ray. Like Ray, Halley was satisfied that Adam had been created rather less than six thousand years ago, but he suspected that the creation of the Earth itself was a much more distant event. In 1715 he suggested to the Royal Society that a true estimate of the Earth's age might be arrived at by examining the waters of the oceans and inland seas over a period of time, to discover the rate at which their salinity increases.[2] Armed with this information, he argued, and assuming the rate of salt accumulation to have been constant, it becomes a simple task to calculate the time necessary for the waters to have attained their present salinity. By this method, Halley claimed, 'the World may be found much older than many have hitherto imagined'. In the event, even he would have been astonished at the result of the calculation, because in 1899, when Joly adopted a method very similar to that suggested by Halley, he arrived at the conclusion that the Earth was between 80 and 90 million years old.[3] Gross underestimate though it is, Halley could only have boggled at such figures.

Concern for the Earth-sciences had already largely evaporated by 1715, and Halley's paper raised no stir. Not until the revival of interest in Earth-history during the 1770s, did scientific attention again come to focus upon the problem of the Earth's age. The issue was then first raised by a somewhat oblique reference to the subject contained in Patrick Brydone's account of his travels in Sicily and Malta. This popular book was first published in 1773, and in it Brydone briefly describes some studies carried out near Mt Etna by one Canonico Recupero.[4] Near the volcano, Brydone relates, Recupero had found a lava-flow which could be identified as having come from the volcano about 210 B.C., during the Second Punic War. The surface of this flow was still unweathered, but only a short distance away, near Catania, Recupero had discovered a section revealing seven lava-flows, separated from each other by weathered horizons containing thick soils. Now, Recupero had reasoned, if it takes more than two thousand years for a lava to weather into a soil, then the Catania section, with its seven soils, must represent a time-interval of at least fourteen thousand years. This deduction, Brydone observed, alarmed Recupero because it seemed to be such a flagrant contradiction of the Mosaic chronology.

Brydone describes Recupero's studies without adding any personal comment, but many of his contemporaries held him guilty of propagating a report which was subversive of Scripture, and therefore morally dangerous. According to a correspondent of the Bishop of Llandaff in 1791, the report had encouraged a great deal of apostasy,[5] and British apologists certainly made repeated reference to Recupero's studies during the closing years of the century. In the 1770s the Bishop of Llandaff himself was at some pains both to refute the conclusions that had been drawn from the Catania section, and to vindicate the Mosaic record.[6] The Bishop was nevertheless possessed of a more liberal outlook than that of most of his predecessors, and he did reluctantly

99

concede that the days of the hexaëmeron might not be days in a literal sense, but, rather, periods of unknown duration. For the first time, the novity of the Earth was being seriously questioned in ecclesiastical circles.

By no means all late eighteenth-century clerics were as liberal as the Bishop of Llandaff, but during the closing decades of the century, the possibility of an allegorical interpretation of the days of the hexaëmeron was widely admitted, and scarcely less widely adopted. The grip of Moses had now lessened somewhat, just as the grip of the ancients had lessened a century earlier. There was still no question of rejecting *Genesis*, but in the face of evidence such as that discovered by Recupero, there seemed little alternative but to reinterpret the days of Creation as representing periods of considerable duration. It was even possible to find Scriptural justification for such reinterpretation. Halley pointed out as early as 1715 that according to Scripture 'one day is with the Lord as a thousand years', and, he reasoned, the five pre-Adamic days may therefore have been God's days rather than normal sidereal days. Others, too, found this a very convenient *modus vivendi* whereby Moses and science could be reconciled. Thus the allegorical interpretation of *Genesis* which Ray and Lhwyd had discussed guardedly in their correspondence, and which had been Burnet's undoing in 1692, became the generally accepted view among the late eighteenth-century geologists. As William Cowper observed ironically in 1785 in Book III of *The Task*:

> ... *Some drill and bore*
> *The solid earth, and from the strata there*
> *Extract a register, by which we learn*
> *That he who made it, and reveal'd its date*
> *To Moses, was mistaken in its age.*

It must nevertheless be emphasised that those who adopted an allegorical interpretation of the first chapter of *Genesis* were still far from proposing a vast extension of the terrestrial time-scale. The days of the hexaëmeron might be reinterpreted, but there was still no escaping the conclusion that mankind had existed for rather less than six thousand normal, sidereal years. Only the date of the beginning of the Creation was open to adjustment, and the interval between the first and sixth days of the Creation could not be opened too far for fear of implying that God had needed an aeon in which to prepare the Earth for human habitation. The days of the hexaëmeron could therefore be reinterpreted as each representing centuries, or perhaps even millennia, but their reinterpretation as representing periods of millions of years was out of the question. Thus, even in the late eighteenth century, an evolutionary view of Nature was still inhibited because of a seeming time-deficiency. A mere abandonment of the literal interpretation of *Genesis* did not suffice to give geomorphology a fresh perspective; scholars also had to shed their anthropocentric view of Earth-history, and this was not achieved until the nineteenth century.

De Luc is pre-eminent among the eighteenth-century geologists who grappled with the problem of Earth-chronology. He accepted that the days of the hexaëmeron represented periods of unknown duration, and that the date of the Earth's creation was beyond human cognisance. He therefore concentrated his attention upon discovering the age of the Earth's present continents, which he believed to be very much younger than the Earth itself. He arrived at the conclusion that the continents had all been formed during the Flood, and although he fixed no precise date for this event, he did adduce the evidence of what he termed 'natural chronometers' in proof of the continental novity.[7] The chronometers are too numerous to receive individual mention, but five examples will serve to illustrate their general nature.

1. *Lake Basins*. He regarded all lake basins as pristine features of the continental topography, and he claimed that throughout their existence, the basins had been receptacles into which rivers were steadily sweeping the loose debris left on the continents at the time of their late diluvial emergence. In every case he had investigated, however, the volume of infill in the basins was quite insignificant, and he regarded this fact as perhaps the most striking proof of the youth of the continents.

2. *The Rhine Gravels*. At Koblenz, in 1778, he saw a section exposed in the Rhine gravels, and he concluded that the deposit represented the sum total of the river's aggradational work in post-diluvial time. Exposed in the section, 8 feet below the surface, was the remains of a Roman cemetery, and in view of the amount of post-Roman aggradation, he concluded that the total (unspecified) thickness of the gravels could not represent a time interval of more than a few thousand years.

3. *The North-German Tumuli*. De Luc believed the north-German

tumuli to have been built by the region's earliest post-diluvial inhabitants. He therefore examined the tumuli to discover how the thickness of the 'vegetable layer' formed on them since their construction compared with the depth of the same layer on the adjacent undisturbed ground. Finding little difference, he concluded that the continents had emerged only shortly before the tumuli were built.

4. *The Alpine Glaciers.* De Luc lived towards the end of the so-called 'Little Ice Age' which had begun in the sixteenth century, and, being Genevese, he was doubtless familiar with tales of Alpine villages being overwhelmed by the advancing glaciers. He believed that all mountain glaciers had formed as a result of refrigeration caused by the raising of some areas to a high level during the late diluvial emergence. The continued growth of the glaciers therefore seemed to place the emergence at no great distance in time. Similarly, he had observed that glaciers transport boulders, and he claimed that the fewness of these erratic blocks around the glacial snouts was further proof of the youth of the glaciers and, therefore, of the novity of the continents.

5. *Talus Sheets.* He perceived that steep Alpine slopes are rapidly being broken down by the freeze-thaw processes, and he recognised the mountain-foot talus sheets as the waste-product of this activity. He offered the widespread survival of steep mountain slopes, despite Nature's ravages, and the smallness of many of the talus slopes, as a further indication of the youth of the continents.

Through his natural chronometers, de Luc employed the observational techniques of the modern field-geologist to support the Mosaic chronology. He was a scientist of international repute, and his claim to have discovered ample evidence showing Nature to be in entire accord with the Scriptures, carried great weight in the late eighteenth century.

In eighteenth-century Britain only one very small group of individuals had any appreciation of the true magnitude of the terrestrial timescale – the deists. Deism had not been unknown in earlier centuries, but eighteenth-century deism had its roots in the seventeenth-century belief that God could be approached through both Nature and the Scriptures. Most eighteenth-century scholars retained this dual approach, but a few independent minds rejected the Bible because they claimed to find it full of inconsistencies and enigmas, and henceforth they relied upon Nature as their sole revelation of the Deity. From the trickle of deistical works which began to appear after the lapse of the Press Licensing Act in 1695, and, more especially, from the works of their Christian opponents, it emerges that having rejected the Mosaic chronology along with the Scriptural revelation, most deists adopted the Aristotelian view that the universe is eternal. This reversion to Aristotle was merely a reaction against the limited Christian conception of time, but the development is nevertheless of some importance because one of the deists – James Hutton – introduced the deistical time-scale into the earth-sciences. He recognised that such a time-scale could form a back-cloth to an evolutionary view of Nature, and his brilliant exposition of this theme, dating from 1785, will engage our attention in a later chapter. As a deist-geologist, however, Hutton was almost unique; his geological contemporaries were orthodox Christians who adhered to a time-scale which was only one degree removed from that of Ussher.

Primitive Rocks and Primitive Topography

We have already seen that the early seventeenth-century belief in the existence of primitive topography gradually faded as the wide extent of fossiliferous strata began to be appreciated. By 1700 it seems to have been widely accepted that all the Earth's original features must have been destroyed either during the Flood, or as a result of seismic and volcanic catastrophes. A belief in the existence of primitive topography nevertheless does occasionally crop up in early eighteenth-century litereature. The idea persisted, for example, in the eighteenth-century English translations of *Geographia Generalis* by Varenius, and in the writings of William Stukeley and John Strachey.

Stukeley, a somewhat gullible antiquary, claimed in 1724 that most topography is asymmetrical, with gentle eastern slopes, and steep, westward-facing escarpments. It was doubtless the form of the English scarplands which gave him this idea, and he believed all such topography to be primitive. He explained its mode of formation in the following passage.

> [When] the body of the earth was in a mixt state between solid and fluid, before its present form of land and sea was perfectly determin'd, the almighty Artist gave it its great diurnal motion. By this means the elevated parts or mountainous tracts, as they consolidated whilst yet soft and yielding, flew somewhat westwards and spread forth a long declivity

to the east; the same is to be said of the plains, their natural descent trending that way, and as I doubt not, of the superfice of the earth below the ocean.[8]

In the following year the same idea was developed by Strachey,[9] one of the very few early eighteenth-century scholars deserving of the title 'geologist'. He suggested that during the creation the Earth was stationary, and that successive strata had been laid down, not as concentric skins parallel to the globe's circumference, but as a series of horizontal beds forming chords drawn across its diameter. Then, when the Earth began to rotate, he pictured it as converted into a kind of gigantic Catherine-wheel. One end of each stratum was drawn down to the centre of the Earth, while the cther end outcropped at the surface where, because of the eastward rotation, it formed the asymmetrical topography which Stukeley had described (Plate III).

Later in the century wide currency was accorded to this idea that all topography is asymmetrical, with its steepest slopes to the west. Indeed, it was accepted as a general, universal law. 'In all parts of the world', Edinburgh students were informed in the 1780s, 'the Mountains are high and precipitous towards the Western side, and they decline and shelve away towards the East.'[10] Only Stukeley and Strachey, however, seem to have claimed the asymmetry as primitive. How they proposed to explain the presence of fossils in the rocks forming the primitive topography of the continents is not clear, but it was evidently this problem which in 1758 prompted William Borlase, a Cornish cleric, to divide mountains into two types. On the one hand he recognised 'natural mountains' dating back to the Creation, and formed as a result of precipitation in a chaotic fluid which then enveloped the Earth, and on the other hand he recognised 'factitious mountains' produced by seismic and volcanic catastrophes.[11]

Borlase was the precursor in Britain of a new school of thought upon the subject of primitive topography. Until the middle years of the eighteenth century, naturalists throughout Europe had possessed only the vaguest notions about the structure of the Earth's surface, but now field-investigations carried out by continental geologists showed that rocks could be differentiated into two broad groups. Basically, the distinction recognised was that between the igneous and metamorphic rocks on the one hand, and the sedimentary rocks on the other, but the two groups were known to eighteenth-century geologists as the Primitive and Secondary rocks respectively. The Primitive rocks were regarded as part of the Earth's original crust, formed by chemical precipitation in a chaotic fluid, while the Secondary rocks were interpreted as much younger, clastic sediments, which had been deposited upon the bed of some former ocean.

This eighteenth-century distinction between Primitive and Secondary rocks was of great geomorphic significance because there was believed to be a close correlation between the age of the rocks in any region, and the age of the region's topography. Mountains developed on Primitive rocks, for example, were regarded as primeval features, formed, as Newton had suggested long before, by the uneven precipitation of material in the chaotic fluid. Indeed, so close was the correlation of geology and topography supposed to be, that the terms 'Primitive rocks' and 'Primitive mountains' were regarded as synonymous, and most authors preferred the latter term even though they were concerned solely with geological problems. Similarly, the Secondary rocks were believed to form their own distinctive landscape, containing features that were essentially sedimentary in origin, and of the same age as the rocks themselves. Not until the last decade of the eighteenth century did a distinction begin to be drawn between the age of a topography and the age of the rocks beneath.

In Edinburgh, towards the end of the century, John Walker offered his students the following criteria by which they might differentiate between Primitive mountains and younger topography.[12]

1. Primitive mountains are built of rocks such as quartzite, porphyry, serpentine, granite, and basalt, whereas Secondary mountains are formed of rocks such as limestone, gypsum, sandstone, and coal.
2. The rocks forming Primitive mountains are extensively obscured beneath younger deposits but are themselves nowhere superincumbent upon Secondary strata.
3. Primitive mountains are invariably much higher than Secondary mountains.
4. The rocks in Primitive mountains dip steeply at angles of between 60° and 90°, whereas the rocks of Secondary mountains are commonly horizontal, and never display dips in excess of 60°. As a result of their steeper geological dips, Primitive mountains are usually the more precipitous.
5. Primitive mountains have a greater geological homogeneity

than Secondary mountains, which commonly exhibit a wide variety of strata within a small area.

6. Primitive mountains contain numerous mineral veins, whereas such veins are rare in Secondary mountains.

7. The rocks of Primitive mountains never incorporate fragments of Secondary strata, whereas fragments of the Primitive rocks are frequently encountered amidst Secondary mountains.

8. Primitive mountains are not fossiliferous, whereas fossils are abundant in most Secondary strata.

Thus by the second half of the eighteenth century the wheel had turned full cycle. Belief in primitive topography had been largely abandoned during the seventeenth century because of the seeming ubiquity of fossiliferous strata, but now, a century later, the belief was revived because a closer acquaintance with Nature revealed that fossiliferous, clastic sediments were by no means the sole rocks present in the continents. The Italian geologist Antonio Lazzaro Moro, in 1740, was apparently the first to draw the distinction between Primitive and Secondary rocks, but the concept was refined by the three German geologists Lehmann, Pallas, and Werner. British scientists were so tardy in returning to the study of Earth-history that it was very late in the century before the revived continental belief in primitive topography gained much currency in Britain. The belief was imported into Britain chiefly through the writings of the late eighteenth-century theorists, and in that context the belief will be encountered again.

Topography and the Subterranean Fires

The seventeenth-century belief that volcanoes and earthquakes have played a major part in shaping the Earth's topography, persisted throughout the eighteenth century, and as late as the 1770s Goldsmith, Whitehurst, and John Wesley were still quoting Pliny as an authority upon the subject. In 1773 William Worthington wrote of seismic shocks shattering the Earth at the time of man's Fall, so that 'fertile plains instantly started up into bleak and dreary mountains'.[13] Three years later, that painstaking student of Vesuvius, Sir William Hamilton (he is better remembered as the husband of Nelson's Emma), urged that the significance of the subterranean fires in the raising of mountains and islands was still insufficiently appreciated.[14] Nearer home, Arthur Young became convinced of the volcanic origin of the southern Irish mountains after mistaking their deep corries for volcanic craters,[15] while William G. Maton, later physician to the infant Queen Victoria, claimed the Cheddar Gorge as an earthquake fissure.[16] Erasmus Darwin, grandfather of the more famous Charles, followed his friend Whitehurst in a bold assertion that almost every feature of the Earth's surface owed its origin to seismic shocks. He believed that during the Earth's youth, the heat of the internal fires had caused the crust to expand and crack, whereupon sea-water descended into the vaults and was suddenly vapourised in a gigantic explosion which left the continents scarred with mountains. Only such an explosion, he claimed, could explain the fact that in a mountain, the strata invariably dip parallel to the mountain flanks, leaving the oldest rocks exposed at the mountain top. A mountain might be likened, he suggested, to the burr produced when a bodkin is thrust through a pile of papers.[17]

The persistence of a belief in the topographical significance of volcanoes and earthquakes is easily understood. The Earth was still believed to be riddled with caverns full of flames and noxious vapours, and it still seemed reasonable to suppose that periodic explosions would shatter the cavern roofs, and throw the Earth's surface into new disarray. Perhaps the influence of the exaggerated classical accounts of seismic and volcanic phenomena had waned with the passage of time, but a series of mid-eighteenth-century earth-tremors came as a vivid reminder of the forces pent up within the Earth. In February and March 1750 London itself was shaken, and on All Saints Day 1755, the whole of south-western Europe felt the tremor which destroyed Lisbon and cost more than 10 000 lives. Such events doubtless stimulated the belief in the unlimited potential of earthquakes, but even late eighteenth-century writing on the seismic and volcanic origin of topography differs so little from the theories being expounded a century earlier, that further discussion of the subject here is unnecessary.

Topography and the Flood

Eighteenth-century writers on Earth-history, like their predecessors of Burnet's generation, were obsessed with the Flood. Neither the emergence of an allegorical interpretation of the early chapters of *Genesis*, nor the late eighteenth-century revival of the earth-sciences, did anything to impair faith in the literal truth of the Flood story. Throughout the period covered by the present chapter, the Deluge continued to be regarded as an event of the utmost geological importance. Primitive, seismic, and volcanic topography might exist, but most authors believed that the greater part of the Earth owed its form to a diluvial metamorphosis. Features varying in scale from the continents and ocean basins, down to such minutiae of the landscape as tors and sinkholes, were all claimed as part of the diluvial legacy. The Flood figured prominently in most of the late eighteenth-century theories of the Earth to be considered later, and here reference to the works of a few typical eighteenth-century scholars will suffice to illustrate the diluvialism of the period.

The Bishop of Clogher, in 1754, claimed that all the world's ocean basins had been scooped-out during the Deluge, and that the resultant debris had been heaped onto the adjacent continents to form mountain ranges. For this reason, he observed, 'the Range of the Hills near any Sea, is generally guided by the Disposition and Extent of the next adjoining Sea out of which they were taken'.[18] In the West Indies, the Rev. Griffith Hughes, rector of St. Lucy's, Barbados, observed that the east to west movement of the diluvial waters under the influence of the Trade Winds, had shattered the eastern slopes of the island's hills, leaving the western slopes undamaged.[19] Philip Howard, writing in the 1790s, suggested that the diluvial shock had first tilted the entire Scandinavian peninsula, and then shivered its high western seaboard to form the Norwegian fiords.[20] Even volcanoes were attributed to the Flood: in 1756 Dr. Edward Wright claimed they were the result of the Flood waters sweeping animal and vegetable debris into immense heaps, which then fermented, putrefied, and caught fire.[21]

Various eighteenth-century authors devoted entire books to the history of the Flood, but of these only Alexander Catcott's *Treatise on the Deluge* of 1761 is of any significance in the history of geomorphology. Catcott, a Bristol cleric, took a Woodwardian view of the Deluge. He believed that the Flood waters had first dissolved the old world, and that a new world had then been shaped as the solute was precipitated to form expansive sheets of sedimentary strata. Like Woodward, Catcott evidently had some field experience, and he recognised that these supposedly diluvial strata are today discontinuous in their distribution (Plate IV). He linked their present condition to that of a ruined building, and he observed:

> ... if a person was to see the broken walls of a palace or castle that had been in part demolished, he would certainly conclude that the breaches or vacant spaces in those walls were once filled up with similar sub-

stances, and in conjunction with the rest of the walls, and could easily with his eye see the lines in which the walls were carried, and in thought fill up the breaches and re-unite the whole: And in the same manner if a person was to view the naked ends or broken edges of the strata in a mountain on one side of a valley and compare them with their correspondent ends in the mountain on the other side of the valley, he would manifestly perceive that the space between each was once filled up, and the strata continued from mountain to mountain.[22]

In seeking an explanation of this ruinous condition of the Earth's surface, he dismissed the seismic theory, and claimed that all breaks in the continuity of strata were the result of the erosive work of moving water. This was proved, he argued, by the serpentine plan of most valleys, and by the fact that they are arranged not haphazardly, but in integrated systems consisting of major valleys fed by smaller tributaries. The moving water which he invoked, however, was not that of a normal river. He believed that all the erosion had been effected by the torrents which drained off the continents as the Flood waters retreated into their subterranean reservoirs through great vents in the sea-bed. He tested his theory in an experiment described in the following passage:

> I provided a large vessel of Glass, had several holes of different sizes bored in the sides about six inches from the bottom, and stopped each with cork: I then filled the vessel with water; and having pulverized beforehand certain portions of the various strata of which the earth consist, as Stone, Coal, Clay, Chalk, &c. I permitted these substances to subside one after another through the water, 'till the terrestrial mass reached about two inches above the level of the holes: and the whole settled in regular layers one upon another, just according to the disposition of things in the earth. I then (with the assistance of another) pulled the corks out of each hole as nearly at the same time as possible. The water immediately began to drive the earthy parts through the holes, and scooped or tore the surface of the earthy mass....[23]

The result of the experiment, he claimed, was a replica of the Earth's topography, complete with miniature mountains, valleys, and plains.

Catcott might have been one of the outstanding figures in the history of eighteenth-century geomorphology. His field-studies had given him an understanding of the circumdenudational origin of most topography, and repeatedly in his writing he seems to be on the verge of advancing a fluvialistic doctrine of landscape development. In the event, however, he invokes nothing more than his diluvial torrents. In this respect it may be that he was a victim of his environment. He passed the greater part of his life in Bristol, and many of his observations were evidently made in the near-by limestone country of the Mendip Hills. There he doubtless became familiar with the numerous dry valleys. These clearly owed nothing to present fluvial action, and it must have seemed reasonable to attribute them to late diluvial torrents. Dry valleys were certainly one of the phenomena which led many later, and much more distinguished geologists, to embrace diluvialism and to invoke those very diluvial torrents which Catcott had tried to reproduce in his glass tank.

Landforms and the new Theology

Although most eighteenth-century scholars regarded the Earth's topography as an amalgam of diluvial, primitive, seismic, and volcanic features, they admitted, as had their seventeenth-century predecessors, that all landforms owe something of their present character to the work of subaerial denudation. Before examining the eighteenth-century understanding of the exogenetic forces, however, we must pause to consider an important theological development which took place just before 1700. This development was touched upon earlier in discussing Woodward's theory, but the change profoundly influenced the eighteenth-century view of denudation.

Late in the seventeenth century, English Calvinism was brusquely swept aside by a Restoration clergy which was predominantly Arminian.[24] As a result, there emerged a completely fresh conception of the Deity. The Calvinists had seen God as an awesome, wrathful Being, inflicting terrible punishments upon a sinful mankind, but the faithful of the new era saw Him as the benign and merciful Being who is reflected in the enlightened teaching of the Cambridge Platonists and the Latitudinarians. Thus by 1700, God the angry judge of mankind, had given way to God the gracious architect of a magnificent creation.

This changed conception of the Deity was doubtless a theological advance, but it had a twofold adverse effect upon geomorphology. Firstly, it played a major part in killing the old belief in a decadent universe. The notion of a cruel, insidious, but divinely ordained cancer spreading slowly through the entire Creation merely because of the frailty of one woman in Eden, was hardly reconcilable with the new image of a compassionate God. Thus the new theology removed one of the foundation stones upon which the seventeenth-century belief in denudation had rested. Secondly, the new theology intensified the conviction that the divine architect was fully revealed in the splendour of the Creation. The plenitude of Nature, and the argument from design, became the two chief weapons of the Christian apologist, and among naturalists, the new theology is reflected in a teleological approach to their subject. Here John Ray set the fashion with his popular book *The Wisdom of God Manifested in the Works of the Creation*, first published in 1691. He took as his text the Psalmist's words 'O Lord how manifold are thy works! in wisdom hast thou made them all', and he sought to show how each component of the Creation is a masterpiece, designed to serve some specific purpose in the divine plan for human happiness. He explained, for instance, that the Moon was fashioned to illuminate the night, to help man to measure time, and to cause the tides which prevent the oceans putrefying; that the air was created merely to allow man and the animals to breathe; and that fire, water, soil, rocks, and minerals were included in the world solely because God recognised that they would be needed by mankind.

Ray and his many imitators approached topography in the same vein. Throughout most of the seventeenth century, hills and mountains were regarded as ugly blemishes upon the Earth's skin, but slowly, during the second half of the century, this view gradually gave way to a new appreciation of the beauty of mountain landscapes. Henry More writing in 1662 managed to combine the two views when he wrote:

> . . . even those rudely-scattered Mountains, that seem but so many Wens and unnatural Protuberancies upon the face of the Earth, if you consider but of what consequence they are, thus reconciled you may deem them ornaments as well.[25]

The same juxtaposition of the old and the new is seen in Burnet's observation that whereas the orator may proclaim the beauty of the globe, the philosopher can only view it as a ruin. This regard of the Earth as a ruin lingered on far into the eighteenth century, but by about 1720 the majority of scholars seem to have accepted mountains not merely as aesthetically satisfying, but as integral and valuable parts of the Creation.

In his well-known teleological discussion of mountains, Ray himself claimed that they exist to serve the following eight divinely ordained ends.[26]

1. They give rise to springs and are the source of all rivers.
2. They are the place where mineral ores grow best, and where, being elevated, the veins are most accessible to miners.
3. They keep off cold winds and provide shelter for settlement.
4. They are themselves ornamental, and at the same time they are a viewpoint from which an observer may appreciate the beauty of the adjacent lowlands.
5. They are clothed with a wide variety of soils, and therefore support a diversity of valuable plants and trees.
6. They provide a suitable habitat for upland dwelling animals.
7. They cause condensation, and therefore give rise to refreshing showers of rain.
8. They form convenient political boundaries.

Ray's discussion of mountains was plagiarised by many other teleologists, and his views on the subject must have been very widely disseminated because Ray's book alone ran through sixteen English editions during the eighteenth century. Even in the last decade of the century, John Walker was still using Ray's work as the basis for a teleological discussion of mountains which he included in his Edinburgh lectures on geology.[27]

The teleological approach to landforms brought a perplexing problem in its train. If, as was now claimed, mountains were essential to the terrestrial economy, was it reasonable to suppose that they might be destroyed by denudation? Earlier we saw Varenius suggesting that the decay of mountains proved them to be no part of the divinely conceived Creation, while others had hailed denudation as a beneficial process which was slowly removing the Earth's mountainous blemishes and restoring the continents to their pristine regularity. Now these arguments were put into reverse. If, as was now believed, mountains were an integral part of the Creation, then clearly their destruction by so insidious a force as denudation could have no place within the divine order. Even limited denudation, it seemed, could only blur the divine masterpiece. On the other hand, denudation, as evinced by turbid rivers, landslides, and marine erosion seemed a reality. This was the rub: how could a belief in the destructiveness of denudation be reconciled with a teleology of Nature which demanded that landforms be regarded as immutable features of the Creation? This problem, which sorely perplexed most eighteenth-century naturalists, has been termed 'the denudation dilemma'.[28]

The denudation dilemma was aggravated by another theological development of the late seventeenth century. The Parousia had for long been regarded as imminent, and it had therefore been unnecessary to consider the effects of indefinitely prolonged denudation, but by 1700 the end of the present order was being predicted with much less confidence. For the first time it now became necessary to face the unpleasant fact that if denudation was a reality, then its extended action could result only in the obliteration of man's continental homes. Clearly, such an event could have no place in the plans of a benign deity, and in any case, as the Bishop of Clogher pointed out in 1754,[29] such destruction of the continents would obviously contravene God's covenant with Noah mentioned in *Genesis*: 'neither shall all flesh be cut off any more by the waters of a flood; neither shall there any more be a flood to destroy the earth'. The bishop found the idea of denudation so repugnant that he dismissed it as an atheistical fallacy.

The bishop was by no means unique in his views on denudation. During the previous century 'adherence unto antiquity', and the doctrine of the macrocosm and the microcosm had both helped to reinforce belief in the reality of denudation, but by 1700 these two influences were fast dying. Gone too was the strong support which had been afforded by the belief in a decadent universe, and in the face of the denudation dilemma, faith in the reality of denudation was seriously shaken. True, bibliolatry was not yet dead, and there were still texts that could be adduced to support a belief in denudation, but as soon as the reality of denudation was called into question, scholars speedily discovered that other texts were available seeming to support the view that topography is immutable. As a result of all these developments, eighteenth-century scholars were left with an understanding of the exogenetic forces which was much inferior to that possessed by their seventeenth-century predecessors.

Resolutions of the Denudation Dilemma

The denudation dilemma was an eighteenth-century problem, but its roots can be traced back into the previous century. It was never then a problem of any significance, but a few of the more perceptive

113

seventeenth-century scholars did have an inkling of the nature of the dilemma. Their references to the subject were not of sufficient importance to detain our attention earlier, but now it is proposed to transgress the time-bounds of the present chapter by taking a backward glance into the seventeenth century in order to examine the earliest glimmerings of the denudation dilemma. The purpose of such a backward glance is to set the eighteenth-century dilemma in its perspective, and to present a synoptic account of the dilemma from its birth in the 1620s, down to the early years of the nineteenth century. There is need of such a synoptic account because the existence of the dilemma has escaped the notice of historians of science.

George Hakewill in his book *An Apologie of the Power and Providence of God in the Government of the World*, first published in 1627, was evidently the earliest British scholar to grapple with the denudation dilemma. The object of the book was to refute the doctrine of a decadent universe – Hakewill regarded the idea as insulting to the Creator – and he sought to show that although the world is constantly changing, everything remains in balance so that there is neither improvement nor deterioration. In the human field, for example, he admitted that the ancients had far surpassed his contemporaries as architects, sculptors, and orators, but he believed that his contemporaries were unequalled as geographers, printers, and gunners. Similarly, he held that as one empire declines, another rises, and that in Nature, periods of cold and of flooding are balanced respectively by periods of heat and of drought. His interpretation of Nature was thus identical to that expounded by Hooke forty years later. Hakewill claimed to have derived his philosophy from Philo, Ovid, and Plato, and he termed it a belief in 'circular progress' or 'reciprocall compensation'.

When it came to discussing denudation, Hakewill was too wise to deny its reality. Instead, he tried to show that denudation plays an important role in the terrestrial economy, and he suggested three ways of reconciling a belief in denudation with his notion of a permanency in Nature. Firstly, he urged that a limited amount of denudation, far from being harmful to the globe, might actually improve it as a home for mankind, and he suggested, for example, that in the course of time denudation might convert barren mountain slopes into cultivable land. Secondly, he claimed that even if denudation did completely destroy a mountain, the debris would merely be deposited in neighbouring valleys, resulting eventually in a kind of inversion of relief.

. . . and consequently that in the whole globe of the earth nothing is lost, but onely removed from one place to another so that in processe of time the highest mountaines may be humbled into valleyes, and againe the lowest valleyes exalted into mountaines.[30]

Similarly, he claimed that what the sea gains by erosion in one place, it loses in another as a result of coastal progradation, so that the globe's total land area ever remains constant. Finally, and perhaps a little inconsistently in view of his second argument, he claimed that Nature contains some built-in mechanism whereby most of the damage effected by denudation is repaired almost instantaneously.

The notion that rocks vegetate was of course widespread in Hakewill's day, but his belief in a renovating process involved much more than the mere growth of new rocks to replace those which had decayed. He suggested, as had Antonio Galateo and Bernard Palissy long before,[31] that the material removed by denudation and carried off to the sea by rivers, is in some way restored to the continents by means of a return atmospheric circulation. Violent storms in mountain regions, he wrote, 'doe not reave more from them at some times, then moderate gentle showres dewes & mists at other times repay home to them againe'.[32] He was vague as to the precise working of the restorative process, but he observed:

> My opinion then is, that all this earth, some sooner some latter is by agitation turned into water, and this water partly drayned out by rivers, and partly drawne up in vapous by the Sunne beames and carried by winds into those places from whence this earth was taken, and so thickened into water and this water by degrees condensate into earth, and from thence a sufficient recompence made though not in all places alike, yet enough for the preservation of the whole.[33]

It was this process of regrowth, he claimed, which made it possible for Habakkuk (3, verse 6) to refer to mountains as 'everlasting', and to hills as 'perpetual'.

Views on denudation similar to those expressed by Hakewill, are to be found in the works of a few other seventeenth-century scholars. Sir Matthew Hale, for example, accepted that the ravages of denudation are repaired by 'terrene fæces' deposited along with the rain,[34] and the notion that the sea's gains in one place are balanced by its losses in another, crops up in many texts, including those of Carpenter and Ray.

Ray was Hakewill's chief disciple. He lived at a time when the denudation dilemma was beginning to be generally recognised, and in his discussion of denudation he freely confessed his indebtedness to Hakewill's work. Two of Hakewill's ideas found their way into Ray's writing. Firstly, Ray adopted the view that denudation is valuable because the debris worn from the mountains replenishes the soil of neighbouring lowlands, and secondly, he accepted the probable existence of a mysterious renovating agent. In borrowing the latter idea from Hakewill, however, Ray was not really concerned with that terrestrial *status quo* which was the crux of the denudation dilemma; he was more concerned with explaining why secular rates of degradation were negligible, when, as evinced by the debris mantling the world's lowlands, denudation had evidently accomplished so much during the six millennia of the Mosaic chronology.

Ray was one of a tiny minority; among most seventeenth-century authors there was far too thorough-going a belief in denudation for much heed to be paid to Hakewill's unorthodox views on the subject. It was not until the appearance of Woodward's theory of the Earth in 1695 that the denudation dilemma really impinged upon the history of geomorphology. It will be remembered that Woodward viewed the Flood as an act of divine re-creation, and he believed that the world which emerged from the diluvial metamorphosis was a world perfectly adapted to the needs of fallen man. He refused to accept that this carefully planned world could have undergone any important modification as a result of post-diluvial seismic or volcanic activity, and he would equally have liked to dispense with the idea of topography slowly being defaced by denudation. The modifications which previous writers had attributed to agents such as denudation, Woodward noted in a passage quoted earlier, 'are without use, and have no end at all, or, which is worse than none, a bad one: and tend to the damage and detriment of the Earth and its Productions'.[35] The age of teleology had arrived; a geomorphic process was now acceptable only if it could be shown that it contributed to the welfare of mankind. Woodward's attitude towards terrestrial changes is epitomised in the following passage:

> If that same Power be yet at the Helm: if it preside in the Government of the Natural World: and hath still the same peculiar Care of Mankind, and, for their Sake, of the Earth, as heretofore . . . then may we very reasonably conclude 'twill also continue to preserve this Earth, to be a

convenient Habitation for the future Races of Mankind, and to furnish forth all things necessary for their use, Animals, Vegetables, and Minerals, as long as Mankind it self shall endure; that is, till the Design and Reason of its Preservation shall cease; and till then, so steady are the Purposes of Almighty Wisdom, so firm, establish'd, and constant the Laws, whereby it supports and rules the Universe; the Earth, Sea, and all natural things will continue in the state wherein they now are, without the least Senescence or Decay, without jarring, disorder, or invasion of one another, without inversion or variation of the ordinary Periods, Revolutions, and Successions of things: and we have the highest security imaginable, that While the Earth remaineth, Seed-time and Harvest, and Cold and Heat, and Summer and Winter, and Day and Night shall not cease.[36]

Although Woodward's teleological view of nature led him to regard the continents as immutable, his considerable field experience had taught him that rocks do indeed show signs of decay. He sought escape from the denudation dilemma in two ways. Firstly, he joined Hakewill in claiming that Nature contains a built-in process capable of repairing most of the damage inflicted by denudation. The material worn off the continents, Woodward held, is swept down to the sea by rivers, but once there, it is prevented from sinking by the 'greater crassitude and gravity' of the sea-water. Instead, the debris remains at the surface of the sea whence it is raised into the atmosphere along with water vapour, and then returned to the continents by means of 'fruitful showers'. In this way he supposed the effects of most denudation to be nullified. Secondly, and again in a manner reminiscent of Hakewill, Woodward admitted that a limited amount of unrepaired denudation might have its place in the divine plan for human happiness. The reasoning that elicited this concession is outlined in the following passage. As mankind multiplied in post-diluvial time, Woodward observed,

. . . the Hills and higher Grounds began to be needed, those Rocks and Mountains which in the first Ages were high, steep, and craggy, and consequently then inconvenient and unfit for Habitation, were by this continual Deterration brought to a lower pitch, rendered more plain and even, and reduced nearer to the ordinary Level of the Earth; by which means they were made habitable by such time as there was occasion for them.[37]

It was nevertheless only with the greatest reluctance that Woodward made these concessions to a belief in denudation. So convinced was he that Nature is designed to preserve a *status quo* – a 'just Æquilibrium' as he termed it – that he firmly denied the possibility of both marine erosion and coastal progradation. Indeed, so strong was his conviction, that he put the clock back several millennia by claiming that Herodotus and Aristotle had both been mistaken in supposing the Nile capable of extending Egypt northwards. The second edition of Woodward's *Essay* was published in 1702 and the third in 1723, and throughout the eighteenth century the book remained one of the clearest illustrations of the nature of the denudation dilemma.

During the eighteenth century the denudation dilemma became a general problem. Numerous writers sought to resolve the dilemma, although the nature of the problem is usually implicit in their works, rather than explicit. Like Woodward, most of these scholars sought escape from the dilemma by invoking a restorative process or by claiming that limited denudation was beneficial to the Earth, but slowly, as the century passed, a new, third solution crept upon the scene. Woodward had denied the reality of marine erosion, but half a century later some scholars were seriously claiming that the whole concept of subaerial denudation was an absurd fiction. Some quotations will serve to illustrate these three resolutions of the dilemma as they appear in eighteenth-century literature.

The mysterious renovating process was the least popular solution. It nevertheless found expression in Dr. John Clarke's 1723 translation of Jacques Rohault's *System of Natural Philosophy*:

> The Earth therefore, which has so long withstood the Force of the subtle Matter of its Vortex, must long since have been entirely worn out and destroy'd, or at least, very much changed to the worse from what it once was, unless it had been continually supplied and repaired from somewhere else. But since we are sure that it does subsist still, and that it does not appear at all different to us from what the Antients describe it, this is sufficient Proof that it is repaired as fast as it wastes.[38]

Later in the century this particular solution to the dilemma fell from favour doubtless as a result of the death of the ancient belief in rock-growth, but, as we will see later, it made its final appearance in British geological literature as late as 1831. The notion that a *status quo* is maintained along the world's coastlines as a result of erosion being balanced by progradation proved rather more popular, and it too survived into the nineteenth century. Ralph Sneyd in 1783 likened the changes resulting from the advance of the sea in one place and its recession in another, to the change which takes place in a gentleman's appearance when his wig slips slightly, exposing one cheek and covering the other.[39]

The second suggested solution to the denudation dilemma – the notion that limited denudation was beneficial to the earth – proved very popular. Goldsmith was one who subscribed to this view. He observed in 1774:

> If mountains, therefore, were of such great utility as some philosophers make them to mankind, it would be a very melancholy consideration that such benefits were diminishing every day. But the truth is, the valleys are fertilized by that earth which is washed from their sides; and the plains become richer, in proportion as the mountains decay.[40]

John Williams expressed a similar view in 1789. He claimed that none of the material worn from the mountains is ever deposited in the ocean, because there it would serve no useful purpose. The debris is, he observed,

> . . . all well and wisely disposed of for the benefit and advantage of the present earth, and the inhabitants of it. Part of it is lodged in lakes, and in deep unseemly gulphs, in the course of the rivers, which are improved thereby into rich and pleasant valleys and plains, and the residue is carried along by the floods, to the borders of the ocean, where it is very happily disposed of to form new land, which in fact enlarges the bounds of our habitations, and in time becomes the most useful, the richest, and most convenient parts of the earth for society and commerce. Great numbers of magnificent friths, extensive bays, long inlets and arms of the sea, have been filled up by the waste of the mountains, which are now improved into rich and plentiful countries; and upon which are built many flourishing towns and cities, which enrich those countries by the extensive commerce carried on in them.[41]

Williams even claimed that the lowering of some of the world's higher mountains would be advantageous to mankind.

> The high and inaccessible mountains, which are immersed in the clouds, and in the cold and frozen regions of the atmosphere, are penetrated and decomposed by the changes of the air and weather, and washed down

by the rains and melted snows; and the matter carried down by the floods is formed into new land, more level, useful, and commodious for man and beast.[42]

He nevertheless refused to believe that mountains could ever be completely destroyed by denudation. He argued that as the topography is lowered, harder rocks are exposed, and the rivers flow less rapidly, with the result that denudation is brought to a halt.

The third proposed solution to the denudation dilemma – a comlete denial of the possibility of denudation on the grand scale – was adopted by some of the most eminent British geologists of the second half of the eighteenth century. No man was more strenuous in his denunciation of the concept of prodigious denudation than was de Luc. He admitted that some limited denudation does take place, and he recognised talus sheets and alluvial fans in a region such as the Alps, as clear evidence of mountain decay, but he vigorously denied that such decay is perpetual. He claimed, rather, that once a slope has been reduced to a particular angle, it becomes stable and clothed with vegetation which 'produces an everlasting security against further demolitions by wind and rain'.[43] Indeed, he maintained that once a surface has become vegetated, it not only undergoes no further denudation, but it is likely to receive accretion from decaying vegetable matter. He refused to believe that any debris from the continents could ever be lost into the ocean basins, and his complete denial of the reality of marine erosion, obviated even the necessity of postulating a balance between coastal losses in one place and progradation in another. The impotence of the sea, he claimed, is proved by the profusion o seaweed and barnacles which grows on even the most exposed of headlands.

Many of de Luc's contemporaries adopted similar views in their attempt to resolve the dilemma. Philip Howard, for example, admitted that a limited amount of denudation was beneficial, but he refused to believe that it was a perpetual process.

Our highest mountains, become very gently-sloping hills, will be defended by grasses, plants, and woods, from any further considerable depredation, whilst the great causes of degradation will be diminished. The contracted surface of the sea will furnish less to evaporation to charge the air with superabundant moisture: the clouds, no longer attracted and broken by the aspiring summits of high mountains, will give less frequent and less violent rains. The boisterous torrent and the impetuous river become gentle streams will carry little to the sea to raise new lands, and still less to the deep to raise its bed. What may yet be washed away from one part of the land will only change place to raise it in another. Hence, from the unerring testimony of the course of nature, I will conclude with the scriptures that the waters of the ocean shall never rise again to inundate the earth.[44]

John Gough, the blind botanist and mathematician, wrote similarly of a river in 1790:

Hence it appears, that the lower parts of the channel are continually rising from the accession of fresh materials; and the upper end is gradually depressed by the removal of the same, till the whole becomes a gentle declivity, down which the current will glide, no longer capable of disturbing the impediments lying in its way.[45]

He thus had an inkling of the modern concept of the graded river, and he claimed that once such a balanced condition had been attained, a thick growth of water-plants would protect the stream-bed from any further erosion. The Irish geologist, William Richardson, was another who denied the possibility of extensive denudation. In 1803 he claimed that soil, far from being the product of rock decay, is nothing more than animal and vegetable debris, and, he continued, soil serves 'as a suit of armour, with which nature, in her wisdom, clothes the world, to protect its loose, moveable materials, and to prevent their being carried off by the rain and winds'.[46] Richardson's compatriot, Richard Kirwan, admitted the possibility of some denudation, but, like de Luc, he jibbed at the idea that valuable continental material could ever be lost into the ocean basins. Kirwan satisfied himself as to the impotency of the sea

. . . by inspecting the basaltic pillars on the coast of Antrim; the angles of such of these as are and have been exposed to the waves, perhaps for some thousand years, are just as sharp as those of such pillars as are placed far beyond their reach.[47]

During the seventeenth century, Carpenter, Varenius, Burnet, Bentley, and even Ray, had freely admitted that prolonged denudation could result only in the destruction of the Earth's mountains, and perhaps even in the elimination of the continents themselves. Now, a century later, some of the leading British geologists were denying denudation even the power to etch a few lines upon the landscape. Thus did a religio-scientific problem cause a major relapse in geomorphic thought, and the long-forgotten denudation dilemma is the vital key to the understanding of eighteenth-century British geomorphology. A few vestiges of the dilemma lingered on into the nineteenth century, but the dilemma ceased to be a major factor in geomorphology soon after 1800 as a result of the emergence of the new scientific approach to the history of the Earth's surface.

The Survival of a Concept

Despite the denudation dilemma, the concept of denudation never entirely disappeared into limbo. It persisted throughout the eighteenth century for three reasons. Firstly, those who resolved the dilemma by claiming that some limited denudation was beneficial to the Earth, were free to discern the work of weathering and erosion in present landscapes. Goldsmith, for example, having satisfied himself that mountains decay in order to replenish the lowland soils, went on to describe the destructive processes in the following passage:

> ... time is every day, and every hour, making depredations; and huge fragments are seen tumbling down the precipice, either loosened from the summit by frost or rains, or struck down by lightening ... and sometimes undermined by rains; but the most usual manner in which they are disunited from the mountains, is by frost; the rains insinuating between the interstices of the mountain, continue there until there comes a frost, and then, when converted into ice, the water swells with an irresistible force, and produces the same effect as gun-powder, splitting the most solid rocks, and thus shattering the summits of the mountain.[48]

Secondly, the concept of denudation persisted among a wide range of authors who accepted it as a common-sense belief, but who never thought sufficiently deeply about the subject to encounter the denudation dilemma. Daniel Defoe belongs to this category. His *Tour*, published between 1724 and 1726, contains mention of water-worn gullies in the Pennines, numerous examples of marine erosion, and a reference to the famous 'shivering' Derbyshire mountain of Mam Tor which is 'continually falling down in small quantities, as the force of hasty showers, or solid heavy rains, loosens and washes it off, or as frosts and thaws operate upon it in common with other parts of the earth'.[49] Similarly, some twenty years later, the Irish topographer, Charles Smith, claimed the precipices and scree slopes of the Monavullagh Mountains of Co. Waterford as clear evidence of wasting caused 'from their being exposed to the vast quantities of hail and snow, which fall on them'.[50] It is strange that men such as Defoe and Smith should have been so ready to invoke denudation in explanation of topographical features, when such eminent eighteenth-century geologists as de Luc and Kirwan could only dismiss belief in Nature's destructive forces as an absurd fiction. Paradoxically, it was those who thought most deeply about denudation, and about its place in the natural order, who were most likely to minimise its importance.

The third reason for the persistence of a belief in denudation is that many of the participants in the late eighteenth-century revival of geology had shed some of the earlier preconceptions about Nature. In particular, they had largely abandoned the teleological approach to landforms. As a result, they were untroubled by the denudation dilemma, and were free to accept the field-evidence of Nature's decay. Joseph Black freely discussed the work of denudation in his Edinburgh lectures as early as the 1760s, and in his later years (he died in 1799) he advised his students to visit the Alps if they wished to see a landscape resulting from denudation on the grand scale.[51] Black's Edinburgh colleague, John Walker, lectured similarly upon 'the graduall Attrition and Degradation' of mountains, and he informed his class:

> It is reasonable to think that in the progress of time mountains may be gradually worn down, because there is such a quantity of terrestrial matter thrown down from them by the rains. . . .[52]

Another Edinburgh scholar – Robert Jameson – was equally convinced of the potency of denudation. In 1798 he described the view from the summit of Ceum na Caillich in Arran in the following passage.

> Here a wonderful and most tremendous scene presents itself to our view. An immense hollow, many hundred feet deep, dreadfully rugged and broken, almost entirely surrounded with mountains, whose serrated summits are covered with immense tumuli of granite, exhibits to us, in very legible characters, the vast operations of nature, in the formation and decomposition of our globe.[53]

Seven years later, Jameson observed of the mountains of Dumfries:

... from the nature of the rocks of which they are composed, we have good reason for expecting, by the continual alteration produced on their surface by the action of frost, torrents, &c. that many metalliferous repositories, at present hid from us, will by these great natural mining operations be brought to light. At the end of every year the surface of the country is in a very different state from what it was twelve months before.[54]

At Cambridge the syllabus of John Hailstone, the Woodwardian professor, included a lecture on the terrestrial changes resulting from the decay of rocks[55] (the lecture was never delivered; in thirty years as Woodwardian professor, Hailstone apparently never gave so much as a single lecture), while in Ireland the Rev. William Hamilton became convinced of the reality of denudation after a careful study of the Antrim coast. In 1784 he observed

... that the slow but certain operations of heat and cold, together with the continued action of the air and storms, are capable of breaking and changing the most firm bodies, even the hardest rocks; while the numerous rivers on the earth's surface, and the waves which wash its shores, perpetually labour to bear all these substances into the bottom of the ocean, and thereby to reduce all things to a level situation.[56]

Hamilton believed that the persistence of mountains in the world despite the ravages of denudation, was striking proof that the Earth was only a few millennia old. Thus by the close of the eighteenth century, the concept of denudation was again sufficiently well established to allow a reversion to one of the favourite seventeenth-century proofs of the Earth's novity.

To say that eighteenth-century writers recognised the reality and potential of denudation, however, is not the same as saying that they recognised the role of denudation in shaping the Earth's present landscapes. It was possible to believe denudation capable of remodelling the Earth's surface, without at the same time seeing the Earth's present topography as representing one stage in the remodelling process. This point was noted earlier in a seventeenth-century context, but the distinction between the supposed past work of denudation, as compared with its future potential, emerges clearly from the writings of the late eighteenth-century geologists. There were then a few scholars of vision such as Black, Hutton, and Playfair (the work of Hutton and Playfair will be discussed in detail in Chapter VI), who recognised the Earth's topography as the residue left after prolonged denudation, but most of their contemporaries, while admitting the potential of denudation, denied it any importance in the shaping of present landscapes. They preferred to regard the Earth's configuration as either primitive, or else the result of seismic, volcanic, and diluvial catastrophes. Walker, for example, in one breath proclaimed that denudation was capable of destroying entire mountain ranges, while in the next breath he denied the fluvial origin of river valleys. Again, after advising proprietors to survey their estates annually in order to discover mineral-veins newly laid bare by denudation, Jameson went on to argue that much of the Earth's surface owed its form to uneven precipitation in the chaotic fluid. Finally, Hamilton, another of our exemplars of the late eighteenth-century belief in denudation, held that Nature's destructive forces were capable of base-levelling the continents, but at the same time he invoked 'some violent shock' to explain the configuration of the north Antrim coast.

This tacit rejection of a fluvialistic interpretation of present topography, in favour of an essentially catastrophic interpretation, can only be attributed to the very limited time-scale current even in an age when the hexaëmeron was being interpreted allegorically. The potential of denudation was never in question among those who had escaped from the shackles of the denudation dilemma, but the wearing down of the continents was clearly a very slow process, and it seemed that during the few millennia since the Creation, denudation could have proceeded no distance towards its eventual remodelling of the Earth's surface. A greater familiarity with Nature in the field had perhaps made the late eighteenth-century geologists more aware than their seventeenth-century predecessors, of the slowness with which Nature conducts her operations. For this reason, and despite their extended time-scale, the late eighteenth-century geologists were perhaps more reluctant to see the work of denudation in present landscapes, than had been the much less experienced scholars of Ray's generation.

Four features stand out in the history of British geomorphology between 1705 and 1807. Firstly, there is the long period of the subject's neglect, lasting down until the 1770s. Secondly, there is the persistence of catastrophism because of the limited conception of time that prevailed even after the allegory of the hexaëmeron had been conceded. Thirdly, there is the unfortunate resurrection of a belief in the existence of primitive topography. Finally, there is the denudation dilemma which brought the concept of denudation into such serious disrepute. As a result of these four factors, and despite the vigorous late eighteenth-century revival of the earth-sciences, geomorphology in 1800 had barely moved beyond the position which it had already attained a century earlier. The great promise of the late seventeenth century had remained unfulfilled. Only one really significant advance can be attributed to the eighteenth-century scholars: late in the century they came to a realisation that any history of the Earth's surface must be based upon detailed field observations. Geological maps and cross-sections, tables of strata, and landscape sketches, began to feature in geological texts after 1778, and for the first time, naturalists started to pay serious attention to landforms in the field. Here Whitehurst and de Luc were the pioneers, but unfortunately they and those who followed them, commonly misunderstood the field evidence. This was partly because of the preconceived notions which they carried into the field with them, but equally important was their failure to appreciate the complexity of the Earth's history. To this point we will return later.

REFERENCES

1. *Philos. Trans.*, XXXVI (1729–30), No. 415, pp. 397–424.
2. *Philos. Trans.*, XXIX (1714–16), No. 344, pp. 296–300.
3. *Sci. Proc. R. Dublin Soc.*, Series II, VII (1898–1902), pp. 23–66.
4. *A Tour Through Sicily and Malta*, pp. 123 & 124; 140 & 141 (London 1775).
5. Richard Watson, *Anecdotes of the Life of Richard Watson, Bishop of Landaff*, p. 248 (London 1817).
6. Richard Watson, *An Apology for Christianity*, pp. 169–174 (Dublin 1777).
7. Jean A. de Luc, *Lettres Physiques et Morales sur l'Histoire de la Terre et de l'Homme* (Paris 1779); *The Monthly Review*, V (1791), pp. 564–585; *The British Critic*, II (1793), pp. 231–238, 351–358; IV, pp. 447–459, 569–578; *Geological Travels* (London 1810–1811).
8. *Itinerarium Curiosum*, p. 4 (London 1724).
9. *Philos. Trans.*, XXXIII (1724–25), No. 391, pp. 395–398.
10. John Walker, *Lectures on Geology*, edited by H. W. Scott, p. 168 (Univ. Chicago Press 1966).
11. *The Natural History of Cornwall*, p. 80 (Oxford 1758).
12. Walker, *op. cit.* (1966), pp. 173–175.
13. *The Scripture-Theory of the Earth*, p. 88 (London 1773).
14. *Campi Phlegræi* (Naples 1776).
15. *Arthur Young's Tour in Ireland (1776–1779)*, edited by A. W. Hutton, I, pp. 362 & 466 (London 1892).

16. *Observations relative chiefly to the Natural History, Picturesque Scenery, and Antiquities, of the Western Counties of England*, II, pp. 122-125 (Salisbury 1797).
17. *Philos. Trans.*, LXXV(1) (1785), pp. 1-7; *The Botanic Garden*, second edition, Pt. I, p. 65f (London 1791).
18. Robert Clayton, *A Vindication of the Histories of the Old and New Testament*, Pt. II, pp. 167 & 168 (Dublin 1754).
19. *The Natural History of Barbados*, pp. 3 & 4 (London 1750).
20. Philip Howard, *The Scriptural History of the Earth and of Mankind*, p. 540 (London 1797).
21. *Philos. Trans.*, XLIX(2) (1756), pp. 672-682.
22. *A Treatise on the Deluge*, p. 163 (London 1761).
23. *Ibid.*, pp. 170 & 171.
24. Gerald R. Cragg, *From Puritanism to the Age of Reason* (Cambridge 1950).
25. *An Antidote against Atheism*, Bk. II, Chap. III (London 1662).
26. *The Wisdom of God Manifested in the Works of the Creation*, pp. 200-206 (London 1692).
27. Walker, *op. cit.* (1966), p. 176.
28. G. L. Davies, 'The Eighteenth-Century Denudation Dilemma and the Huttonian Theory of the Earth', *Ann. Sci.*, XXII (1966), pp. 129-138.
29. Clayton, *op. cit.* (1754), Pt. II, pp. 17 & 18.
30. George Hakewill, *An Apologie of the Power and Providence of God in the Government of the World*, Bk. II, p. 129 (Oxford 1627).
31. Lynn Thorndike, *Science and Thought in the Fifteenth Century*, p. 213 (New York 1929); Bernard Palissy, *Recepte Véritable* (La Rochelle 1563).
32. Hakewill, *op. cit.*, third edition, Bk. V, p. 62 (Oxford 1635).
33. *Ibid.*, Bk. V, p. 63.
34. *The Primitive Origination of Mankind*, p. 96 (London 1677).
35. John Woodward, *An Essay Toward a Natural History of the Earth*, p. 227 (London 1695).
36. *Ibid.*, pp. 61 & 62.
37. *Ibid.*, pp. 238 & 239.
38. *Rohault's System of Natural Philosophy*, II, p. 123 (London 1723).
39. *A Letter to Dr. Toulmin, M.D., Relative to his Book on the Antiquity of the World* (Lewes 1783).
40. Oliver Goldsmith, *An History of the Earth, and Animated Nature*, I, pp. 162 & 163 (London 1774).
41. *The Natural History of the Mineral Kingdom*, II, pp. 105 & 106 (Edinburgh 1789).
42. *Ibid.*, II, pp. 110 & 111.
43. *The Monthly Review*, II (1790), p. 590.
44. Howard, *op. cit.* (1797), p. 540.
45. *Mem. Proc. Manchester, Lit. Phil. Soc.*, IV(1) (1793), pp. 13 & 14.
46. *Trans. R. Ir. Acad.*, IX (1803), p. 443.
47. *Ibid.*, n.d., V. p. 55.
48. Goldsmith, *op. cit.* (1774), I, pp. 155 & 156.
49. *A Tour Thro' the Whole Island of Great Britain*, edited by G. D. H. Cole, II, p. 578 (London 1927).

50. *The Antient and Present State of the County and City of Waterford,* p. 207 (Dublin 1746).

51. Various sets of notes taken by students who attended Black's lectures have survived. See National Library of Scotland MSS. 3533, 3534, and 5725. Another set of notes, taken by Thomas Cochrane, is in the Andersonian Library of the University of Strathclyde, and was published in 1966 for private circulation under the editorship of Professor Douglas McKie. See also John Robison, *Lectures on the Elements of Chemistry . . . by the Late Joseph Black, M.D.,* II, p. 6 *et seq.* (Edinburgh 1803).

52. Walker, *op. cit.* (1966), p. 173. See also Walker's *Institutes of Natural History* (Edinburgh 1792).

53. *An Outline of the Mineralogy of the Shetland Islands, and of the Island of Arran,* pp. 106 & 107 (Edinburgh 1798).

54. *A Mineralogical Description of the County of Dumfries,* pp. 75 & 76 (Edinburgh 1805).

55. *A Plan of a Course of Lectures on Mineralogy* (Cambridge 1792).

56. *Letters Concerning the Northern Coast of the County of Antrim,* p. 190 (Dublin 1786).

Chapter Five
Theorists Again
1778-1808

A detailed account of the speculations of philosophers concerning the original formation of the earth, or of the successive changes to which it has been subjected, might afford some amusement to the reader, and might not perhaps be altogether devoid of instruction, as it would exhibit, in a striking light, the rashness, folly, and presumption of the human mind, in overleaping the bounds of sober investigation and calm inquiry.

James Millar
editor of the 1810 edition of 'The
Natural History of the Mineral
Kingdom' by John Williams.

AFTER the prolonged relapse in the history of the Earth-sciences, British geologists returned in the 1770s to the task of devising theories of the Earth. The first of the new generation of theories, and the work which marks the beginning of the revival of interest in the Earth-sciences generally, was John Whitehurst's *Inquiry into the Original State and Formation of the Earth*, published in 1778. During the following thirty years other theories came from de Luc, Hutton, Williams, Kirwan, and Jameson. The present chapter is concerned with all these theories save that of Hutton, which is a work of such moment that it must stand by itself, and it will be reserved for later examination.

The work of Hutton and Jameson is well known, but the other theorists of the period have received much less attention. Whitehurst, de Luc, Williams, and Kirwan were among the leading geologists of their day, but their theories, unlike those of their seventeenth-century predecessors, have been allowed to lapse into limbo. The reasons for this are clear. The four theories are bizarre works in the tradition of Woodward, and coming a century later, they lack the interest which attaches to the seventeenth-century theories as the earliest gropings in historical geology. These four theories were in fact the last of their breed. By invoking catastrophes, their authors sought to marry *Genesis* to a rudimentary knowledge of Nature gleaned through field experience, and they claimed the results as rational, comprehensive Earth-histories. Instead of examining such freakish works, historians have preferred to focus attention upon the gentle contemporary simmerings which marked the first stages in the emergence of geology as a modern science. The four theories stand like the gigantic, slow-witted Mesozoic reptiles on the eve of their extinction, while beneath them new, rapidly evolving life-forms of enormous potential began to scurry busily to and fro.

Whitehurst and a Shivered Earth

John Whitehurst[1] was born at Congleton in Cheshire, in 1713. His father was the local clock-maker, and after receiving a rudimentary education, Whitehurst went into his father's workshop where he soon showed outstanding ability as a craftsman. He found his recreation in exploring the near-by Pennines, and the broad, wind-swept upland vistas doubtless provided a welcome relief from the microscopic work of his craft. His interest, however, soon began to focus upon the region's geology. Here he received paternal encouragement, and together father and son began to seek an understanding of the Earth's structure by exploring some of the many caves which seam the Derbyshire limestone. On some occasions, we are told, Whitehurst's geologising was sufficiently strenuous to impair his health if not his life.

About 1736, having completed his apprenticeship, he opened his own business in Derby, making clocks and scientific instruments, and his reputation grew steadily. Within a year of his arrival in Derby he made a fine new clock for the town-hall, and a grateful town-council promptly elected him a burgess. Another ingenious clock, made for the Duke of Newcastle's seat at Clumber Park in the Dukeries, earned him the Duke's patronage. Whitehurst's interest in geology survived the removal to Derby, and during his years there he joined the Lunar Society of Birmingham. He was the first to introduce geology into the ambit of that brilliant circle, and through the society he came into close contact with men such as Matthew Boulton, James Watt, and

Josiah Wedgwood, all of whom had a keen, practical interest in the Earth-sciences.

In 1775, thanks to the influence of his patron, Whitehurst was appointed Stamper of the Money-Weights, after the passage of an act to regulate the standard of the gold coinage. This appointment necessitated a removal to London, where four years later he was elected a Fellow of the Royal Society. He lived in Bolt Court, off Fleet Street, a modest man who 'never affected, after the manner of some, to know what he did not know', and there he died in February 1788.

Today Whitehurst is remembered in two contexts. Firstly, he is remembered as an ingenious maker of turret-clocks and as the inventor of a time-clock which was later manufactured and sold by Boulton. Secondly, he is remembered as a confirmed geologist of the Vulcanist school who recognised the igneous origin of the Derbyshire toad-stones. Here, however, we are concerned with his forgotten theory. He was at work upon the theory as early as 1763, and it was then that Benjamin Franklin wrote to him saying: 'Your new Theory of the Earth is very sensible, and in most particulars quite satisfactory'.[2] The work nevertheless had a prolonged gestation period, and a further fifteen years elapsed before it was finally published. Boulton, Erasmus Darwin, Joseph Priestley, and Wedgwood were among the many subscribers. A second edition, incorporating only minor changes in the theory, was published in 1786.

Whitehurst adopted a very much freer interpretation of *Genesis* than had the seventeenth-century theorists. He certainly had no qualms about accepting the allegorical interpretation of the hexaëmeron and he was even prepared to admit that the Earth might conceivably be eternal. (In this admission we can perhaps detect the influence of his friend Erasmus Darwin, who was one of the deists.) Wedgwood, however, had not expected even the mild bibliolatry which permeates the work. He knew Whitehurst as 'the free philosopher of Derby', and confessed himself 'astonish'd beyond all measure at the labour'd & repeated efforts to bring in & justify the mosaic account beyond all rhime or reason'.[3] Evidently Whitehurst was more outspoken in the Lunar circle than he dared to be in print.

In the theory, Whitehurst maintained that the Earth had originated as a liquid ball, and that this was the chaos referred to by Moses as being 'without form, and void'. Slowly, as the elements within the ball separated out, a tripartite division became clear: at the centre there was a solid nucleus which became the globe; next came a continuous envelope of water; and then finally there was an outer zone consisting of the atmosphere. Like Woodward, he believed that the rocks forming the nucleus had settled according to their relative densities, with the heaviest rocks lying deepest in the crust, but he denied that the precipitation had everywhere been equal. He claimed, rather, that the Sun and Moon had attracted the elements as they settled, so that in some places great mounds of material formed on the bed of the primordial sea, while in other places, areas of minimum precipitation were represented by large hollows. Thus the Earth acquired its primitive topography. Next, sandbanks accumulated atop the primitive seamounts, and eventually, continued deposition allowed the sandbanks to emerge to form a series of islands, which became the site of man's earliest, paradisal home.

After a time, and for reasons which Whitehurst confessed not to understand, fires broke out within the Earth and caused the crust to rise as a result of heat-expansion. This uplift was uneven; it chiefly affected the floor of the primordial oceans because elsewhere expansion was held in check by the superincumbent weight of the primitive islands. In consequence of this differential uplift, the former ocean beds emerged, and as they did so, the resultant wash swept over and drowned the primitive islands. In this way he explained the Flood, and he regarded our present continents as fashioned from the primitive ocean floors, while he saw the modern oceans as marking the watery graves of the primitive islands.

Whitehurst was a thorough-going catastrophist. He envisaged the raising of the continents as a violent process during which the strata were so broken that sea-water was able to descend into the Earth's vaults. There the water was suddenly vaporised in a series of violent explosions, and it was to these explosions that he attributed all the Earth's present topography. (This is the idea which was later elaborated by Erasmus Darwin.) He quoted contemporary accounts of earthquakes, and stories from Hooke, Burnet, Varenius, and even from Pliny, in order to demonstrate the power of earthquakes, and he regarded the Earth's topography as a ruin 'burst into millions of fragments'. He saw mountains as 'heaps of ruins'; valleys as fissures; caves as cavities amidst the fallen debris; glacial erratics as blocks hurled about by the explosions; and igneous rocks as molten material poured out during the catastrophe. His interpretation of topography therefore differs little from that offered by Burnet a century earlier, but Whitehurst, in characteristic eighteenth-century fashion, added a dash of teleology; he claimed that the innumerable fissures in the ruined strata exist to allow the dissipation of terrestrial heat which otherwise would accumulate and cause a second Flood.

To turn from the seventeenth-century theories of the Earth, to Whitehurst's theory, provides a striking illustration of the lack of progress made in the Earth-sciences during the first half of the eighteenth century. Whitehurst's theory is entirely in the seventeenth-century tradition, even to the extent of his including chapters on ante-diluvian longevity, climatic changes caused by the Flood, and the first, post-diluvian rainbow. Whitehurst's book is nevertheless not concerned exclusively with such fanciful Earth-history. The latter part of the work is devoted to a detailed discussion of the geology of Derbyshire, and in the second edition he added a similar discussion of north Antrim, together with a note on the geology of Flintshire. These regional essays show him to have been an enlightened and very competent field-geologist. He had worked out a rough succession in the Carboniferous rocks of Derbyshire, he had gained an understanding of the structure of north Antrim, and while he offered no geological maps, he did make full use of geological cross-sections to illustrate his conclusions. No previous British author had discussed the geology of any region in such detail. Whitehurst was at once both the author of the first of a new generation of fantastic theories of the Earth, and the earliest British representative of that new scientific school of geology which was already flourishing on the continent. His book therefore looks two ways. His theory carries us back to the days of Burnet, Woodward, and Whiston, but his accounts of regional geology inaugurated a new age and set a pattern for the innumerable other geologists who were soon making the British countryside ring with their hammer-blows. Whitehurst's discussions of regional geology contain scant reference to landforms, and here too he unfortunately established a precedent. He was the earliest of a horde of scientific geologists who were so obsessed with what lay beneath the Earth's surface, that they lost all sight of the form of the surface itself.

De Luc, the new Steno

Jean André de Luc[4] was born in 1727 at Geneva, where his family had resided for some three centuries. After receiving a sound education, he chose a career in commerce, and he was soon one of the leading figures in Genevese life. He devoted his spare time to scientific pursuits, and geology became his special interest. Indeed, he was the precursor of the much more famous de Saussure, because he travelled widely in the Alps, observing, collecting geological specimens, and measuring heights barometrically. The turning point in his life came in 1773 when his business failed. He quitted Geneva and moved to England where he was welcomed into intellectual life. James Watt became one of his closest friends, and Watt found de Luc 'one of the most amiable and entertaining of men'.[5] The expatriate was elected to Fellowship of the Royal Society, and he was appointed Reader to Queen Charlotte, the German consort of King George III. His duties in this office required him to be in regular attendance at court, and he therefore took up residence at Windsor where he lived for the remainder of his long life.

As a scientist he won a considerable European reputation. He made significant contributions to chemistry, meteorology, and physics, and he was a prolific writer in many other fields. Geology nevertheless remained his prime interest, and after his appointment at court, he periodically obtained leave of absence in order to make long geological excursions. The most important of these were a series of excursions made in England and on the Continent between 1797 and 1809, and they formed the basis of his *Geological Travels* published between 1810 and 1813. Considering that these particular excursions were made when de Luc was in his seventies and eighties, they represent a remarkable achievement. He emerges from his journals as a field-geologist whose enthusiasm was quite undiminished with age, and he spared himself no effort in the making of observations. He travelled long distances on foot, ascended innumerable towers and church-spires in order to obtain general prospects of a region, and when proceeding by coach, he invariably took an outside seat so as to get a better view of the topography. He was a deeply religious man, and his field-studies had only one objective – to prove that *Genesis* and Nature are in entire accord. This he achieved to the satisfaction of himself and at least some of his contemporaries, and when he died in 1817 after a long illness, he was still convinced that he had formulated the ultimate theory of the Earth.

De Luc's views on Earth-history are scattered through many of his works published between 1778 and 1813,[6] but although these writings span more than thirty momentous years in the history of geology, his views show little progressive development. He evidently formulated his theory relatively early in his geological career, and thereafter his excursions were designed to do nothing more than provide further evidence in the theory's support. As we have seen already, de Luc accepted that the six Mosaic days of Creation each represented a period of unknown duration, but otherwise his theory follows *Genesis* as closely as possible. He divided the Earth's early history into the following six periods, each period corresponding to one day of the hexaëmeron.

Period I. At the divine command 'Let there be light', a series of chemical reactions took place within the primordial chaos, and as a result the Earth's solid nucleus was formed. The remainder of the terrestrial material became a heavy, turbid fluid – the Universal Liquid – enveloping the nucleus, and holding all the constituents of the Earth's present rocks in solution.

Period II. Precipitation from the Universal Liquid began. First a thick bed of slime and mud was laid down upon the nucleus, and then there followed various chemical substances which immediately congealed to form a thick granite crust over the sludge. He claimed that granite commonly displays bedding as proof of its precipitative origin.

Period III. Precipitation from the liquid continued, and formed horizontal beds of such Primitive rocks as gneiss and schist, while inside the Earth, great cavities developed as the sludge layer dried out and became compacted. By the end of the period large areas of the crust had been undermined as a result of the cavern development, and crustal collapse occurred on a grand scale. The remaining liquid then gathered over the sunken areas, leaving the more stable regions to form the globe's earliest continents. This was the first of the Earth's general revolutions. Many of the gaping fissures formed during the collapse were filled with liquid from the ocean above, and by precipitation, were converted into mineral veins and dykes.

Period IV. This was a time of tranquillity on Earth because the divine creative hand was employed in the heavens where the Sun, Moon, and stars now came into being. Chemical precipitation nevertheless continued in the terrestrial ocean basins, and many of the rocks formed were breccias and conglomerates containing debris shaken off the continents during the preceding general revolution.

Period V. The first important event of this period was the creation of living creatures, and unlike the older, Primitive rocks, the rocks formed during this period contain great quantities of fossils. These rocks constitute the Secondary strata. The second important event was another general revolution, this time affecting only the floor of the oceans, and caused by renewed crustal collapse into great cavities beneath. The catastrophe shattered all the rocks laid down since the earlier revolution, and differential movement left the sea-bed diversified with a confusion of mountain ranges, hills, and plains. Again fissures were filled by percolating solutions from above, and new dykes and veins were formed.

Period VI. Now the Universal Liquid had almost completed its work of laying down the Earth's rocks. The final precipitates formed such unconsolidated deposits as sand and clay, and then, having lost its turbidity, the remains of the liquid became ordinary sea-water.

After the sixth 'day', the Earth's primitive, paradisal continents enjoyed a long period of tranquillity, but deep within the Earth an insidious decay was opening up a fresh generation of caverns. Perhaps de Luc pictured this decay as the macrocosmic equivalent of the moral decline of post-Adamic mankind, and he certainly saw the two trends as leading to one common end – the Flood. This was the third of the Earth's general revolutions – a revolution on such a scale that the relative positions of land and sea were completely reversed. At the divine command, the primitive continents collapsed into the voids beneath and were drowned by the inrushing sea, while the floors of the primeval oceans were left high and dry to become the Earth's present continents. The evidence of his natural chronometers satisfied him that this emergence of the modern continents was a very recent event, and that Moses was correct in placing the Deluge somewhere about four thousand years ago. De Luc seems to have imagined that his interpretation of the Flood had some claim to originality; he was evidently unaware that Hooke had advocated precisely such a diluvial interchange of land and sea more than a century earlier.

Having resolved the denudation dilemma through a denial of the potency of denudation, de Luc arrived at the conclusion that the Earth's continental topography was almost entirely the result of differential collapse occurring during the second of the Earth's general revolutions. (The third general revolution affected only the primitive continents, and did not remodel the relief already extant on the floors of the primeval oceans.) He regarded mountains as blocks which had undergone the least collapse, and hills, plains, and lake basins as a scale of features representing increasing degrees of crustal foundering. He interpreted all folding and faulting in the continental rocks as a result of movement during the revolution, and he claimed erratic boulders as contemporaneous debris which had been brought to the surface by compressed air escaping from the Earth's vaults with explosive force. Even relatively small landforms were explained as the consequence of differential subsidence, and during his English tours he recognised all the following features as collapse structures: the Avon Gorge at Clifton, the Cheddar Gorge, the Solent and Spithead, Poole Harbour, Lulworth Cove, the Dart estuary, the Hamoaze and the Tamar estuary at Plymouth, and the cliffs at Land's End.

All the present continental topography of course lay on the ocean bed from the time of the second general revolution until the diluvial metamorphosis, and de Luc regarded all the drift mantling the topography as the precipitates laid down from the Universal Liquid during the sixth 'day'. This was clearly the case, he claimed, because in most areas the character of the drifts is entirely different from the character of the underlying rocks, and the drifts therefore cannot be the product of weathering *in situ*. So convinced was he of the validity of his theory, and of the impotency of denudation, that he regarded the growan mantling the Dartmoor granite, and even the thinnest of the Earth's soils, as the final, hard-wrung deposits laid down by the remarkable primeval liquid.

De Luc was the most widely experienced British geologist of his day, and it is sad that his prolonged and strenuous efforts in the field should have yielded nothing more than this fanciful theory. He deserves credit for his recognition that any account of Earth-history must be founded upon a painstaking examination of the field evidence, but, like Woodward before him, de Luc never escaped from his bibliolatry, and as a result he viewed every facet of the landscape through Mosaic spectacles. Again we must note the failure of the late eighteenth-century theorists to advance beyond the scholars of Burnet's generation, and de

Luc's theory, with its notion of topography resulting from the collapse of the crust into expanding caverns beneath, is little more than an amplified version of Steno's scheme. Indeed, in two respects de Luc's theory is actually inferior to Steno's. Firstly, de Luc was at rather greater pains than Steno to effect a reconciliation between *Genesis* and Nature. Secondly, Steno admitted that some of his second generation rocks were clastic sediments, formed out of the plentiful debris worn from the continents by denudation. De Luc, on the other hand, minimised the work of denudation, and denied that in the ordinary course of events any valuable continental material is ever lost into the ocean basins. His Secondary rocks were essentially chemical precipitates, and their only clastic inclusions were beds containing debris broken off the continents during one or other of the general revolutions. De Luc thus affords a splendid example of the eighteenth-century relapse from sense to nonsense.

Williams the Diluvialist

Comparatively little is known about John Williams, the third of our eighteenth-century theorists, and the only two published accounts of his life are sharply contradictory at many points. He was evidently born in Glamorganshire about 1730,[7] but as a boy he lived for some time in Anglesey close to the great Parys copper mine, and this gave him an interest in mining. Later he travelled to Scotland, and there he was employed for many years in the survey of some of the large Highland properties which were then in the hands of the Commissioners for Forfeited Estates. About 1770 he took a lease of the coal-workings at Brora in Sutherland, but after four lean years there he moved on to try his luck first in Dumfries, at the lead-mines at Wanlockhead, and then in Peebles, at the silver mines near West Linton. Nowhere was he any more successful than at Brora, and having lost his capital, he became a coal-mine manager at Blackburn in West Lothian and, later, at Gilmerton near Edinburgh.

These were the years when the Industrial Revolution was beginning and providing an immense stimulus to the mining industry. Landowners were increasingly anxious to be informed of the mineral potential of their estates, and to this end Williams was frequently consulted during his years at Blackburn and Gilmerton. In consequence he was able to broaden his knowledge of Scottish geology, and early in the 1790s he enlarged his horizons still further by making an extended

tour of the English mining districts. Shortly afterwards he was persuaded to go to Italy to oversee some mining enterprise, and he died
there of typhoid in 1797.

During the Industrial Revolution, geology ceased to be a purely
academic study and became, in some of its aspects at least, a severely
practical science of value to canal and road builders, mineral prospectors and the like. Williams was the earliest of these practical,
'mechanic' geologists to launch into print, and the nature of his interest in the subject is evident from his writing. The book containing his
theory of the Earth[8] – first published in 1789 – displays on the one hand
a sound, miner's knowledge of the structure of the Earth's crust, but on
the other hand it reveals a weakness in drawing out general principles,
a confusion in presentation which perhaps reflects the poverty of his early
education, and above all, a woeful ignorance of geological literature.

If de Luc was a new Steno, then Williams was a new Woodward.
Williams believed that during the Deluge the pre-diluvial world had
been entirely dismantled, that its remains had been dissolved by the
Flood, and that then, as the waters abated, the debris had been precipitated in carefully ordered sequence. First the granites were deposited as
a foundation, and then horizontal beds of successively younger strata
were precipitated, until, near the top of the succession, the coal
measures were laid down, formed from the remains of the antediluvian
forests. Williams nevertheless differed from Woodward in one important respect: unlike Woodward, he insisted that the precipitation had
been uneven, and for this he held a diluvial tidal system responsible. He
believed that the normal celestial influences had thrown the universal,
diluvial menstruum into two antipodal tidal bulges of such size that
their passage submerged even the highest of the Earth's present mountains. The movement of the bulges around the Earth, however, was
irregular. Because of their momentum, the bulges tended to advance
ahead of the solar-lunar pull, so that every twelve hours the bulges
became stationary for a time until the celestial influences caught up
and allowed the tidal advance to be resumed. This tidal still-stand
occurred repeatedly over the same portions of the Earth's surface,
and as a result the Afro-Eurasian land complex was built up beneath
one bulge, while the Americas were precipitated beneath the antipodal
bulge. Between these two stadial positions, the tides ran freely, so that
precipitation was at a minimum, and these gaps are today represented
by the ocean basins.

Having explained the continents and ocean basins – the first order landforms – as the result of uneven precipitation, Williams went on to offer a similar explanation for such second order features as the world's major mountain chains. These, he claimed, were formed by intense local precipitation, and the highest mountains lie in the tropics because there the tidal bulges were best developed. Similarly, he insisted that mountain ranges, and the 'course or bearing' of strata (the 'strike' in modern terminology), all trend parallel to the axes of the diluvial bulges. These moved fastest in low latitudes, and mountains and strata therefore strike south-eastwards in the southern hemisphere, and north-eastwards in the northern hemisphere. The north-eastward striking Caledonian structures of Scotland and his native Wales had clearly not escaped Williams's notice.

Although his theory is in many respects similar to Woodward's, Williams rejected the idea that the expansive sheets of diluvial strata had been shattered by earthquakes. He preferred to believe that most of the disruption of the rocks was the work of diluvial torrents draining off the continents every twelve hours as the tidal bulges passed by. The damage inflicted on the tender young rocks by these powerful torrents, he claimed, is still strikingly evident in our landscapes.

> I have frequently stood upon the summit of a high mountain in a clear day, and taken a general view of the outlines of this great subject; and from such elevated situations, and indeed every where else, I see deep and evident marks of the amazing effects of the high diluvian tides, even after the higher mountains and other elevated parts of the earth were formed. All the long and deep channels through chains of lofty mountains, were cut out by the running of these weighty tides. All the gulfs and deep bays upon the face of the globe, were scooped out by the prodigious weight and force of the violent agitation and progress of the tides of the ocean from east to west.[9]

Among the features he specifically mentioned as having been produced by the diluvial torrents are the Mediterranean, the Red Sea, the Gulf of Mexico, most valleys, the lake basins of Scotland, and the Minch lying between the Scottish mainland and the Outer Hebrides.

Williams's was the least original of the late eighteenth-century theories. Consciously or unconsciously he had done nothing more than present an amalgam of elements drawn from the works of Woodward,

Catcott and Whitehurst. When a second edition of Williams's book was published in 1810, its editor (James Millar, the editor of the fourth edition of *Encyclopædia Britannica*) offered apology for the theory, but claimed – with some justification – that republication of the volume was worth while because of the valuable account of the coal measures contained therein. Lyell considered the account 'a work of great merit for that day',[10] and had Williams devoted himself to Carboniferous stratigraphy, rather than to theorising, he would doubtless today be a very much better remembered figure in the history of geology.

Kirwan and the Crystalline Mountains

Ireland has made little contribution to the evolution of British geomorphology, and only one Irishman deserves any prominence in our story – Richard Kirwan.[11] He was born in 1733, the second son of a wealthy Co. Galway family. As a child he was precocious, but, being a Roman Catholic, the penal laws of the day made it virtually impossible for him to enter a university in the British Isles. He therefore went to France to complete his education, and in 1754 he entered upon a Jesuit noviciate. At this juncture, news arrived from Dublin that his elder brother had been killed in a duel, leaving Richard as the heir to the family estates and an annual income of more than £3000. The temporal triumphed over the spiritual; Kirwan abandoned his noviciate and returned to Ireland where he later renounced his faith in favour of Protestantism.

Apart from two years spent as a member of the Irish Bar, Kirwan now devoted himself exclusively to science. Chemistry and mineralogy were his particular fortes, but his published work ranges from meteorology to music, and from philology to philosophy. He was one of the last of the polymaths. At first he lived in Dublin, but in 1769 he moved to London where he became associated with Sir Joseph Banks, de Luc, Joseph Priestley, and other leading figures in English intellectual life. Strangely, it was not until 1780 that he became a Fellow of the Royal Society, but if the Society at first overlooked his scientific work, amends were soon made, because in 1782 he received the Society's Copley Medal.

Kirwan was a hypochondriac; about 1787 he decided that his supposedly delicate frame could no longer withstand the rigours of the English climate, and he returned to Dublin. Ireland lacked the stimulating atmosphere and general scientific glitter of London, but Kirwan's reputation as a scientist of international importance was already firmly established, and honours were showered upon him. In 1807 his name was in the first list of honorary members elected to the newly founded Geological Society of London, and he was similarly honoured by other learned societies throughout Europe. Lord Castlereagh offered him a baronetage; he became Inspector General of His Majesty's Mines in Ireland; a Dublin scientific society founded in 1812 named itself The Kirwanian Society in his honour; and from 1799 until his death, he was the President of the Royal Irish Academy. In Dublin society he was noted for his many eccentricities, and one of these proved to be his undoing. He believed that illness could be cured by starving the patient, and when, at the age of seventy-nine, complications followed his eating of an ill-baked apple-dumpling, the self-inflicted starvation remedy was more than his constitution could stand. He died in June 1812.

Despite the reputation he enjoyed among his contemporaries, Kirwan in fact made no lasting contributions to the advancement of science. He was one of the last defenders of the phlogiston theory, and as one of his biographers has justly remarked: 'Scarcely ever did he advocate a theory, which was not almost immediately discovered to be unfounded'.[12] Although as a geologist Kirwan's chief interest lay in mineralogy, he late in life turned to the problem of Earth-history. His views on the subject are contained in his *Geological Essays* of 1799, and in a series of papers which he presented to the Royal Irish Academy between 1793 and 1800.[13] These writings reveal him as a diluvial catastrophist, thoroughly conversant with the literature of geology, but sadly deficient in field experience. He is said to have made numerous geological excursions in Ireland, but scarcely ever does he adduce his own observations in support of his beliefs.

Like his fellow theorists, Kirwan believed that all the rocks of the Earth's crust had been precipitated from some primordial fluid, and he agreed with Williams that the Earth's topography owed much of its present form to the unevenness of the precipitation. He dispensed, however, with Williams's tidal mechanism, and claimed that the precipitation of the Primitive rocks was irregular merely because it took place around random local centres, just as precipitation does in a chemist's retort. Kirwan joined Newton in his regard of primitive mountains as gigantic crystals or crystal agglomerations, and conversely he saw plains as areas of minimum precipitation.

The formation of plains is easily understood, in the wide intervals of distant mountains, after the first crystallized masses had been deposited, the solid particles still contained in the chaotic fluid, but too distant from each others sphere of attraction to concrete into crystals, and particularly those that are known to be least disposed to crystallize, and also to have least affinity to water, were gradually and uniformly deposited.[14]

After the formation of this primitive topography, the level of the chaotic fluid sank, partly, Kirwan claimed, as a result of volcanoes scooping-out the ocean basins of the southern hemisphere (he offered no further enlightenment upon this remarkable event), and partly as a result of some of the fluid sinking into the vaults already prepared for its reception. As the fluid subsided, the primitive continents emerged, dried out, and became consolidated. Next the fish were created. This event took place when the fluid still stood 9000 feet above present sea-level, because, according to Kirwan, fossiliferous Secondary strata are widespread below that level, but are never found at a higher altitude. When a traveller later claimed to have discovered fossiliferous rocks at over 13,000 feet in the Peruvian Andes, Kirwan was unabashed; he immediately wrote a paper claiming that the man was incompetent and had miscalculated the altitude in question.[15]

After the creation of the fish, the lowering of the fluid continued for several centuries, and during this time the Secondary strata were formed. The material composing these new deposits came from such sources as submarine volcanoes, the destruction of the pristine continents by earthquakes, and from some slight continued precipitation from the chaotic fluid. Again the deposition was uneven, and as a result Secondary mountains were constructed in many places. In particular, such mountains were formed around the flanks of the Primitive mountains, and Kirwan, like many of his contemporaries, pictured the world's major mountain chains as consisting of a core of Primitive peaks surrounded by Secondary ridges. In explaining the uneven distribution of the Secondary rocks he resorted to a tidal theory akin to that invoked by Williams. According to Kirwan the chaotic fluid had been stirred by two sets of tidal movements. Firstly, there was a normal east to west tidal system resulting from celestial influences, and secondly, there was a vigorous north to south movement caused by 'the water trending to those vast abysses then formed in the vicinity of the south pole'. These two tidal streams, he held, were responsible for giving mountains their widely observed asymmetry, because the tides caused precipitation on the northern and eastern faces of mountains, which were therefore made gentle, while the other faces received no such accretion and remained steep.

Throughout the period when the Secondary rocks were forming, the chaotic fluid continued its retreat into the Earth's cavernous interior, until eventually the present continents were fully exposed. Then, after a suitable interval to allow the youngest of the Secondary strata to become hard, the globe entered the maelstrom of the Deluge. Kirwan believed that even the highest of the Primitive mountains had been submerged during the catastrophe, and his account of the event is scarcely less graphic than that penned by Burnet. He disputed de Luc's claim that the period had witnessed a wholesale interchange of land and sea, and he trotted out all the old arguments – some of them identical to those used by Carpenter in 1625 – for believing that the primitive continents had survived the Flood, albeit in a somewhat shattered condition.

According to Kirwan, the Flood originated in the southern hemisphere and swept northwards with 'resistless impetuosity'. Its surging waters reshaped the continents to give them their southward taper, shivered a primitive land-mass in the north Pacific to leave only a few islands, and then swept across Asia and North America, in some regions dashing mountains to pieces, and in others scouring the terrain to leave barren, soil-less deserts such as the Gobi. Further south a westerly branch of the main Pacific deluge excavated the gulfs of Tonkin and Siam, the Bay of Bengal, the Arabian, Red, and Caspian Seas, and converted Persia, Arabia, and north Africa into sandy wastes. Another, less powerful branch of the Deluge surged northwards up the Atlantic, and there, somewhere to the west of Ireland, it met the other torrents which had swept across Eurasia from the Pacific. Of this event, Kirwan observed:

> The effect of the encounter of such enormous masses of water, rushing in opposite directions, must have been stupendous, it was such as appears to have shaken and shattered some of the solid vaults that supported the subjacent strata of the globe. To this concussion I ascribe the formation of the bed of the Atlantic from latitude 20° south up to the north pole.[16]

Similarly, but on a smaller scale, he claimed that Galway Bay, where he had often played as a child, was the site of a Primitive granite mountain which had been swallowed during the catastrophe, and he attributed the folding and faulting of rocks throughout the British Isles to the same diluvial shocks. Even the columnar basalts at Staffa and the Giant's Causeway, he claimed, were 'rent into pillars' by diluvial blows.

The Flood left much of the Earth's crust unstable, and there followed a series of violent post-diluvial earthquakes as the crustal blocks settled. These shocks continued until about 2000 B.C., and they were responsible for forming the Irish Sea, the Straits of Dover and Gibraltar, the Dardanelles, the Kattegat, and the Bering Straits.

This theory in a sense epitomises the geomorphic thought of the late eighteenth century because Kirwan employs all the topography-moulding processes then in vogue: uneven precipitation in the chaotic fluid, diluvial shocks, and recent earthquakes. Kirwan believed that his theory was in entire accord with *Genesis*, but de Luc thought otherwise; in 1809 he accused Kirwan not only of being deficient in field experience, but of superficiality in his reading of Moses.[17] It may be that Kirwan took this latter point to heart because we are told that during the closing years of his life he spent long periods immersed in the Scriptures.

Jameson the Neptunist

Each of the theories of the Earth published in Britain between Steno's theory in 1671 and Kirwan's theory in 1799, has its own individual character, but one feature they all possess in common: they all claim that a large proportion of the Earth's rocks are precipitates laid down in some chaotic fluid. Admittedly, in some of the theories the fluid is primitive, while in others it is diluvial, but the employment of such a fluid, whatever its age, entitles the theories to be described broadly as 'Neptunian'. This designation, however, is normally reserved for the particular theory of the Earth devised and popularised by the German geologist Abraham Gottlob Werner, and later imported into Britain by Robert Jameson.

Werner without question was the most remarkable figure in eighteenth-century geology.[18] He came of a family which had a long connection with the mines in the Saxon Erzgebirge, and after being educated at the Freiberg Mining Academy and the University of Leipzig, he returned to Freiberg in 1775 as Professor of Mining and Mineralogy. There he proved himself to be one of the greatest science teachers of all time, and under his direction the Freiberg academy became the focus of the world's geological attention. His brilliant discourses, his painstaking practical instruction, and his immense personal charm, brought students flocking to Freiberg from all over Europe and North America.

More than any other single individual, Werner was responsible both for establishing geology as a science in its own right, and for placing the subject firmly upon the basis of field investigation. It is therefore surprising to find that he was himself a man of somewhat limited field experience. The reason for this is nevertheless clear: he was satisfied that his native state contained a full range of geological phenomena, and that it afforded the observant geologist with an ample education. Saxony was to Werner, as Tuscany had been to Steno; he saw it as a microcosm of the globe, and as a key to the whole of Earth-history. Thus, despite his insistence upon the field examination of rocks, Werner was guilty of theorising upon far too narrow an observational basis, and in consequence, his theory is just as defective as the theories of his contemporaries. So ably did he expound his theory, however, that even the most discerning members of his classes were soon convinced of the truth of their master's words. At Werner's feet students were all too readily converted into disciples, and as such they carried his ill-found doctrines throughout the world. Many geologists shed their Mosaic spectacles only to don scarcely more satisfactory lenses of Freiberg manufacture.

Werner earned his immense reputation almost entirely through his teaching; his publications were few, and they were mostly slight. He evidently had an antipathy to the physical act of writing, and there is a story that when his sister sent a messenger from Dresden to Freiberg to secure Werner's signature upon some legal document, the emissary was kept waiting for two months before Werner could be persuaded to do so much as write his name. Throughout his years at Freiberg, he was full of promises to publish detailed accounts of all his geological work, but these promises were still unfulfilled when he died in 1817. We therefore have to turn to the writings of his students in order to discover the nature of Werner's teaching, and among these students, Robert Jameson is one of the most important.

During the last decade of the eighteenth century a steady stream of British scholars made their way to Freiberg. Some went to join Werner's classes, while others, like de Luc in 1798, made the pilgrimage merely to meet the oracle and to inspect his magnificent collections. Through these personal contacts, and through the writings of his continental disciples, Werner's work became familiar to British geologists. Whitehurst's theory was completed before Werner had made much impact upon geological thought, and we are expressly informed that Williams knew nothing of Werner's particular brand of Neptunism.[19] Kirwan's theory, however, and the later writings of de Luc, doubtless owed something to the growing Wernerian influence. After 1800 this influence became strongly marked in British literature, and in 1802 John Murray published a brief English summary of Werner's Neptunian theory.[20] Not until the publication of Jameson's *Elements of Geognosy* in 1808,[21] however, did a detailed exposition of the Wernerian doctrines become available to British readers. Geognosy was the Wernerian title for what we now know as geology, and Jameson's book was the text for his geognostical lectures in Edinburgh. Reading the book, replete with Saxon detail which must have been meaningless to Edinburgh students, it is easy for one to understand why pupils found his discourses so tiresome. The young Charles Darwin was a member of Jameson's class in 1826–27, and he found the lectures 'incredibly dull. The sole effect they produced on me was the determination never as long as I lived to read a book on Geology, or in any way to study the science.'[22]

Jameson was born in 1774, the son of a wealthy soap-manufacturer in Leith.[23] He early developed a taste for natural history, and as a student in Edinburgh University under John Walker, his interest came to focus upon geology. In the 1790s, financed by his father, he made a series of geological tours in Scotland, and these provided the material for two major works upon the geology of the Scottish islands.[24] These volumes, published in 1798 and 1800, were pioneer regional studies in the tradition established by Whitehurst in Derbyshire, and from them we learn that Jameson was already a convinced Wernerian. It was therefore only natural that in 1800 he should have made his way to Freiberg. There he fell completely under Werner's sway, and he soon became one of the master's favourite pupils. In 1802 the illness of his father necessitated Jameson's return to Scotland, but his two years at Freiberg had been sufficient to convert him into Neptunism's most devoted and ardent disciple. Two years later, when he succeeded Walker in the Edinburgh chair of Natural History, Jameson set about establishing a Scottish school of Neptunian geology. Towards this end in

1808, the same year as he published his *Elements of Geognosy*, he founded the Wernerian Natural History Society of Edinburgh, with Werner, Kirwan, and Sir Joseph Banks as its first three honorary members.

In his *Geognosy*, Jameson followed Werner very closely. His Neptunian creed contained the same two fundamental tenets that we have already encountered in the other eighteenth-century theories of the Earth. Firstly, he believed that almost all the Earth's rocks – including granite and basalt, dykes, and mineral veins – had been precipitated from a chaotic fluid which enveloped the Earth during its early history. Secondly, he claimed that the passage of time had brought about gradual changes in the nature of the fluid, so that each successive rock formation differed in character from its predecessor. Those formations which were world-wide in their distribution were designated 'Universal Formations', while those of lesser extent were termed 'Partial Formations'.

According to the Wernerian system, the following three main categories of rock could be recognised, each class containing a number of formations and representing a particular stage in the life of the chaotic fluid.

I. *The Primitive Class* (Urgebirge). These are the oldest rocks. They were formed before both the creation of life and the emergence of the continents, and the rocks are therefore crystalline chemical precipitates, devoid of all extraneous material. The chemical substances were laid down upon the uneven surface of the Earth's primeval nucleus (the cause of the unevenness was not explained), not as horizontal beds, but as successive strata conforming to the underlying irregularities, just as precipitation in the laboratory takes place on both the bottom and sides of a retort. The steeply dipping bedding sometimes observed in the rocks of this class is therefore original to them. Among the rocks of the class are such Universal Formations as granite, gneiss, mica-slate, clay-slate, Primitive limestone, Primitive trap (the term 'trap' was applied to any dark, fine-grained rock such as basalt), and serpentine-rock.

II. *The Transition Class* (Übergangs-Gebirge). After the formation of the Primitive rocks, the level of the fluid fell slightly to allow the highest of the primitive peaks to emerge. The denudation of these emergent lands fed debris into the fluid, and the rocks formed during this period are therefore partly chemical precipitates, and partly normal clastic sediments. Life now existed on a small scale and some of the rocks are fossiliferous. The rocks of the class are therefore transitional in the sense that they span the interval between Primitive and Secondary times. Among the formations of the class are the Transition limestone, the Transition trap, the Greywacke, the Transition flinty-slate, and the Transition gypsum. Again, these formations are Universal.

III. *The Floetz Class* (Flötz-Gebirge). By this stage the level of the fluid had fallen sufficiently far to expose extensive land surfaces, and the volume of detrital material fed into the fluid increased proportionally. The rocks are therefore chiefly clastic sediments, and they contain abundant fossils. They are broadly equivalent to the Secondary strata of earlier writers. The Floetz formations include the Floetz limestone, the Floetz gypsum, coal, basalt, Old Red Sandstone, and chalk. Because of the comparatively low level of the fluid at the time of their deposition, the Floetz rocks rarely rise to any altitude; some of their formations are only Partial, and the rocks normally form the lowlands around the Primitive mountains. Late in the period, however, the fluid suddenly rose, inundated the continents, and precipitated the transgressive Newest Floetz Trap Formation. This is the formation which Jameson supposed to form the Castle Rock and Arthur's Seat in Edinburgh, the Pentland Hills, north Antrim, and most of Iceland. With the precipitation of the Newest Floetz Trap, the work of the fluid was almost complete; it resumed its subsidence, and the remains of the fluid became the water of the modern oceans.

Jameson in 1808[25] was convinced that the broad pattern of the Earth's relief was the result of uneven precipitation within the chaotic fluid. He believed that the Earth's highest mountains and uplands were primitive features, formed as a result of the Primitive rocks being laid down over the irregular surface of the primeval nucleus and reflecting its every contour. He pictured the remainder of the Earth's topography as the result of the Transition and Floetz rocks filling in the intervals between the primitive features, with each of the younger formations giving rise to its own distinctive type of terrain. He did nevertheless admit that the topography formed by precipitation had later been modified in detail by other processes. To two of these processes he attached particular importance. Firstly, in a manner reminiscent of Williams's theory, he claimed that many valleys and other features had been scooped-out by currents in the retreating waters of the Newest

Floetz Trap transgression. Secondly, he admitted, as we saw earlier, that many present landscapes owed their minutiae to subaerial denudation effected since the continental emergence. As a result of the activities of these currents within the fluid, and of more recent denudation, he claimed, the Universal Formations are today much less extensive than at the time of their deposition. He recognised the products of this rock destruction – sand, gravel, clay, and alluvium – as forming a fourth, albeit relatively insignificant class of rocks – the Alluvial Class (Aüfgeschwemmte-Gebirge).

Jameson had shed the bibliolatry which had so hampered his predecessors, and he made no attempt to relate the Neptunian theory to *Genesis*. The theory was nevertheless attractive to those who were still anxious to reconcile geology with the Scriptures. It represented the continents as the planned end-product of a now extinguished creative process, and the theory therefore accorded well with the linear Christian interpretation of history. With a little imagination it was possible to fit the theory to the allegorical days of the hexaëmeron, and in the Newest Floetz Trap transgression the theory even offered an event to correspond to the Noachian deluge. Neptunism represents the adolescent stage in the evolution of modern geology. The first fruits of the new scientific approach to the subject had become available, and in their naïve ebullience, Werner and his followers assembled their rudimentary knowledge of Nature into a theory which accorded with their preconceived notion of Earth-history as a steady linear progression. Only later did sufficient knowledge accumulate to show that the pattern of Earth-history is in fact not a straight line, but a series of complex cycles within cycles.

Jameson was of course aware that the Neptunian theory left certain puzzling questions unanswered. What, for example, was the trigger which caused the chaotic fluid to begin its remarkable series of precipitations? How was the fluid able to contain sufficient material to form all the Earth's rocks? Above all there was the problem of the fate of the fluid. Where had it disappeared to as it completed its work? Jameson, like Werner, and in a manner reminiscent of Whiston a century earlier, suspected that the fluid had gradually been drawn off into space. In view of these difficulties, and in view of his considerable field experience, it is strange that Jameson should have remained loyal to this extravagant theory long after he had escaped from the compulsive atmosphere of Freiberg. More especially, it is difficult to understand how he could still believe basalt to be a chemical precipitate decades after the Vulcanists had convincingly demonstrated its true igneous origin. Yet we have Darwin's assurance that as late as 1827 Jameson was still teaching the Neptunian origin of basalts, dykes and veins. Jameson never declared his renunciation of the Neptunian theory in print, but it seems that he did publicly admit the error of his former beliefs at a meeting of the Royal Society of Edinburgh sometime before 1839.[26] He himself lived on until 1854 when the University of Edinburgh had the misfortune to lose two Professors of Natural History in the same year; Jameson, an anachronistic survival from a bygone age, died in April, and his successor, the youthful and much respected Edward Forbes, died in November.

The seventeenth-century theories of Burnet, Woodward, and Whiston stimulated enormous interest at the time of their publication, and they continued to be widely read down to the closing years of the following century. The theories of Whitehurst, de Luc, Williams, and Kirwan, on the other hand, fell still-born from the presses, and they were soon forgotten amidst the general excitement which accompanied the forward surge of geology during the early years of the nineteenth century. The Neptunian theory was more successful. Jameson was not alone among British geologists in his devotion to the Wernerian system, and despite the dryness of his lectures, and a rather forbidding personality, he did succeed in establishing a thriving Edinburgh school of Neptunian geology. In due course, however, this theory too met the fate it deserved. Today it stands together with the earlier theories in a dusty and little frequented corner of the museum of geological curiosities. Very different is the story of the theory devised by one of Jameson's Edinburgh contemporaries – James Hutton. In their day, the Neptunian and Huttonian theories were bitter rivals in seeking general acceptance, but whereas the Neptunian theory now lies in limbo, the Huttonian theory lies securely at the core of modern geology. To the Huttonian theory we must now turn our attention.

REFERENCES

1. *The Works of John Whitehurst, F.R.S.*, edited by C. Hutton, pp. 6–20 (London 1792); J. Challinor, 'From Whitehurst's "Inquiry" to Farey's "Derbyshire": A Chapter in the History of English Geology', *Trans. N. Staffs. Fld. Cl.*, LXXXI (1946–47), pp. 52–88; Robert E. Schofield, *The Lunar Society of Birmingham* (Oxford 1963), *passim*.

2. Schofield, *op. cit.* (1963), p. 176.

3. *Ibid.*, p. 177.

4. Despite his interest, de Luc has never found an English biographer. The best English accounts of his life are contained in the *Dictionary of National Biography*, and in *Phil. Mag.*, L (1817), pp. 392–394.

5. James P. Muirhead, *The Origin and Progress of the Mechanical Inventions of James Watt*, II, p. 189 (London 1854).

6. See especially de Luc's letters to Professor Blumenbach in *The British Critic*, II (1793), pp. 231–238, 351–358; III (1794), pp. 110–118, 226–237, 467–478, 589–598; IV (1794), pp. 212–218, 328–336, 447–459, 569–578; V (1795), pp. 197–207, 316–326. The letters, edited by Henry de la Fite, were republished in London in 1831 as *Letters on the Physical History of the Earth, addressed to Professor Blumenbach*. Other works by de Luc are included in the bibliography of the present volume.

7. John Williams, *The Natural History of the Mineral Kingdom*, second edition, edited by James Millar, I, pp. vii and viii (Edinburgh 1810); P. Neill, 'Biographical Account of Mr. Williams, the Mineralogist', *Annals of Philosophy*, IV (1814), pp. 81–83.

8. *The Natural History of the Mineral Kingdom* (Edinburgh 1789).

9. *Ibid.*, II, pp. 213 & 214.

10. Charles Lyell, *Principles of Geology*, I, p. 67, (London 1830).

11. J. O'Reardon, 'The Life and Works of Richard Kirwan', *The National Magazine*, I (1830), pp. 330–342; M. Donovan, 'Biographical Account of the Late Richard Kirwan, Esq.', *Proc. R. Irish Acad.*, IV (1847–1850), pp. lxxxi–cxviii; J. Reilly and N. O'Flynn, 'Richard Kirwan, an Irish Chemist of the Eighteenth Century', *Isis*, XIII (1930), pp. 298–319; P. J. McLaughlin, 'Richard Kirwan: 1733–1812', *Studies*, XXVIII (1939), pp. 461–474 and 593–605, XXIX (1940), pp. 71–83 and 281–300.

12. Richard Ryan, *Biographia Hibernica*, II, p. 358 (London 1821).

13. 'Examination of the Supposed Igneous Origin of Stony Substances', *Trans. R. Ir. Acad.*, V, N.D., pp. 51–81; 'On the Primitive State of the Globe and its Subsequent Catastrophe', *ibid.*, VI (1797), pp. 233–308; 'An Essay on the Declivities of Mountains', *ibid.*, VIII (1802), pp. 35–52; *Geological Essays* (London 1799). The first three essays in the 1799 volume were a reprint of the paper of 1797 on the primitive state of the globe.

14. *Trans. R. Ir. Acad.*, VI (1797), p. 248.

15. *Ibid.*, VIII (1802), pp. 29–34.

16. *Ibid.*, VI (1797), p. 288.

17. *An Elementary Treatise on Geology*, pp. 368 & 383, (London 1809).

18. G. Cuvier, 'Historical Eloge of Abraham Gottlob Werner', *Edinb. Philos. Journ.*, IV (1820–21), pp. 1–16; Frank D. Adams, *The Birth and Development of the Geological Sciences*, pp. 209–227 (Baltimore 1938).

19. Neill, *op. cit.* (1814), p. 82.

20. *A Comparative View of the Huttonian and Neptunian Systems of Geology* (Edinburgh 1802).

21. The work is volume III of Jameson's *System of Mineralogy* (Edinburgh 1804–1808).

22. Francis Darwin, *The Life and Letters of Charles Darwin*, I, p. 41 (London 1887).

23. L. Jameson, 'Biographical Memoir of the late Professor Jameson', *Edinb. New Philos. Journ.*, LVII (1854), pp. 1–49; V. A. Eyles, 'Robert Jameson and the Royal Scottish Museum', *Discovery*, XV (1954), pp. 155–162; J. Ritchie, 'A Double Centenary – Two Notable Naturalists, Robert Jameson and Edward Forbes', *Proc. roy. Soc. Edinb.*, LXVI B (1955–57), pp. 29–58.
24. *An Outline of the Mineralogy of the Shetland Islands, and of the Island of Arran* (Edinburgh 1798); *Mineralogy of the Scottish Isles* (Edinburgh 1800).
25. See also Jameson's *A Mineralogical Description of the County of Dumfries*, pp. 12–24 (Edinburgh 1805).
26. Archibald Geikie, *Life of Sir Roderick I. Murchison*, I, pp. 108 and 109 f (London 1875); *Edinb. Rev.*, LXIX (1839), p. 455 f.

Chapter Six

The Huttonian
Earth-Machine 1785-1802

A constitution of the world which did not maintain itself without a miracle, has not the character of that stability which is the mark of the choice of God.

Immanuel Kant
in his 'Allgemeine Naturgeschichte
und Theorie des Himmels' of 1755.

No figure in the history of the Earth-sciences has received more posthumous adulation than has James Hutton; a tablet placed over his grave in a secluded corner of Edinburgh's historic Greyfriars Churchyard in 1947, hails him as 'The Founder of Modern Geology'. Geomorphologists have joined the geologists in their homage, and for more than a century Hutton has been acclaimed as the father of modern geomorphic thought. He is therefore deserving of our especial attention, even though we may conclude that in some respects the praise lavished upon him has been misplaced.

Hutton was born in 1726, the son of a respected Edinburgh merchant.[1] He was educated at the city's High School, and at the age of 14 he moved on to Edinburgh University as a student of the humanities. There, in the course of a class in logic, an allusion to a simple experiment with acids aroused his interest in chemistry, and that subject became his lifelong passion. Chemistry of course found no place among the supposedly humanising subjects of Hutton's curriculum, and he therefore had to be self-taught in the subject. His earliest tutor was *Lexicon Technicum* by the same John Harris who had been a protagonist in the controversy over the Woodwardian theory of the earth, and the *Lexicon* contains a discussion of the theories of Burnet, Woodward, and Whiston. It was thus perhaps while thumbing through the pages of the *Lexicon* in search of chemical instruction, that Hutton first encountered the problem of Earth-history.

In 1743 he left the University and was apprenticed to a Writer to the Signet, but it soon became clear that he was not destined to make his mark in the legal profession; instead of copying papers and studying court proceedings, he was frequently found amusing both himself and his fellow apprentices with chemical experiments. His master, no doubt eager to rid the office of so diverting an influence, speedily released him from his engagement, and Hutton now turned to medicine as a field where his chemical interests might find some more legitimate outlet. After three years as a medical student in Edinburgh, he moved in 1747 to Paris, and then to Leiden, to complete his studies. At Leiden, in September 1749, he received the degree of M.D. for a dissertation entitled *De Sanguine et Circulatione Microcosmi*. It was exactly one hundred years since Bernhard Varenius had received the same Leiden degree.

Although he was now a qualified physician, Hutton had no desire to enter practice; henceforth all attempts to solicit his medical advice were jokingly turned aside. At this juncture the works of Jethro Tull, and a chance Edinburgh encounter with a Norfolk farmer, suddenly aroused his interest in agriculture, and he immediately resolved to apply himself to the study of the subject. Perhaps as a gentleman-farmer he had hopes of finding the leisure and independence he needed to develop his scientific interests. Be that as it may, he certainly took his new career very seriously. He went to live with farming families in East Anglia to study the revolutionary agrarian techniques being developed there, and from the eastern counties he made a number of pedestrian tours into neighbouring parts of England in order to investigate their farming-practice. Then, late in 1754, after a brief agrarian sally into the Low Countries, he decided that his novitiate was over; he purchased one of the light Norfolk ploughs, hired a Norfolk ploughman, and returned to Scotland to a small property near Duns, in Berwickshire, which he had inherited from his father.

For the next fourteen years Hutton led the life of a country gentleman, but he had returned from England with much more than the novel agrarian techniques which at first so diverted his neighbours – he had brought back an interest in geology. In a letter written at Yarmouth as early as 1753, he confessed that he was searching eagerly for rock-exposures in quarries, ditches, and river-beds, and while he lived at Duns he continued to make regular geological excursions. From a few surviving Huttonian letters it emerges that by about 1770 his travels had given him a sound, if rudimentary understanding of the stratigraphy of both England and Scotland, and in one of his letters he promised to make a map of 'the mineral geography' of Scotland for his correspondent.[2] If this ambitious promise was kept, the map must have been the earliest geological map of Scotland.

About 1768, after some differences of opinion with his Norfolk ploughman, Hutton abandoned agriculture and moved to Edinburgh, where for the remainder of his life he lived as a gentleman of leisure, devoted to science and scholarship. Of the many notable figures then resident in the Scottish capital, his closest companions were Joseph Black, the chemist, Adam Smith, the author of *The Wealth of Nations* (Hutton and Black edited Smith's posthumous essays), and John Clerk of Eldin, an amateur naval tactician who had never been to sea, but whose doctrines found favour with Nelson himself. One other of Hutton's friends deserves mention, although he was not an habitué of the Edinburgh circle – James Watt the engineer. Watt was doubtless introduced into Hutton's circle by Black, because before coming to Edinburgh in 1766, Black had been a professor in Glasgow where Watt was a mathematical-instrument-maker, and the two men had developed a high regard of each other. Hutton and Watt were firm friends as early as 1767, and in 1774, soon after Watt's removal from Glasgow to Matthew Boulton's factory at Soho, near Birmingham, Hutton paid Watt a visit. There he met other members of the Lunar circle, and afterwards he and Watt together toured the Cheshire salt-mines. Watt's departure from Scotland did nothing to break the friendship, because in 1784 we find him inviting Hutton to Soho to meet Jean de Luc, and a letter preserved in the Black Papers in Edinburgh University reveals that four years later Watt presented Hutton with one of the multifarious products of Boulton's factory – a plated tea urn. This association of Hutton with Watt is of some interest because it will be shown later that Hutton may have found geological inspiration in Watt's steam-engine.

Hutton's earliest publication – a pamphlet on the difference between coal and culm – appeared in Edinburgh in 1777, and during the next twenty years he published works in the fields of chemistry, meteorology, physics, and metaphysics, while at the time of his death he was writing a treatise on agriculture. His reputation with posterity nevertheless rests entirely upon his geological writings, and more especially upon his famous theory of the Earth which was first made public in 1785.

It has recently been suggested that in devising his theory Hutton may have drawn upon the slightly earlier writings of an English physician named George Hoggart Toulmin,[3] who between 1780 and 1789 published four books containing the shadowy sketch of a theory of Earth-history very similar to that now associated with Hutton.[4] The two men may well have known each other because Toulmin was a medical student in Edinburgh for the three sessions between 1776 and 1779, but it seems that Hutton's theory was formulated before 1776 and certainly long before the appearance of the earliest of Toulmin's books in 1780. Evidence bearing upon the date of the theory comes from three sources. Firstly there is Hutton himself; in 1795 he observed:

> When first I conceived my theory, naturalists were far from suspecting that basaltic rocks were of volcanic origin.[5]

He was well-read in the continental literature, and this statement must surely place the formulation of his theory earlier than the appearance of Desmarest's great memoir on the basalts of Auvergne in 1771 and 1773.[6] Secondly, there is the less decisive assurance of his biographer, John Playfair, that the theory was completed and communicated to Black and Clerk of Eldin 'several years' before 1785. Playfair adds the explanation that Hutton was in no haste to publish the theory because 'he was one of those who are much more delighted with the contemplation of truth, than with the praise of having discovered it'. Finally, there is the important testimony of Black himself, who in a letter written in 1787 affirmed that the principal elements of the Huttonian theory had been completed more than twenty years earlier.[7] At least the outline of the theory must therefore have existed before Hutton moved to Edinburgh about 1768, and at a time when Toulmin (he died in 1817, received his M.D. at Edinburgh in 1779, and must have been born in the 1750s) was still little more than a child. If, as McIntyre has suggested, textual parallels exist between the 1783 version of Toulmin's book and the account of the Huttonian theory published five years later, then one must suspect that Toulmin was plagiarising a manuscript version of the theory to which he had been given access during his three-year sojourn in Edinburgh.[8] At least one early draft of the theory certainly did exist, because it was used by Playfair after Hutton's death, and in the Edinburgh circle of that day it was evidently common for works to circulate in manuscript long before they were published.

After fermenting in Hutton's mind for perhaps twenty years, the theory might well have died with him had the Royal Society of Edinburgh not been founded in 1783. Always eager to support the cause of learning, Hutton offered to fill out the Society's lecture-programme by communicating his theory. The first part of his paper was read on 7th March 1785 by Black, Hutton himself being indisposed, and the reading was completed by Hutton on the 4th of the following month. A 32-page summary of the theory was printed and circulated in the same year,[9] but the full paper, running to 96 quarto pages, was not published until 1788, when it appeared in the first volume of the *Transactions of the Royal Society of Edinburgh* under the title 'Theory of the Earth; or an Investigation of the Laws observable in the Composition, Dissolution, and Restoration of Land upon the Globe'.

Hutton's theory, like many of its predecessors, was acclaimed in some quarters and denounced in others. His disciples urged him to publish a more extended account of the theory than that contained in the 1788 paper, but he procrastinated, and it seemed likely that the work would never be completed. The strictures of one of his critics, however, succeeded where the encouragement of his friends had failed. In February 1793 Richard Kirwan presented a critical analysis of the theory to the Royal Irish Academy, and Hutton believed some of Kirwan's arguments to be so specious as to make a rejoinder imperative. It is strange that he should have been so aggrieved by Kirwan's censure because some earlier critics had been equally damning, but the very day after he received a copy of Kirwan's paper, Hutton began to prepare the manuscript of his expanded work for the press. He intended that the new work should be completed in three volumes, and the first two of these were published in Edinburgh in 1795 under the title *Theory of the Earth, with Proofs and Illustrations*, and they sold at 14 shillings for the pair. The first chapter of volume I is merely a reprint of the 1788 essay incorporating a few minor changes, and in the remainder of the two bulky octavo volumes Hutton develops his theory, replies to some of his critics, and adduces new evidence gleaned during his geological excursions since 1785. Eyles suggests that only 400 or 500 sets of the 1795 volumes were printed, and today the work is certainly a valuable rarity; the most recent set to come on to the market was sold in 1964 at a price of £410. Fortunately, the work is also now available in a facsimile-reprint published in 1959.[10]

Hutton died at his Edinburgh home on St. John's Hill in March 1797 after a prolonged and painful illness. He left a vast quantity of manuscripts, but unfortunately very few of these have survived. Among them was the completed text of the third volume of the *Theory of the Earth*, and after passing through various hands, six chapters from this manuscript (chapter IV to IX) came into the possession of the Geological Society of London. They were published in 1899, under the editorship of Sir Archibald Geikie, as volume III of the *Theory*, but all attempts to find the other missing chapters have proved fruitless.

The Huttonian Theory

Hutton, like his friend Adam Smith, was one of the small band of eighteenth-century British deists. He tacitly rejected all notion of divine revelation through Scripture, but at the same time he claimed that the wonders of Nature afford ample proof of the existence of a beneficent and omnipotent deity. Like so many before him, he was convinced that this deity had designed and created the world to serve as a magnificent home for mankind, and this belief is the basic premise from which the entire Huttonian theory stems.

Metaphysics led him to geology through a simple chain of reasoning. If, he argued, the Earth has been designed as a home for man, then it follows that one of its prime functions must be to provide him with nourishment in the form of plant and animal foods. Now both plants and animals depend directly or indirectly upon the existence of deep, fertile soils, and such soils are therefore basic to human existence. As a keen-eyed farmer and traveller, however, Hutton had noticed firstly that soil gradually moves downslope towards rivers, and secondly that after storms the rivers themselves are often turbid because of the volume of soil being swept seawards. At first he must have been puzzled at these observations; did the removal of soil mean that the Earth's soil resources are steadily being depleted, and that the Earth is losing its capacity to support mankind? Clearly such notions were difficult to reconcile with a belief that the Earth is the consummate creation of a divine intelligence. He claimed, therefore, that despite soil erosion, no depletion of the soil mantle occurs, because the wise Creator has furnished the Earth with rocks which steadily decay to form fresh soils.

He thus allied himself with those earlier naturalists who had viewed denudation as a beneficial process which replenishes the Earth's soil-resources, but at this point Hutton, like his predecessors, came face to face with the denudation dilemma. The formation of new soil through the continual decay of solid rocks could result only in the eventual destruction of the continents and the extinction of mankind for want of a suitable habitat. Such a fate was unthinkable, but whereas others had sought escape from the dilemma by invoking some mysterious renovating process, or by minimising the work of denudation, Hutton tried to show that denudation, far from being an evil, destructive force, is really only the first stage in the formation of a fresh generation of continents which will be ready to receive settlers as soon as the older continents become too decayed for habitation.

Hutton stated the object of his theory in the following passage taken from the exposition of 1795:

> Therefore, a proper system of the earth should lead us to see that wise construction, by which this earth is made to answer the purpose of its intention, and to preserve itself from every accident by which the design of this living world might be frustrated. For, as this world is an active scene or a material machine moving in all its parts, we must see how this machine is so contrived, as either to have those parts to move without wearing and decay, or to have those parts, which are wasting and decaying, again repaired.[11]

Apart from its expression of Hutton's intention in formulating the theory, this passage is of some interest because of its reference to the Earth as 'a material machine moving in all its parts'. This notion that the Earth is a gigantic piece of machinery crops up time and again in his geological writing, and he clearly regarded the Earth as a form of self-repairing, perpetual-motion engine. Unlike most of the earlier theorists, Hutton made no attempt to trace the origin of the Earth; he concerned himself solely with the terrestrial economy. His theory is a manual on the day-to-day running and repair of the Earth-machine, and not a treatise on its original construction.

According to Hutton, the Earth-machine is a three-phase engine. The first phase is one of denudation during which the continental rocks moulder into soil in order to maintain the Earth's fertile mantle. The second phase – a phase which overlaps the first – sees the continental debris swept oceanwards, firstly by rivers and then by marine currents, until finally it comes to rest on the ocean-floor. There the detritus is converted into new sedimentary strata as a result of fusion caused by the terrestrial heat. The third and final phase witnesses the elimination of the old continents as a result either of prolonged denudation or the collapse of their subterranean supports, and the formation of a fresh generation of continents as the newly consolidated strata are raised from the ocean floor. This elevation Hutton, like Whitehurst, attributed to the heating and expansion of material at depth in the crust, and he regarded all folding, faulting, and other displacement of the strata as the result of differential movement during the uplift. He nevertheless confessed some difficulty in understanding why the continents should remain elevated once all the local heat had been dissipated. Thus, according to Hutton's theory, a new world is shaped from the ruins of the old so that the globe is never deficient in habitable lands.

With the emergence of a new generation of continents, the whole cycle of course recommences, and continues until these continents are in their turn lowered beneath the waves. Hutton, like Hooke, recognised that after its submergence a former continent might receive a blanket of fresh sedimentary strata and then be resuscitated. These lands, composed partly of ancient rocks and partly of more recent superincumbent strata, he termed 'compound masses', and in a passage which displays his inkling of the concept of metamorphism, he observed of such a mass:

> In that case, the inferior mass must have undergone a double course of mineral changes and displacement; consequently, the effect of subterranean heat or fusion must be more apparent in this mass, and the marks of its original formation more and more obliterated.[12]

It was important for Hutton to discover some actual examples of compound masses, because they were the proof he needed to confirm his theory that portions of the Earth's surface have undergone successive periods of denudation, sedimentation, lapidification, uplift, and renewed denudation. It must therefore have been with the greatest satisfaction that in the course of his Scottish travels he discovered three localities where the contorted rocks of an ancient world plainly lie beneath the much less disturbed strata of a very much younger world. He found the first of these compound masses – unconformities in modern terminology – in 1787 at Loch Ranza, in Arran, where he saw what are now known to be Upper Old Red Sandstone beds dipping northwards and resting upon southward-dipping Dalradian rocks. He discovered his second unconformity in the same year in the Tweed basin, where he observed the Old Red Sandstone lying atop the steeply inclined Silurian greywackes, and a drawing of this section by Clerk of Eldin was included in the *Theory* of 1795 (Plate V). Finally, in the course of a boating expedition along the Berwickshire coast in 1788, he lighted upon his third unconformity at Siccar Point, where the gently dipping Old Red Sandstone again rests upon the nearly vertical Silurian rocks. Because of their association with Hutton, and their importance as confirmatory evidence of his theory, these three unconformities are today numbered among the classical geological sections of the British Isles.

Hutton recognised that the three-phase cycle must have been completed innumerable times during the life of the Earth-machine, and he regarded our present continents as merely the most recent in a whole succession of similar land masses. All former continents, he claimed, were composed of materials identical to those forming the modern continents, and similarly he believed that the processes which had operated upon the earlier continents had differed neither in their nature nor their intensity from those slow-acting processes still current today. He admitted that during a human life-span the effect of such processes may seem infinitesimal, but he recognised that if given sufficient time, even the most sluggish of those processes may effect prodigious changes. He summarised his views on the nature of the Earth-processes – and more especially his views on the violent processes so often invoked by his predecessors and contemporaries – in an important passage in the final chapter of his 1795 *Theory*:

> Not only are no powers to be employed that are not natural to the globe, no action to be admitted of except those of which we know the principle, and no extraordinary events to be alledged in order to explain a common appearance, the powers of nature are not to be employed in order to destroy the very object of those powers; we are not to make nature act in violation to that order which we actually observe, and in subversion of that end which is to be perceived in the system of created things. In whatever manner, therefore, we are to employ the great agents, fire and water, for producing those things which appear, it ought to be in such a way as is consistent with the propagation of plants and life of animals upon the surface of the earth. Chaos and confusion are not to be introduced into the order of nature, because

certain things appear to our partial views as being in some disorder. Nor are we to proceed in feigning causes, when those seem insufficient which occur in our experience.[13]

He thus banished all catastrophes from his theory. Even the Noachian deluge was excluded because, he observed, 'general deluges form no part of the theory of the earth; for, the purpose of this earth is evidently to maintain vegetable and animal life, and not to destroy them'.[14] To Hutton must go the credit for being the first British naturalist to make a complete break with Moses.

Hutton's belief that the present continents are very much younger than the Earth itself, and that they are largely composed of clastic sediments, left no room for a belief in primitive rocks which date back to the Creation and which are genetically different from all younger strata. The majority of rocks which others claimed as primitive, were for Hutton merely the remains of ancient worlds which had experienced two or more periods of mineral change. He and his disciples were able to adduce field-evidence to support this belief because they found fossils in some of the supposedly primitive rocks of Wales, the English Lake District, and the Scottish Southern Uplands, and they thereby demonstrated that these rocks were in fact marine sediments.

Granite presented a more difficult problem. It was clearly not a normal marine sediment, and if the Neptunists were correct in their regard of granite and granite topography as primitive, then the entire Huttonian theory was in jeopardy. As early as his 1785 paper, Hutton sought escape from this particular dilemma by claiming that all granites had been formed by the solidification of molten material intruded into the crust from the Earth's hot interior. Thus, in contrast to the Neptunists, he regarded granite masses as younger than the surrounding strata. He invoked a similar igneous origin for basalt, and he likewise claimed that all sills, dykes, and mineral veins have been filled with molten material rising from deep inside the Earth. Indeed, Hutton relied so heavily upon the power of subterranean heat to form rocks and to raise continents, that his system was soon christened – in derision at first – 'the Plutonic theory'.

Hutton later confessed that at the time of reading his paper in the spring of 1785, he had seen granite *in situ* in only one locality (between Peterhead and Aberdeen). Not until the summer of 1785 did he set out to examine one of the Scottish granite bodies in the hope of finding an intrusive contact which would support his granite hypothesis and, in turn, his cyclic interpretation of Earth-history. His search was soon rewarded; near Forest Lodge in Glen Tilt, Perthshire, he made his famous discovery of a series of granite veins set in the local schists, and, in his own words, 'breaking and displacing the strata in every conceivable manner . . .' These exposures in the bed of the Tilt were just as important as the three unconformities in confirming the validity of the Huttonian theory, and Hutton himself was so excited at his discovery of the granite veins that those accompanying him were convinced that he must have lighted upon nothing less than a lode of silver or gold.

There remains to be considered one final aspect of Hutton's theory: his demands upon the bank of time. The Huttonian conception of continent succeeding continent, and of vast changes being wrought by slow-acting natural processes, clearly involved a belief that the Earth is extremely ancient, and Hutton, in common with the other deists, had no hesitation in regarding time as almost unlimited. 'Time, which measures every thing in our idea, and is often deficient to our schemes,' he wrote, 'is to nature endless and as nothing . . .'[15] In tracing the history of the Earth, he noted,

> . . . we come to a period in which we cannot see any farther. This, however, is not the beginning of those operations which proceed in time and according to the wise oeconomy of this world; nor is it the establishing of that, which, in the course of time, had no beginning; it is only the limit of our retrospective view of those operations which have come to pass in time, and have been conducted by supreme intelligence.[16]

The end of the world, too, he observed, is hidden from our sight, and he summarised his views on time and Earth-history in the now famous terse phrase with which he ended his 1788 essay: '. . . we find no vestige of a beginning, – no prospect of an end.' He was nevertheless anxious to avoid offending the susceptibilities of his orthodox Christian readers. He for example emphasised that his inability to discover a beginning to Earth-history was not to be interpreted as a denial of the act of creation; he meant merely that the Creation was now so distant as to lie far beyond the limit of human perception. Similarly, he explained that while he claimed a great antiquity for the Earth, for most strata, and for the fossil life-forms contained therein, he had no quarrel with Moses over the novity of mankind.

> The Mosaic history places this beginning of man at no great distance; and there has not been found, in natural history, any document by which a high antiquity might be attributed to the human race.[17]

This nevertheless seems to have been a somewhat grudging concession to convention, and in spite of his protestations to the contrary, it is probable that he inclined to the usual deistical view that the universe is eternal.

Huttonian Geomorphology

Denudation is only one phase in the cyclic operation of the Huttonian Earth-machine, but, being a subaerial activity, it is much more in evidence than the other two phases which take place chiefly in the ocean-deeps. Hutton in fact found it impossible to discuss the submarine aspects of the theory in anything but the most general of terms, and as a result he devoted a quite disproportionate amount of space – 11 chapters out of 22 in the 1795 presentation – to the study of the work of denudation as it is reflected in topography. No previous British author had ever deemed landforms worthy of such a weight of words.

Hutton regarded topography as a function of three sets of processes: those of sedimentation, earth-movement, and denudation. He claimed that sedimentation gives rise to an initial relief on the ocean floor, and that this relief is then modified, firstly by folding and faulting as the newly formed strata are elevated, and secondly by the processes of denudation which come into play as soon as a continent rears above the waves. Important though the topographical effects of sedimentation and earth-movement might be locally, it was to the third set of processes – the processes of denudation – that Hutton attached by far the greatest significance. Having resolved the denudation dilemma and shown, to his own satisfaction at least, that denudation is merely the first stage in a carefully ordered reconstructive cycle, he felt free to give full rein to Nature's destructive forces in his explanatory discussion of topography. His sound understanding of the extent of denudation is revealed in the following passage from the *Theory* of 1795:

> Whether we examine the mountain or the plain; whether we consider the degradation of the rocks, or the softer strata of the earth; whether we contemplate nature, and the operations of time, upon the shores of the sea, or in the middle of the continent, in fertile countries, or in barren

165

deserts, we shall find the evidence of a general dissolution on the surface of the earth, and of decay among the hard and solid bodies of the globe; and we shall be convinced, by a careful examination, that there is a gradual destruction of every thing which comes to the view of man, and of every thing that might serve as a resting place for animals above the surface of the sea.[18]

He fully appreciated that the Earth's present relief is largely the result of prolonged circumdenudation, and 'that what we see remaining is but a specimen of what had been removed . . .' Mountains, he observed, 'are parts which either from their situation had been less exposed to those injuries of what is called time, or from the solidity of their constitution have been able to resist them better,' and, he asserted, 'we have but to consider the mountains as formed by the hollowing out of the valleys . . .'[19] He envisaged denudation on so grand a scale that even the highest of a region's modern mountain peaks might be far below the level of the initial land-surface existing before denudation began its attack. The Scottish Southern Uplands were offered as an example of one such region; he claimed that the sandstone which he had seen resting unconformably upon the greywackes in the Tweed basin and at Siccar Point, was the remains of a sandstone cover which formerly had blanketed the greywackes throughout the Southern Uplands, with the base of the sandstone standing at least as high as the present mountain tops.

In adducing evidence to support his belief in prodigious denudation, Hutton drew his readers' attention firstly to the great volume of detritus contained in sedimentary strata, secondly to those compound masses where, as in the Tweed basin, the lowest beds of the overmass are clearly derived from the rocks below the plane of unconformity, and finally to those areas where the strata are displaced by faults possessed of a large throw but yet devoid of topographical expression. He regarded an area of almost horizontal strata near Newcastle-upon-Tyne as affording evidence of this latter type, and he wrote:

> But in those strata there is a slip, or hitch, which runs from north-east to south-west, for 17 or 18 miles in a straight line; the surface on each side of this line is perfectly equal, and nothing distinguishable in the soil above; but, in sinking mines, the same strata are found at the distance of 70 fathoms from each other. Here therefore is a demonstration, that there had been worn away, and removed into the sea, 70 fathoms more

from the country on the one side of this line, than from that on the other. It is far from having given us all the height of country which has been washed away, but it gives us a minimum of that quantity.[20]

One proof of the reality of denudation eluded Hutton: he could find no secular evidence of decay during recent or historic time. Even on English hill-tops, he noted, the roads built by the Romans, and the neighbouring pits from which they dug their materials, are scarcely less fresh today than they were almost two thousand years ago. Such facts nevertheless did nothing to impair his belief in denudation; they merely further convinced him of the slowness with which Nature works, and of the vastness of the time-scale against which she conducts her operations.

According to Hutton, the agents of denudation are the Sun, the atmosphere, alternating moisture and drought, frost, running-water, and the sea, although it is not always clear exactly how he supposed each of these agents to operate. As a chemist he certainly understood the nature of chemical weathering, but he evidently regarded running-water as by far the most potent of the topography-moulding processes. The fluvial origin of valleys seemed so completely self-evident to him that he scarcely thought the matter necessary of proof. Of the then popular diluvial hypothesis of valley formation, he remarked:

> To suppose the currents of the ocean to have formed that system of hill and dale, of branching rivers and rivulets, divided almost *ad infinitum*, which assemble together the water poured at large upon the surface of the earth . . . would be to suppose a systematic order in the currents of the ocean, an order which, with as much reason, we might look for, in the wind.[21]

Although Hutton offered no logical proof of the fluvial origin of valleys, there are two places in the *Theory* where he does linger on the subject of valley-formation. The first of these occasions is in the second volume of the 1795 pair where he outlines the evolution of a drainage system on a granite body in the following terms:

> We are to suppose our mass of granite without any structure except that of the veins and cutters, formed by the contraction of the solid mass in cooling. Now, those separations will naturally give direction to the operation of the wasting causes, whether we consider these as chymical or mechanical. Hollow tracts would thus be formed in the solid mass; in those hollow ways would flow the water, carrying the detached

portions of the rock; and those hard materials, by their attrition upon the solid mass, would more and more increase the channels in which they move. Thus there would be early formed a system of valleys in this rock, and among those valleys a number of central points, or summits over which no running water would carry hard materials to operate upon the solid rock over which it flows.[22]

The other passage of interest in this context is in the third volume of the *Theory* where Hutton discusses the form of the Tay valley in Perthshire. Just below Dunkeld the Tay traverses a narrow mountain-ridge, and Hutton suggests, reasonably enough, that the difference in height here between the valley floor and the adjacent mountain peaks (a difference of more than 1000 feet) affords a minimum value for the amount of degradation performed by the river. He also draws attention to a series of sand and gravel terraces near Dunkeld standing up to 200 feet above the Tay. These are now known to be of glaciofluvial origin, but Hutton interprets them as ancient valley-floor deposits marking levels at which the Tay formerly flowed, and he accepts them as conclusive proof that the whole valley has indeed been excavated by the river itself. He concludes his discussion of the Tay valley by summarising his views on fluvial erosion in the following passage:

> In short, it is impossible to examine this river, and its many branchings, without being convinced that this district of the globe, as well as every other alpine country, has acquired its present form by the operation of water running upon the surface of the earth; that it has required an indefinite space of time to have hollowed out those valleys; and that the continual tendency of those operations, natural to the surface of the earth, is to diminish the heights of mountains, to form plains below, and to provide soil for the growth of plants.[23]

Although Hutton laid such emphasis upon fluvial processes, and although he rejected the diluvial-current hypothesis of valley formation, he never for a moment doubted the potency of the sea as an agent of erosion. He did admit that in some areas coastal progradation results from the inability of the sea to remove all the debris brought down by rivers, but he was convinced that in most places the sea is steadily devouring the continents. The configuration of any coastline, he suggested, depends firstly upon the geological structure, and secondly upon the sea's local erosive capacity. If the structure is uniform, he claimed, then the coastline will be of simple outline, but if the structure is varied, and the sea powerful, then the weaker rocks will be eaten away to form an irregular coastline such as exists in Norway or western Scotland. Like so many before him, he regarded most offshore islands as the last remnants of former land-masses which had been destroyed by the sea, but unlike his predecessors, he was able to adduce geological evidence in support of the belief. He pointed out that such Scottish islands as St. Kilda, Ailsa Craig, and the Bass Rock are all composed of rocks which must have solidified from igneous-melts beneath a thick covermass. They therefore cannot be pristine features, but are, rather, the resistant relics left after prolonged marine erosion.

One other natural process recognised by Hutton deserves mention – the work of glacier ice. His writing on this subject is tantalisingly vague, but he claimed that at one time the Alps had experienced a cold period when there were 'immense valleys of ice sliding down in all directions towards the lower country . . .'[24] He attributed this glacial period not to any refrigeration of climate, but to the fact that before their lowering by denudation, the Alpine peaks had supported the vast snowfields necessary to nourish extensive glacier systems. Clearly, this kind of reasoning is equally applicable to the other mountain regions of the world, but Hutton discusses only the Alps, and there is nothing to suggest that he recognised the former existence of glaciers throughout much of Europe. Similarly, there is nothing but the following enigmatic sentence to suggest that he appreciated the enormous erosive potential of glacier-ice:

> The motion of things in those icy valleys is commonly exceeding slow, the operation however of protruding bodies, as well as that of fracture and attrition, is extremely powerful.[25]

On the other hand, he was well aware of the ability of glaciers to transport heavy boulders over long distances. Having dismissed all attempts to explain the dispersal of alpine erratics in terms of debacles and explosions, he resorted to the suggestion that all the blocks in question had been transported to their present sites by the glaciers of his postulated alpine glaciation.

Hutton and the Denudation Dilemma

According to Playfair, a belief in the clastic origin of many strata, and in the reality of denudation, were the first two elements of the Huttonian theory to crystallise in its author's mind. If Playfair is correct here, and if Black was correct in supposing the theory to have been completed before Hutton took up residence in Edinburgh, then it follows that Hutton must have accepted the reality of denudation no later than the 1760s, and at a time when the denudation dilemma was still very much a live issue. Indeed, being a deist, Hutton may have felt the dilemma even more acutely than did his orthodox christian contemporaries. Whereas they believed that God was revealed both in the Bible and in Nature, Hutton had to rely upon Nature as his sole revelation of the benign deity, and initially denudation must have seemed a strangely jarring element in his world-picture. From a very early date in his geological career – perhaps even from the time of his pedestrian tours in England in the early 1750s – he must have struggled to reconcile his belief in denudation with his essentially teleological view of Nature, and it was in the course of this struggle that the Huttonian theory was born. This point has not been appreciated by historians of science; in their obsession with Hutton as the progenitor of modern geology, they have paid scant attention to the heritage to which he was heir, and their consequent failure to appreciate the existence of the denudation dilemma has resulted in a misunderstanding of the true nature and purpose of the Huttonian theory.

The theory was evidently intended as a novel resolution of the denudation dilemma. As we have seen, Hutton, like many an earlier writer, found a place for denudation within a teleological system by showing that incessant rock-decay is necessary to maintain the Earth's soil mantle, but unlike his predecessors he refused to impose any limits upon either the extent or duration of denudation. He freely confessed that prolonged denudation could result only in the destruction of continents, but at the same time he demonstrated through his theory that while denudation is destructive in the narrow context, it is really the first stage in a grand restorative cycle. In a sense Hutton's resolution of the denudation dilemma is directly comparable to that suggested by Woodward. Whereas Woodward envisaged the survival of habitable land being ensured through the regular repair of the present continents by some obscure subaerial process, Hutton pictured the same end being achieved through the slow growth of fresh continents under the influence of submarine processes.

We cannot know whether the denudation dilemma was the sole seminal problem which engendered the Huttonian theory, but the internal evidence of Hutton's writings makes it plain that the dilemma

170

was very much in his mind as he expounded his theory. The title of the original paper presented to the Royal Society of Edinburgh in 1785 is itself noteworthy: 'Examination of the System of the habitable Earth with regard to its Duration & Stability'. Thus at the outset he makes it clear that he will be concerned with that terrestrial *status quo* ('duration' and 'stability') which is the crux of the denudation dilemma, and in the first paragraph of the 1785 abstract of his paper he states that one of the objects of the theory is 'to see how far an end or termination to this system of things may be perceived, from the consideration of that which has already come to pass'. In the 1788 version of the theory Hutton approaches his subject entirely from the direction of the denudation dilemma. He first points out that soil is essential to a habitable world, and then he proceeds to explain that soil is constantly washed away into the oceans, that rocks are subject to decay so that the soil mantle may be replenished, and that this decay can result only in the eventual destruction of our continents. He then continues:

> But is this world to be considered thus merely as a machine, to last no longer than its parts retain their present position, their proper forms and qualities? Or may it not be also considered as an organized body? Such as has a constitution in which the necessary decay of the machine is naturally repaired, in the exertion of those productive powers by which it had been formed.
>
> This is the view in which we are now to examine the globe; to see if there be, in the constitution of this world, a reproductive operation, by which a ruined constitution may be again repaired, and a duration or stability thus procured to the machine, considered as a world sustaining plants and animals.[26]

Similarly, and a few pages later in the essay, he writes:

> In what follows, therefore, we are to examine the construction of the present earth, in order to understand the natural operations of time past; to acquire principles, by which we may conclude with regard to the future course of things, or judge of those operations, by which a world, so wisely ordered, goes into decay; and to learn, by what means such a decayed world may be renovated, or the waste of habitable land upon the globe repaired.[27]

Even in the final version of the theory, the denudation dilemma was still to the fore in his mind. This is revealed by the following three passages taken from the 1795 exposition:[28]

It has already been our business to show that the land is actually wasted universally, and carried away into the sea. Now, What is the final cause of this event? – Is it in order to destroy the system of this living world, that the operations of nature are thus disposed upon the surface of this earth? Or, Is it to perpetuate the progress of that system, which, in other respects, appears to be contrived with so much wisdom? Here are questions which a Theory of the Earth must solve . . .

The object of my theory is to shew, that this decaying nature of the solid earth is the very *perfection* of its constitution, as a living world.

. . . we shall thus be led to admire the wisdom of nature, providing for the continuation of this living world, and employing those very means by which, in a more partial view of things, this beautiful structure of an inhabited earth seems to be necessarily going into destruction.

Turning aside from Hutton's own writings, it is interesting to discover that he was not the only member of the Edinburgh circle to be concerned with the denudation dilemma. Particularly intriguing is the case of David Hume, who in his *Dialogues Concerning Natural Religion* of 1779, discusses the problem of discerning some underlying permanency behind Nature's constant mutations. Although nowhere very explicit on the subject, he clearly believed Nature to contain some built-in system whereby decay is speedily repaired. A study of the influence of Hume upon Hutton – and perhaps of Hutton upon Hume – might well be rewarding, but here it suffices to note that Hume was obviously interested in problems germane to the denudation dilemma at the same time as Hutton was working upon his theory. Joseph Black is another of Hutton's circle who was interested in the dilemma and who recognised the Huttonian theory as a novel and exciting solution to the problem. In 1787, when Black wrote an account of the theory for the Russian Princess Dashkova (she had lived in Edinburgh during the 1770s and was well known in Hutton's circle), he first discussed the work of denudation, and then continued:

Were it [denudation] to go on without being counteracted, the inequalities of the Earth's surface would be levelled in the course of time to a perfect plain, and mostly covered up by the sea or by collections of stagnating water. But there is a Remedy in Nature to prevent such a great change in the constitution of the Globe.[29]

This remedy, Black explained, had first been perceived by Hutton, and it was now the kernel of his theory.

Others outside Hutton's circle took a similar view. G. B. Greenough, for example, the first President of the Geological Society of London, wrote in the journal of his Scottish tour of 1805:

> A man must have some credulity to believe in the Huttonian theory, but unless he believes it he cannot but see that the world is daily approaching towards its dissolution.[30]

About the same date, a lecturer before the Natural History Society in the University of Edinburgh praised Hutton's theory – a 'splendid theory' he claimed – and observed of it:

> The dreary and dismal view of waste and universal ruin is removed, and the mind is presented with the pleasing prospect of a wise and lasting provision for the economy of nature.[31]

Even Hutton's opponents saw the theory as an attempted resolution of the denudation dilemma. Richard Kirwan and William Richardson, for example, having resolved the dilemma to their own satisfaction by denying the possibility of prodigious denudation, both protested that since such denudation is a fiction, then the entire Huttonian theory is rendered superfluous.[32]

The denudation dilemma perplexed scholars for little more than a hundred years, and Hutton was one of the last to be concerned with the problem. By the time of his death, geologists had begun to shed their religious preconceptions, and as they did so, they began to consider the case for and against a belief in denudation solely on its scientific merits. Hutton's amalgam of deism and geology was quite outmoded in the nineteenth century, and while his geology earned him many disciples, the religious undertones of his theory were either overlooked or discarded. Thus when the theory began to be accorded widespread acceptance after 1830, the dilemma which had played so vital a role in the theory's formulation had been almost totally forgotten. In this way the key to the Huttonian theory was lost.

The Earth-Machine, the Macrocosm, and Watt's Steam Engine

Although it was probably a confrontation with the denudation dilemma which led Hutton to formulate his theory, it is equally clear, as Tomkeieff has pointed out,[33] that during the theory's gestation period Hutton's mind felt the influence of two other elements in the intellectual atmosphere of his day. The first of these influences was the doctrine of the macrocosm and the microcosm, which here makes its final appearance in our study, and the second influence was that of the Industrial Revolution.

Hutton undoubtedly retained a vestigial belief in the doctrine of the macrocosm and the microcosm, and it is worth recalling that his doctoral dissertation at Leiden is entitled *De Sanguine et Circulatione Microcosmi*. In his theory he frequently alludes to the Earth as though it were a living creature. At the beginning of the 1788 essay, for example, he suggests that the Earth is 'an organized body', while later he refers to it as 'a living world', and he adopts the old device of likening a drainage system to 'the arteries and veins of the animal body'. As a physician Hutton was thoroughly familiar with the various physiological cycles whereby damage and depletion in the microcosm are made good, and he may therefore have been tempted to seek in the macrocosm for similar cycles which repair the ravages of denudation. He perhaps regarded such terrestrial processes as transportation, sedimentation, lithification, and uplift as the macrocosmic equivalents of circulation, digestion, respiration, and excretion. The ascription of such thoughts to Hutton must not be regarded as too far-fetched, because he himself wrote:

> This earth, like the body of an animal, is wasted at the same time that it is repaired.

Strangely, this idea is exactly paralleled in Part VI of Hume's *Dialogues Concerning Natural Religion*, and here again we may have evidence of the mutual influence of these two scholars.

The Industrial Revolution – the second formative influence apparent in the Huttonian theory – is commonly regarded as having begun with the blowing of the first furnace at the Carron Ironworks near Falkirk – an event which occurred, conveniently enough, on 1st January 1760. Throughout the period when Hutton was musing over his theory, the machines of the new age were the object of both awe and astonishment. Britain's first large factory – the £9,000 works built at Soho by Matthew Boulton in 1764 – was soon regarded as an eighth wonder of the world. A local versifier wrote of

> Soho!—where Genius and the Arts preside
> Europa's wonder and Britannia's pride.[35]

and the factory rapidly became one of the country's chief tourist attractions. Hutton and Dr. Johnson were both visitors in 1774, and two years later Boswell was profoundly impressed by the 'vastness and the contrivance of some of the machinery'.[36] The sight of one of Watt's steam-engines at work in the factory moved Hutton's friend Erasmus Darwin to verse:

> Press'd by the ponderous air the Piston falls
> Resistless, sliding through it's iron walls;
> Quick moves the balanced beams of giant-birth,
> Wields his large limbs, and nodding shakes the earth.[37]

Halévy has remarked that 'the very sight of machinery inclines the mind to seek a mechanical explanation of all natural phenomena',[38] and it was evidently so with Hutton. Living in an age of novel machinery, he began to regard the Earth itself as a gigantic machine whose parts are in perpetual, if slow motion. In the first paragraph of the 1788 essay he refers to the Earth as 'a machine of a peculiar construction', and as we have seen, this notion is reiterated throughout the expositions of his theory. Here, yet again, we may have an example of the inter-action of the minds of Hume and Hutton because long before the Huttonian theory appeared in print, Hume referred to the world as:

> . . . nothing but one great machine, subdivided into an infinite number of lesser machines, which again admit of subdivisions, to a degree beyond what human senses and faculties can trace and explain.[39]

In particular, Hutton seems to have regarded the Earth as some kind of heat-engine, and it is here that we may perhaps detect the influence of his friendship with James Watt. Watt's first full-scale experiments with a steam-engine were carried out in secret near Bo'ness in a glen behind Kinneil House, the home of John Roebuck, who was Watt's patron, the founder of the Carron Ironworks, and a member of Hutton's circle. There, in September 1769, Watt failed to get the engine to work satisfactorily, and there it lay rotting and rusting until 1773 when its metal parts were removed to Soho. Now Kinneil House is less than 20 miles from Edinburgh, and Hutton may well have been one of the select group which attended the abortive trials there. Indeed, it may have been on that occasion that he observed the bed of oyster shells at Kinneil to which he refers in the *Theory*. Be that as it may, Hutton was certainly profoundly impressed by Watt's invention, and in January 1774 he wrote to Watt saying 'I admire so much your reciprocating engine'.[40] When Hutton visited Soho later the same year, he must have seen the Kinneil engine being re-erected in Boulton's factory, and he may well have watched it in action because by the autumn of 1774 the engine was at last working efficiently.

The influence of the steam-engine upon the Huttonian theory must not be exaggerated, because, if we are to believe Black, the theory was formulated before Hutton had had the opportunity of seeing even the unsuccessful trials at Kinneil House. Hutton's close association with the development of the steam-engine is nevertheless interesting because the active agents in Watt's engine and the Huttonian theory are identical – fire and water. Hutton himself observed that his theory is 'a system in which the subterranean power of fire, or heat, co-operates with the action of water upon the surface of the Earth . . .'[41] He perhaps saw the elevation of new continents by subterranean heat, and the lowering of the older continents by atmospheric agencies, as in some way analogous to the rise and fall of the piston in one of Watt's early single-acting pumping engines. One final fact suggesting that he viewed the Earth as a heat-engine is his regard of volcanoes as a kind of safety-valve.

> A volcano is not made on purpose to frighten superstitious people into fits of piety and devotion, nor to overwhelm devoted cities with destruction; a volcano should be considered as a spiracle to the subterranean furnace, in order to prevent the unnecessary elevation of land, and fatal effects of earthquakes.[42]

Such an interpretation of volcanoes has no claim to originality, but it is perhaps further evidence of the impact of Watt's work upon Hutton's geological thought.

The Huttonian theory thus bears trace of curious and varied influences. It evidently resulted from an effort to resolve that peculiarly eighteenth-century problem, the denudation dilemma, and in the course of the theory's formulation the mediæval doctrine of the macrocosm and the microcosm met the steam-engine of the Industrial Revolution.

Hutton's Dialectical Failings

We have it on Playfair's authority that Hutton's verbal expositions of his theory were lucid, precise, and even entertaining, and Hutton's few surviving letters certainly show him to have been a lively and humorous correspondent. On the other hand, as Playfair admitted, the published versions of the theory are tortuous, obscure and extremely dull. Individual passages from his work are lucid enough, but he failed lamentably in trying to order his thoughts into a reasoned and systematic exposition of the theory. The Huttonian quotations offered above can give readers no idea of the confusion and interminable repetition which they will encounter in the original works. The more space Hutton allowed himself, the more glaring do his dialectical failings become, and the two bulky volumes of 1795, running between them to almost 1 200 pages, are sufficient to stall all but the most determined of readers. Quite apart from the prolixity of the two volumes Hutton now clouded the issue by employing quotations on such a scale that almost half of the second volume consists of passages in French taken from the works of continental geologists. In some places Hutton's 'style' degenerates so far as to become a series of footnote comments on the passages quoted. Even the present-day reader, well-versed in the principles of modern geology first enunciated by Hutton, has difficulty in following much of his reasoning, and the task must have been doubly difficult for his contemporaries, whose geology was largely that of the Freiberg school. Richard Kirwan, keen critic though he was of the Huttonian system, evidently never managed to wade through the *Theory* of 1795. His personal copy of the work (now in the Library of the Royal Irish Academy in Dublin) contains marginal comments in volume I but not in volume II, and two leaves in the latter volume remain unopened to the present day (1968). Even so distinguished an Huttonian as Sir Charles Lyell had to confess in 1839 that he had never been able to bear with Hutton throughout his exposition of the theory.[43]

Apart from Hutton's inability to communicate his theory through the written word, there are other defects in his expositions which must have militated seriously against the theory's acceptance. For example, the expositions were mistitled; Hutton's was not a theory of the Earth in the then accepted sense because it neither considers the origin of the Earth, nor does it attempt a systematic and detailed history of the Earth's surface. Again, Hutton clearly alienated some of his readers by his confession that certain aspects of the theory had been formulated and published in advance of the necessary field-observations. His admission in 1795 that when he first proclaimed the igneous origin of granite, he had only once seen the rock *in situ*, was hardly likely to inspire confidence. Similarly, it was not until after the reading of the 1785 paper that he set off in search of the unconformities which are such a vital proof of the theory. As one critic wrote twenty years after Hutton's death:

> ... had he studied nature, and then theorized, his genius would, in all probability, have illustrated many difficult points; but it is obvious, from his own works, that he has frequently reversed this order of proceeding.[44]

The tragedy is that by 1795 he could have offered ample, convincing proof of the theory from his own field-notebooks, but instead, he most unwisely chose to illustrate the theory by reference to the published observations of other naturalists made, in most cases, in regions which he himself, had never even visited. Hutton explained that he adopted this course so as to avoid any suspicion of partiality on the part of the observer, but suspicions of another kind must have remained because in many cases his interpretations of phenomena differ radically from those of the original author. Not until 1899 did Hutton's own ability as a field-geologist become apparent following the publication of the account of his Scottish tours in volume III of the *Theory*.

All told, Hutton's presentation of his theory could hardly have been worse. Mistitled, lacking in form, drowned in words, deficient in field-evidence, and shrouded in an overall obscurity, the theory's chances of finding general acceptance were seriously prejudiced. Many of those who knew of the theory only through Hutton's expositions must have dismissed it as the worthless and indigestible fantasy of a somewhat dated arm-chair geologist.

Hutton's Contribution to Geomorphology

In view of the importance generally ascribed to Hutton's theory, it is surprising that it should have received so little scholarly attention. Some analytical studies of the theory have of course been made, but a neglect of the pre-Huttonian literature has handicapped historians in their attempts to achieve a just evaluation of Hutton's work. Ill-equipped to make their own evaluation, many scholars have been all too willing to accept the conclusions already arrived at by nineteenth-century historians who patently were neither informed nor impartial in their judgments. Among these earlier historians none has been more influential than Sir Archibald Geikie.

Geikie's life was in many respects remarkably similar to Hutton's: they were both born and bred in Edinburgh; they both attended the city's High School and then moved on to the University as students of the humanities; they both made a false start to their careers by going into the office of a Writer to the Signet; and later, they both became keenly interested in geomorphology and igneous geology. Geikie clearly felt some affinity with Hutton, and throughout his long and distinguished career (he retired as Director-General of the Geological Survey of Great Britain in 1901) Hutton's brilliance was a constantly recurring theme both in Geikie's lectures and in his voluminous and eloquent writings.

Perhaps Geikie's most effective plea for Hutton's canonisation dates from 1897, when at Johns Hopkins University, Baltimore, he devoted almost the whole of one of his six Williams Memorial Lectures on the history of geology, to a fluent account of Hutton's life and achievements. To William Smith, in contrast, the scarcely less worthy 'father of English geology' Geikie gave no such prominence; Smith had to share his lecture with Cuvier, Brongniart, and a medley of other, lesser continental geologists. The lectures were later amplified and published as Geikie's well known *Founders of Geology*, which remains one of the most widely read works on the subject. In the book Geikie presents a graphic picture of Hutton as an unsurpassed genius working at the centre of a powerful research-team composed of his scarcely less brilliant friends. We see him in the field working meticulously from section to section, and calling upon the pen of Clerk of Eldin to illustrate each crucial unconformity or intrusive contact. We see him back in Edinburgh poring over his specimens and consulting Black over abstruse problems of geochemistry. Then finally, Geikie shows him emerging from his researches with completely fresh answers to many of Nature's most profound mysteries. Coming as it did from one of the world's leading geologists and, moreover, one who was recognised to be the foremost authority upon the history of British geology, this assessment of Hutton gained universal currency, and to this day it has never been seriously questioned. It is nevertheless clear that Geikie approached Hutton more as a eulogist than as an impartial scholar. He allowed his appreciation of Hutton's work to be influenced by his transparent devotion to the man whom he regarded as the founder of that Scottish School of Geology to which he himself was so proud to belong.

The time is ripe for a reassessment of Hutton's contribution to the Earth-sciences, but to essay such a task here would involve transgression far beyond the confines of the present study. Hutton's contributions to igneous and metamorphic geology, for instance, have little bearing upon the history of geomorphology, and all that is attempted here is an assessment of those aspects of his theory which are germane to our present theme. It so happens that some of these aspects are among the most misappraised elements of his work. Hutton's contribution to geomorphology may conveniently be considered under the following three heads:

I. The Age of the Earth and Huttonian Uniformitarianism
II. The Cyclic Earth-History
III. The Origin of Landforms

I. *The Age of the Earth and Huttonian Uniformitarianism.* It is often asserted that Hutton was the first to appreciate the true magnitude of the terrestrial time-scale, but such a claim is in need of qualification. Many of his predecessors had in fact possessed an equally sound understanding of time. It may be recalled that as early as 1693 John Beaumont had boldly suggested that the only reasonable alternative to the Mosaic chronology was a belief that the Earth is either eternal or else so old that its origin lies far beyond the limits of man's perception. Whitehurst, too, perhaps influenced by his deistical friend Erasmus Darwin, had conceded that the Earth might be eternal. As we have seen, the eternity of the universe was generally accepted by the eighteenth-century deists, and since Hutton himself was a deist (he was also another of Erasmus Darwin's friends), there is nothing remarkable in his adoption of a greatly extended time-scale; he was merely conforming to a long-established tenet of the deistical creed. His contemporaries clearly saw nothing original in his conception of almost limitless time; de Luc, in a critique of the theory, noted that it was a 'common pretence' to suppose that the world is infinitely ancient, and another of Hutton's critics observed that 'the eternity of the world is a favourite atheistical doctrine'.[45]

Hutton's conception of an immensely ancient Earth thus has no claim to originality. What does give him some title to distinction is his application of the deistical time-scale to geology, although even here his priority is by no means uncontested. Very much earlier in the century the French scholar Benoît de Maillet had used the idea of an eternal Earth as the basis for a novel theory of Earth-history. His book had been available in an English translation since 1750,[46] and it is specifically mentioned by Hutton himself. Similarly, G. H. Toulmin, another deist, had drawn attention to some of the geological implications of an eternal Earth five years before Hutton made his theory public, although, as indicated earlier, there is a strong possibility that Toulmin was actually plagiarising some of Hutton's material. Neither de Maillet nor Toulmin, however, was a serious student of Earth-history; they were both theorists in the tradition of Burnet and Whiston, and we must reserve our acclaim for Hutton as the first true geologist to appreciate the magnitude of the terrestrial time-scale. Earlier geologists from Hooke to de Luc had considered extending the Mosaic chronology by perhaps a few millennia, but Hutton was the first geologist to perceive that the age of the Earth was so great as to be almost beyond human comprehension.

Having imported the deistical time-scale into geology, Hutton demonstrated that its use as a back-cloth obviated the need for catastrophism. The catastrophic doctrine had been born of the attempt to compress Earth-history into the narrow compass of the Mosaic chronology, but once the time-shackles were removed, then the way was open for a very different interpretation of Nature. Hutton seized his opportunity and showed that all geological phenomena are readily explicable as the cumulative effect of those same slow-acting processes which are still operative in the modern world. He was thus the first true uniformitarian, and the uniformitarian doctrine, together with the time-scale which made it possible, are two of Hutton's three chief contributions to geomorphology.

II. *The Cyclic Earth-History.* Most of Hutton's contemporaries regarded the Earth as the end-product of now extinguished creative processes, and they saw Earth-history as a linear progression leading through the various stages of creation, down to a relatively static present. Hutton's interpretation of Earth-history is diametrically opposed to such a view. Through his discovery of unconformities he convincingly demonstrated that the Earth's surface bears trace not of one single act of creation, but of innumerable acts of re-creation. His theory, with its vision of an endless succession of continents rising from the ocean only to be destroyed and again submerged, is the earliest comprehensive exposition of a cyclic Earth-history. His was a steady-state theory of the Earth as opposed to the special creation theory held by his contemporaries.

Among pre-Huttonian authors in Britain, only Robert Hooke can rival Hutton in his insight into the true nature of Earth-history. Indeed, as already observed, where the problem of lapidification is concerned, Hooke in 1668 was somewhat in advance of Hutton. By the second half of the eighteenth century, however, Hooke was a forgotten figure, and it was through Hutton's writings, either directly or indirectly, that the world learned of the cyclic history of the continents. Playfair christened the Huttonian cycle 'the great geological cycle', and as 'the geological cycle' or 'the geostrophic cycle' it remains to this day one of the most fundamental tenets of geology and geomorphology. Hutton's cyclic interpretation of Earth-history was probably his greatest achievement.

III. *The Origin of Landforms.* Hutton, having dismissed the belief that some of the Earth's topography was coeval with the globe, and that much of the remainder was the result of the world's diluvial metamorphosis, became the first to state unequivocally that topography is the result of the interaction of the processes of sedimentation, earth-movement, and denudation. So far as the raising of an initial topography is concerned, he was closest in his thinking firstly to Whitehurst, and secondly to those of his predecessors who had regarded landforms as chiefly the product of seismic convulsions, but even here his appreciation of the endogenetic forces represents a marked advance upon previous opinion. Where earlier writers had invoked sudden, gigantic convulsions, Hutton had resort to slow protracted uplift caused by the gradual expansion of material deep in the crust.

Sound though Hutton's understanding of the endogenetic forces may have been, he was careful not to exaggerate their importance in the formation of topography. Topography, he claimed, is essentially the work of denudation. Praise has been lavished upon him for his supposed priority in recognising the potency of denudation. Playfair hailed him as the discoverer of the fact 'that all the hard substances of the mineral kingdom, when elevated into the atmosphere, have a tendency to decay . . .'; Sir Edward Bailey observed that 'Hutton's conception of the connection of landscape with commonplace erosion is strikingly modern in its more essential features . . . there can be no question of the originality of his views . . .'; and as recently as 1964 Hutton was saluted as 'the first great fluvialist'.[47] Such claims on Hutton's behalf are either groundless, or else they are deceptive half-truths. In seeking a true assessment of his views on denudation, two questions must be answered. Firstly, how far was he possessed of an original understanding of the nature and efficacy of the denudational forces? Secondly, was he unique in his understanding of the fluvial and circumdenudational origin of most of the Earth's present topography?

The first question is answered readily. Hutton deserves some credit for his suggestion that former extensive glaciers might have left their legacy in our modern landscapes (he was not the first to make such a suggestion, and he was certainly not the first to appreciate the transporting power of *present* glaciers), but otherwise his writing on the subject of denudation contains little that might not have come from the pen of any seventeenth-century writer upon the subject, if not from the still earlier pen of William Bourne himself. Innumerable earlier scholars had displayed just as sound an understanding of both the nature and efficacy of the denudational forces, and some of the many quotations from the works of earlier writers which Hutton adduces to support his own belief in denudation, should have been sufficient to convince historians that he had no priority of discovery in this field. True, he did have a sounder understanding of denudation than such of his contemporaries as de Luc, Williams, or Kirwan, but this was only because his ingenious resolution of the denudation dilemma allowed him to revert to the thorough-going seventeenth-century belief in the power of Nature's destructive processes.

The question concerning Hutton's interpretation of topography as the work of circumdenudation by fluvial processes is less easily answered. He was certainly not the first to regard topography as the result of circumdenudation; Catcott and others had already amply recognised this fact, although they had invoked diluvial currents as the agent responsible. Similarly, Hutton was by no means the first to recognise that fluvial processes are capable of remodelling the Earth's surface; Burnet and many seventeenth-century scholars had freely confessed that fluvial processes were capable of levelling even the highest of mountain ranges. Here, however, we again encounter the problem that was discussed earlier: how far were pre-Huttonian scholars prepared to see the Earth's *present* landscapes as the result of circumdenudation effected by rain and rivers? They certainly recognised many modern landforms as the result of fluvial erosion, but did they picture denudation as having occurred on the grand scale, or were they content to regard rivers as merely adding the detail to a topo-

183

graphy that was essentially primitive, diluvial, or seismic in origin? No matter what their inclination may have been in this respect, pre-Huttonian scholars were severely restricted in the topographical significance which they could attach to rain and rivers because of the obvious slowness of such processes, and because of the very limited time-scale available. Hutton was certainly the first British author to argue at length that the Earth's topography is merely the residue left after prolonged denudation, but his priority here reflects not so much the brilliance of his geomorphic insight, as the freedom allowed by the deistical time-scale. Had seventeenth-century scholars been less cramped by the Mosaic chronology, they too would doubtless have seen that the Earth's topography owes its form largely to circumdenudation effected by fluvial processes. Indeed, implicit in their belief in the efficacy of the fluvial processes to base-level mountains, there is clearly the notion that if not already in being, a topography produced by circumdenudation must come into existence at some future date, provided the Parousia does not intervene. All that Hutton did was push the Creation back to some remote period, and then claim that the Earth's topography had already attained that stage of development which most of his predecessors had probably regarded as an event reserved for the future. Thus not only does Hutton's understanding of the geomorphic processes have little claim to originality, but there was nothing strikingly novel in his belief in a circumdenudational topography produced by the fluvial processes.

Apart from the factors already mentioned – an ignorance of the pre-Huttonian literature and partisanship on Hutton's behalf – two other factors have probably played a part in causing Hutton's writing on landforms to be falsely assessed. Firstly, Hutton's protracted discussion of denudation has evidently misled historians into thinking that here was an original thesis in need of elaborate proof, when in reality the length of the discussion reflects nothing more than Hutton's prolixity and the fact that denudation, being a subaerial affair, is the only phase of the cycle for which much secular evidence was available. Secondly, it is well known that for long after Hutton's death, geologists were reluctant to admit that fluvial processes could have played any significant part in the shaping of topography. Not until the 1860s did the geomorphic views embodied in the Huttonian theory gain general acceptance and those unfamiliar with the pre-Huttonian literature have therefore come to regard Hutton as a wise prophet crying in the wilderness – a genius living perhaps a century before his time. It has been overlooked that Hutton was not so much the precursor of a late nineteenth-century school of fluvial geomorphology, as the last vestige of a very much earlier school whose approach to the subject of denudation was profoundly influenced by the denudation dilemma.

Hutton's writing on the subject of denudation is nevertheless important, because through his imaginative resolution of the denudation dilemma he showed that denudation has a vital role to play in the terrestrial economy. In this sense his understanding of the work of denudation did far surpass that of his predecessors. His contribution to geomorphology thus lies not at the level of the field-interpretation of topography, but at a more fundamental, philosophical level. Through his new time-scale, his cyclic theory of Earth-history, and the uniformitarian doctrine, he provided the Earth-sciences with their dynamic system. He thus did for geology and geomorphology what Newton had already achieved for astronomy, and what Darwin was destined to do for biology. Indeed Darwin, through Lyell, owed a debt to Hutton, because Darwin's theory of evolution by means of natural selection is merely an application to the organic world of that slow evolutionary process which Hutton had first perceived in Nature.

The Reception of the Huttonian Theory

In his biography of Hutton, Playfair claims that the newly published Huttonian theory passed unnoticed by the world of science at large. Later writers have not only accepted Playfair's statement, but have amplified it by suggesting all manner of reasons to explain the theory's supposed early neglect. This intrenched belief in a time-lag before the first reaction to the theory is nevertheless just another of the myths which have been allowed to grow up around Hutton. From the very moment of the appearance of the paper in 1788 the theory was in fact the subject of widespread comment and criticism. It was discussed in the literary reviews of the day;[48] John Williams, at near-by Gilmerton, immediately added a section to his newly completed *Natural History of the Mineral Kingdom* to refute the theory; de Luc, with the same end in mind, wrote a series of critical letters for *The Monthly Review;* and innumerable other authors seized opportunity to pass judgment upon the theory. The appearance of the two volumes of 1795 further stimulated interest in the theory, and later, during the early years of the nineteenth century, Edinburgh was the scene of the famous battle between the Huttonian Plutonists and the local Neptunists led by Jameson. According to Leopold von Buch,[49] the Revolutionary and Napoleonic wars did prevent the free circulation of the *Theory* in Europe, but even so a speaker (admittedly an Huttonian) before the Edinburgh University Natural History Society in January 1794, affirmed that Hutton had already gained many converts on the continent. Certainly Nicolas Desmarest offered a long summary of the theory in the *Encyclopédie Méthodique*[50] published in Paris in 1794–95, and it seems that even non-scientists in the United States were soon familiar with the theory because it is mentioned in an oration delivered by Ebenezer Grant Marsh at a public commencement at New Haven, Connecticut in 1798.[51] All told one cannot escape the conclusion that despite all its dialectical failings, and the relative obscurity of its author (Hutton, unlike Whitehurst, de Luc, and Kirwan, was never a Fellow of the Royal Society), the Huttonian theory was read just as widely as any of the seventeenth-century theories, and much more widely than the theories of, say, Williams or Kirwan.

By no means all the comment upon the theory was critical, but it was certainly Hutton's opponents who produced the most clamour. The theory was attacked chiefly upon purely geological grounds because it ran counter to the prevailing Neptunian system in almost every respect. The Neptunists protested vigorously at Hutton's conception of rocks being fused by heat, at his idea of mineral-veins being filled by melts rising through the crust, and at the belief that basalt and granite had been formed by igneous activity in what they mockingly termed Hutton's 'subterraneous laboratory'. Although they formed by far the greatest weight in the artillery brought into action against Hutton, these geological objections to the theory are of little consequence in the present context. Instead, attention must be focused upon two lesser skirmishes which took place beneath the main geological barrage. The first of these centred upon the supposed irreligion of the theory, and the second – a very minor affair – was an attack upon Huttonian geomorphology.

It was only to be expected that Hutton's theory would raise religious protests, because quite apart from any inherent heterodoxy, it appeared at a most inopportune moment. The first full version of the theory was published on the eve of the French Revolution, and by the time the 1795 volumes appeared, Britain was deeply embroiled in the Revolutionary War. At first Britain had been sympathetic towards the

Revolution, but its excesses soon alarmed the British upper and middle classes, and during the 1790s there was widespread fear of a similar rising in Britain. Government spies were everywhere, habeas corpus was suspended, and in 1795 Parliament passed the two 'gagging' bills designed to check treasonable practices and seditious meetings. Throughout Britain the French social upheaval soon came to be regarded as the outcome of an international, Voltaire-inspired plot to overthrow the Christian church. Never, it seemed, had there been a more perfect illustration of the truth of Sir Matthew Hales's dictum that Christianity is fundamental to the preservation of law and order. In the face of events, the war with France took on the nature of a religious crusade, while at home the church suddenly found itself regarded as the chief bulwark of the establishment. Out of self-interest, the upper classes became concerned for the religious welfare of the masses, and there prevailed what has been termed 'the policeman theory of religion'. It was this theory which the Rev. John Erskine expounded in a sermon delivered in Greyfriars Church, Edinburgh, before the magistrates of the city in 1792, when he informed his congregation:

> Where there is no religion, the firmest support of government is removed, the surest bond of social union is broken, and a wide door is opened for vice to enter, and to usher in disorder and misery.[52]

In this setting three aspects of Hutton's apostasy drew the ire of his contemporaries. Firstly, there was his implicit rejection of the Mosaic writings and their chronology – a most serious matter because, as one eminent divine put it in a University sermon at Oxford in 1805, 'every thing which relates to revealed Religion depends ultimately on the authenticity of the Mosaic account of the creation and fall of man'.[53] This heterodoxy alone was sufficiently grave to place Hutton in the camp of Voltaire as a dangerous, subversive influence, and to earn him the title of 'atheist'. (Some modern writers have protested at the injustice of Hutton, so patently a deist, being branded as an atheist, but in his day that label was still being attached to any who strayed too far from the narrow path of Christian orthodoxy.) Secondly, Hutton was roundly condemned for his regard of the Earth as a perpetual motion machine capable of running for all time without the attention of even a divine mechanic. John Williams expressed this view forcibly:

If once we entertain a firm persuasion that the world is eternal, and can go on of itself in the reproduction and progressive vicissitude of things, we may then suppose that there is no use for the interposition of a governing power; and because we do not see the Supreme Being with our bodily eyes, we depose the almighty Creator and Governor of the universe from his office, and instead of divine providence, we commit the care of all things to blind chance. Like a mob, who think they can do well enough without legal restraints, depose and slay their Magistrates. But this is rebellion against lawful authority, which must soon end in anarchy, confusion, and misery, and so does our intellectual rebellion.[54]

Finally, there was strong religious objection to Hutton's conception of a cyclic Earth-history. In the traditional Christian view, history was seen as a drama which had a clear beginning in the Creation, and which proceeded steadily through various acts to the eventual denouement of the Second Coming and the Millennium. The Huttonian theory, on the other hand offered neither beginning nor end, but only a seemingly interminable series of cycles of uplift and denudation, and unlike its Neptunian rival, the theory could not be reconciled with a linear interpretation of history. The orthodox therefore regarded it with abhorrence. The Huttonian view of history was for Kirwan 'an abyss from which human reason recoils', and Williams felt equal disquiet. 'Let us turn our eyes from the horrid abyss', he wrote after reviewing the theory, 'and stretch out our hands, and cry, Save, Lord, or we perish!'[55]

Turning now to the criticism of Hutton's geomorphic beliefs, it must be re-emphasised that such attack as there was upon Huttonian geomorphology was a trivial affair as compared with the controversy which raged around his Plutonism. Many historians have overlooked this fact. Seeing a battle raging around the theory, and imagining Huttonian geomorphology to be novel, they have jumped to the conclusion that it was one of the issues at stake. In fact, during the twenty years following the publication of the 1788 essay, only four critics raised any noteworthy protest against Huttonian geomorphology. These four were Williams, de Luc, Kirwan, and Richardson.[56]

Each of these four geologists had sought to resolve the denudation dilemma through a denial of the potency of Nature's destructive forces, and it is therefore hardly surprising that they blenched at Hutton's conception of denudation on the grand scale. Richardson was shocked that anyone could believe the Earth to be a mouldering ruin;

THOMAS BURNET, *from the fifth edition of his* Theory of the Earth (*London* 1722)

WILLIAM WHISTON, *from John Nichols,* Literary Anecdotes of the Eighteenth Century, I (*London* 1812)

RICHARD KIRWAN, *from the* Proceedings of the Royal Irish Academy, IV (1847–1850)

ROBERT JAMESON, *from* The Edinburgh New Philosophical Journal, LVII (1854)

I   Four British Theorists

II   *The frontispiece to both volumes of the English version of Thomas Burnet's* Theory of the Earth *of 1684 and 1690*

The spheres represent the successive phases in the history of the Earth, and they follow in a clockwise sequence from the top right as follows: (1) The chaotic liquid (2) The Earth in its original state as a smooth ball (3) The Earth during the Flood with the Ark floating on the turbulent waters (4) The modern Earth (5) The Earth during the conflagration which will precede the Second Coming (6) The Earth during the Millennium restored to its original perfect form (7) The Earth in its ultimate condition as a star

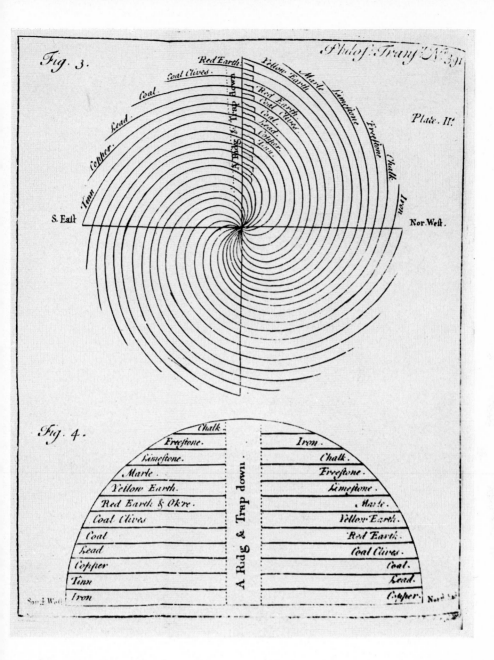

III   *The Earth as a form of Catherine-wheel: John Strachey's attempt to account for the supposed asymmetry of much of the world's topography*

From the Philosophical Transactions of the Royal Society, *No.* 391, *November–December* 1725

An Explanation *of the* COPPER-PLATE,

REPRESENTING

The internal structure of the terraqueous Globe, from the Center to the Circumference, and the Air around it.

D. The *outward Expanse or the open Firmament of Heaven.*
E. A *circular Space* filled with water during the height of the Deluge, but now with the Air that came from the central Hollow of the earth; and at present constitutes what we call our *Atmosphere.*
F. The *shell of the earth* broken into innumerable *apertures* and *fissures,* of various shapes and sizes; the *larger* of which, f. f. f. f. f. being filled with the water that descended from the surface of the earth, form *Seas* and *Lakes;* the *lesser* (which branch from the former, or pass immediately from the under-part of the shell of the earth to the tops of the highest mountains) serve as canals for the water which supplies *Springs* and *Rivers* to run in; the *least* of all (denoted by the *irregular black strokes* in the solid shell of the earth) represent the cracks thro' which *vapours* principally ascend.
G H. The *Great Abyss* of water within the earth; with which all Seas, Lakes, Rivers, &c. communicate; and from whence they receive their supplies. G. H. are divided from each other by a dotted circle, because *one of them* represents the water that, during the Deluge, covered the whole surface of the earth, but which was afterwards forced down, thro' the above-mentioned larger apertures and fissures, to its original place, as the inward Air was forced out thro' the lesser and oblique fissures: and the *other of them* represents that part of the Abyss which, during the Deluge, remained beneath the earth.
I. A *solid Ball* or *Nucleus* of terrestrial matter, formed from what the water in its descent from the surface, and passage through the strata of the earth, tore off, and carried down with it into the Abyss, and reposited at the lowest place, the center of the earth.
☞ So that the Opinion of the Ancients concerning the Earth's resembling an Egg has great propriety in it: for the Central *Nucleus,* (I.) by its innermost situation and shape, may well represent the *Yolk;* the *Abyss* of water, (G. H.) which surrounds it, and is in a middle position, may stand for the *clear Fluid* of the *White;* the *Crust of the Earth* (F.) (allowing only for its breaks and cracks) by its roundness, hardness, uppermost situation, and little inequalities on its surface, is justly analogous to the *Shell.* And on this account the term *the shell of the earth* is frequently used in this treatise.

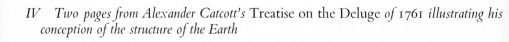

IV   Two pages from Alexander Catcott's Treatise on the Deluge *of* 1761 *illustrating his conception of the structure of the Earth*

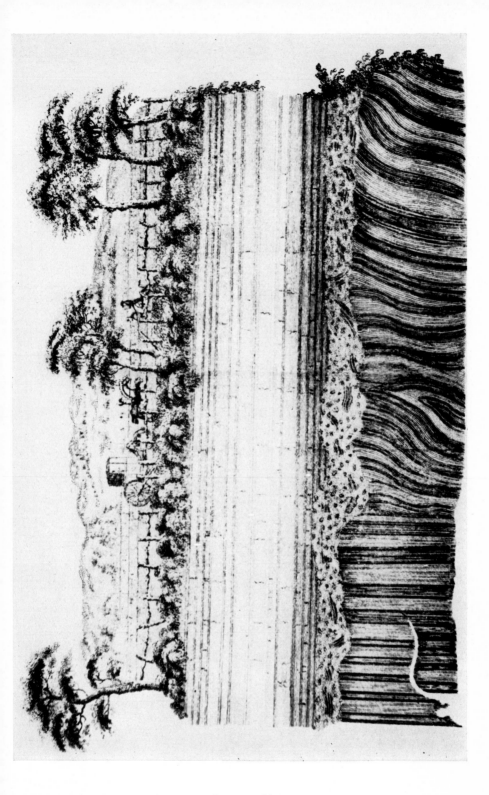

V *James Hutton's Unconformity in the Tweed basin*

This drawing was made by John Clerk of Eldin, and it appeared in the first volume of the 1795 edition of
Hutton's Theory

VIEW OF THE TEMPLE OF SERAPIS AT PUZZUOLI IN 1836.

VI   *Sir Charles Lyell's famous illustration of the so-called Temple of Jupiter Serapis near Pozzuoli on the Bay of Naples*

The three marble columns with marine lamellibranch borings up to a height of 23 feet above the high–water level were regarded as important confirmation of the uniformitarian belief in slow non–violent earth–movements

VII   *Polished and Striated Rocks from Zermatt, Rosenlaui and Landeron*

One of the plates which Louis Agassiz displayed to the members of Section C of the British Association on 22 September 1840. Later in the same year the plate was included in his Études sur les Glaciers *published at Neuchâtel*

JAMES HUTTON, *from John Kay*, A series of Original Portraits, I (*Edinburgh* 1838)

JOSEPH BEETE JUKES, *from C. A. Browne*, Letters and Extracts from the Addresses and Occasional Writings of J. Beete Jukes (*London* 1871)

SIR ANDREW CROMBIE RAMSAY, *from* The Geological Magazine (London) N.S. Dec. II, IX (1882)

SIR ARCHIBALD GEIKIE, *from* The Geological Magazine (London) N.S. Dec. III, VII (1890)

VIII *A Group of Fluvialists*

he called it an 'alarming', 'gloomy', and 'desponding' view. Hutton, it seemed, was pleading for a reversion to the seventeenth-century belief in the decadence of Nature. Richardson and Kirwan both reminded Hutton of the lack of secular evidence of rock decay within historic times, but in doing so they reveal a complete failure to appreciate either the magnitude of Hutton's time-scale, or the slowness with which he supposed Nature to operate. They both still regarded a few centuries as a considerable proportion of Earth-history, and they saw the failure of denudation to inflict significant damage within such a period as ample proof of the immutability of topography. So deeply were they conditioned by the Mosaic time-scale that the Huttonian aeons were evidently quite beyond the comprehension of both men.

Such controversy as there was over the Huttonian belief in denudation naturally came to focus upon valleys. Were they the result of fluvial erosion, were they primitive, or were they features produced by earth-movements and diluvial torrents? Succinctly, were valleys younger or older than their rivers? In Hutton's day there were a number of telling arguments which could be advanced against the fluvialist doctrine of valley formation, and it was de Luc, in his letters to *The Monthly Review* in 1790 and 1791, who taxed Hutton with one of the most serious of these objections. De Luc was of course convinced that all Alpine valleys were collapse-structures, and he pointed out that many of the valleys contain large lakes which serve as natural settling-basins, so that little of the debris swept into them is ever removed. Therefore, he asked Hutton, if the valleys upstream of the lakes have been excavated by river action, as you claim, why have the lakes not been filled in with the resultant debris? He reinforced his very pertinent question by observing that in most cases the volume of the lakes is insignificant as compared with the volume of the valleys further upstream. Had the valleys really been river-cut, they must have yielded sufficient debris to fill the lakes many times over – yet the lakes still survive. Indeed, the infilling of most of the lakes is at such an early stage that de Luc was able to use the rate of infill as one of his natural chronometers proving the Earth's novity.

Hutton clearly felt the force of de Luc's limnological objection to the fluvialist doctrine, and he offered a reply in the *Theory* of 1795. He there suggested that the infilling of the lakes is at an early stage because the lake-basins are late additions to the landscape, and very much younger than the valleys upstream. Here, as we now know, Hutton was essentially correct, but knowing nothing of the glacial excavation of rock-basins, he had difficulty in finding a mechanism which could have formed the lakes at some comparatively recent date. He suggested three possibilities: firstly, some of the lakes might have formed as a result of landslides blocking valleys; secondly, some might be the result of the weathering of soluble rocks exposed on valley-floors; and finally, Hutton came dangerously close to the catastrophism of de Luc himself by suggesting that some of the lake-basins might have been formed by earthquakes.

De Luc's familiarity with the Alps allowed him to dismiss Hutton's first two arguments out of hand, and if the lake-basins were supposed to have been formed by earthquakes, was it not equally reasonable to suppose that the valleys upstream of the lakes were themselves seismic fissures? De Luc was unmoved, and when he reviewed the *Theory* of 1795 for *The British Critic* he again taunted Hutton with the problem of the Alpine lakes. Now, perhaps as a result of the religious fervour which swept Britain at the outbreak of the Revolutionary War, he wrote in much more bitter vein than in his letters to *The Monthly Review* five years earlier. He roundly condemned Hutton's atheism, applied such adjectives as 'extraordinary' and 'futile' to the theory, and wrote sarcastically that its propositions 'will, no doubt, appear like the outlines of an oriental tale: the author, however, seems to be in earnest....' De Luc still regarded Hutton's fluvialism as one of the most extravagant aspects of the theory, but Hutton, now a sick man, offered no rejoinder. He died the following year, and more than sixty years were to elapse before an understanding of the glacial origin of rock-basins finally removed de Luc's limnological objection to the fluvialist doctrine.

John Playfair and the 'Illustrations'

After Hutton's death, John Playfair, the Professor of Mathematics in Edinburgh University, undertook to prepare a biographical memoir of his late friend, and as he read through Hutton's works he realised for the first time the full extent of Hutton's failings as an author. Playfair himself had become convinced of the soundness of the Huttonian theory through its author's lively verbal expositions, but he now perceived that Hutton's published accounts of the theory were unlikely to win him many converts. Playfair therefore resolved to prepare a fresh and more lucid exposition of the theory. He was certainly well equipped to become Hutton's champion; he had discussed the theory regularly with Hutton over a period of several years, but more important, he possessed in abundance those very gifts which are so conspicuously lacking in Hutton's own writing. Playfair had a clear, incisive mind, the ability to marshal arguments to good effect, and a precise and elegant literary style. Work upon the new exposition occupied him for almost five years, and it was published in 1802 as the celebrated *Illustrations of the Huttonian Theory of the Earth*.[57] By skilful distillation he compressed the theory into only a fraction of the number of pages which Hutton had filled to so little purpose in 1795, and after 1802 few troubled to tackle the Huttonian original. It was through Playfair's masterly exposition that the nineteenth century acquainted itself with Hutton's theory.

Playfair accepted Huttonian geomorphology unreservedly; he was no less convinced than his master that the Earth's topography is the result of circumdenudation effected by current geomorphic processes operating throughout vast periods of time. His views follow those of Hutton so closely that a full account of them here would involve a wearisome reiteration of the Huttonian geomorphic teaching already discussed. There is nevertheless one important field where the *Illustrations* represents a marked advance upon Hutton's *Theory*. Despite its significance, Hutton had made little effort to prove the fluvial origin of valleys; he clearly regarded the matter as sufficiently self-evident to render elaborate substantiation unnecessary. Playfair thought otherwise. His logical mind rejected Hutton's peremptory statements and substituted five reasoned proofs of the fluvialist doctrine.[58] Some of Playfair's arguments were directed against those who regarded valleys as crustal fissures opened by earthquakes, while the remainder were intended to counter the belief that valleys had been cut by diluvial or marine currents.

Playfair's first argument arose from his observation that in many areas valleys radiate from some central point, or drain down the two sides of a ridge, and in each case the valleys increase in depth in the downstream direction. The agent that produced such valleys, he maintained, clearly increased in degradational power as it proceeded downslope, and he regarded stream-water as the sole agent capable of meeting this requirement. Such groups of valleys, he wrote, are 'highly unfavourable to the notion that they were produced by any single great torrent, which swept over the surface of the Earth'. His second argument was similar: he pointed out that it was ridiculous to suppose the waters of a single debacle capable of scooping out complex valley systems containing streams flowing in all manner of directions. He derived his third argument from what he described as 'longitudinal valleys, which have the openings by which the water is discharged, not at one extremity, but at the broadside'. Clearly, such valleys arranged at right-angles to each other could hardly be the work of diluvial or marine currents.

Important though the three above arguments may have been in their day, it is Playfair's two final proofs of the fluvial doctrine which are now best remembered. The first of these he drew from the plan of a drainage system, and he observed 'that where a higher valley joins a lower one, of the two angles which it makes with the latter, that which is obtuse is always on the descending side . . .' He thus recognised the dendritic form of most drainage systems, and he pointed out that this regular pattern could hardly have arisen through the opening of fissures in the Earth's crust. (Catcott, it will be remembered, had made a similar point forty years earlier.) His last argument is familiar today as Playfair's Law of Accordant Junctions which he outlined in the following passage:

> Every river appears to consist of a main trunk, fed from a variety of branches, each running in a valley proportioned to its size, and all of them together forming a system of vallies, communicating with one another, and having such a nice adjustment of their declivities, that none of them join the principal valley, either on too high or too low a level; a circumstance which would be infinitely improbable, if each of these vallies were not the work of the stream that flows in it.

Although he was a confirmed fluvialist, Playfair recognised the cogency of de Luc's limnological objection to the doctrine, and he followed Hutton closely in trying to grapple with the problem. He accepted Hutton's suggestion that the lake-basins might have been formed comparatively recently as a result of landslides, solution, or earthquakes, but to these he added a fourth explanation of his own. He suggested that at some time in the past the basins had been completely filled with river-borne debris, but that Nature for some reason had then gone into reverse, cleaned the basins out, and allowed them to become lakes again. This was Playfair's sole geomorphic lapse; elsewhere there is little fault to be found with his views on the origin of topography. Indeed, some portions of the *Illustrations* might still be read with profit by those seeking an introduction to geomorphology, and certainly in its day the book was unrivalled as a geomorphic text.

It was Playfair's intention to publish a second and much expanded edition of the *Illustrations*, and with this in mind he made numerous geological tours in the British Isles. His desire to extend his researches to the continent was long frustrated by the Napoleonic War, but in 1816, aged 68, he set off on an extended tour of France, Switzerland, Austria, and Italy. These were the very districts which Hutton himself had commended to the attention of geologists, but Playfair was not destined to publish an Huttonian interpretation of the Alpine landscapes; he died in July 1819, leaving the second edition of the *Illustrations* unfinished. His failure to complete the book is greatly to be regretted, perhaps more so in a geomorphic context than in any other, because he was clearly especially interested in the geomorphic aspects of the Huttonian theory, and these would doubtless have figured prominently in the projected work. Playfair's long reflection upon the nature of landforms, his wide field experience, and his literary ability could only have resulted in a most valuable treatise, and its appearance in the 1820s might well have saved geomorphology from the renewed neglect which was its fate throughout much of the early nineteenth century.

Some of Playfair's contemporaries deplored his preoccupation with the Huttonian theory. They believed that Hutton had seduced him from the pure fields of mathematics into the treacherous morass of Earth-history. Even Playfair's biographer offered apology for the *Illustrations*, and he suggested that the book was best forgotten in seeking a true evaluation of Playfair's scientific attainments. It is therefore ironic that Playfair is today remembered and acclaimed solely as Hutton's foremost disciple. Such acclaim is entirely justified, but it is important to distinguish the role of the *Illustrations* in the history of geology from its role in the history of geomorphology. This distinction has been insufficiently appreciated. In the geological context the book served as a fan which raised the temperature of the controversy between the Neptunists and the Plutonists to a white heat, but in the geomorphic context the book had no inflammatory effect whatsoever. Playfair must have felt hurt; his carefully reasoned exposition of the fluvial doctrine was neither praised nor condemned – it was merely disregarded. For this there were two chief reasons. Firstly, Huttonian geomorphology was forgotten as the attention of the two rival schools of geology focused upon one single, vital issue – were granite and basalt chemical precipitates or were they igneous melts? Secondly, and for reasons to be explained later, even those geologists who were not committed to the struggle between the Neptunists and the Plutonists, turned away from geomorphology and became increasingly preoccupied with stratigraphical problems. There is, however, another, third reason worthy of mention. While all early nineteenth-century geologists save Playfair rejected the fluvialistic interpretation of present landscapes, they at the same time almost all accepted the reality of subaerial denudation. The days of a teleological approach to landforms were fast departing, and the passing of the denudation dilemma was allowing a reversion to a thorough-going belief in the potency of Nature's destructive forces. Playfair himself exemplifies this new, less inhibited approach to the subject because in his book he presents the Huttonian theory shorn of all its teleology, and he discusses denudation as a scientific fact without pausing to consider whether it is harmful or beneficial to the Earth. Thus while very few of Playfair's readers can have found themselves in agreement with his fluvialistic interpretation of topography, there can have been even fewer who saw anything exceptionable in the idea that the Earth's topography is today mouldering away. Playfair's vociferous Neptunian critics certainly raised no objection to his belief in the reality of denudation.[59] The chief importance of the *Illustrations* therefore lies in the field of geology where its appearance did much to hasten the acceptance of the Plutonic doctrine. In the geomorphic context the book was of much less significance. Indeed, Playfair's exposition of the fluvial doctrine might be adjudged a failure in the sense that it won the doctrine few, if any converts.

Hutton's theory represents an enormous advance upon the fanciful, bibliolatrous theories of most of his contemporaries. Any reassessment of his geological work must leave him as one of the outstanding figures in the history of the science, but in the geomorphic context it seems that much of the praise lavished upon him has been misplaced. His chief contribution to geomorphology was not the fluvialism for which he has been so widely acclaimed, but rather his deistical time-scale, his cyclic Earth-history, and his uniformitarianism. Instead of viewing him in his traditional guise as the founder of modern geomorphology – the precursor of the fluvialists of the 1860s – it is reasonable, and certainly refreshing, to cast him in a new role. He may with justice be regarded not as the vanguard of modern fluvialism, but rather as the last representative of that early lineage of British geomorphologists which has

194

William Bourne at its head. Within little more than ten years of Hutton's death in 1797, geomorphology's geological parent underwent the transformation which converted it from a speculative subject based upon a minimum of field-evidence, into a precise observational science, and Hutton seems to belong in the period before the subject underwent this momentous transformation. Four elements in his work support the conclusion that he should be grouped with Hooke, Ray, and Woodward, rather than with such nineteenth-century figures as Jukes, Ramsay, and Geikie. Firstly, there is Hutton's deism, his teleology, and his concern with final causes. The intrusion of such matters into science was common enough during the seventeenth and eighteenth centuries, but such a mixing of science and religion was utterly foreign to most nineteenth-century geologists. Secondly, there is Hutton's concern with the denudation dilemma – an eighteenth-century problem which was almost unknown to his successors. Thirdly, there is the antiquated presentation of his theory. His heavy reliance upon literary sources, and his reluctance to discuss his own field-observations in detail seem to place him squarely alongside such eighteenth-century figures as Williams and Kirwan. When his friend Sir James Hall, imbued with the spirit of modern geology, sought to test the truth of Hutton's plutonism experimentally in the laboratory, Hutton so discouraged him that Hall postponed his experiments until after Hutton's death. Equally significant is Hutton's neglect of even such elementary techniques of modern geology as maps and sections. The volumes of 1795 contain not a single map and only two rather crude geological sections, whereas many of his more progressive contemporaries, from Whitehurst in 1778 onwards, had already demonstrated the value of detailed field-studies supported by graphical material. Perhaps here we have further evidence that Hutton had indeed written his theory more than twenty years before it was published. He did go some way towards modernising his presentation in the third volume of the *Theory*, but so far as the work published during his lifetime is concerned, one must sympathise with Lyell's judgment that Hutton's writing shows little advance upon that of Hooke or Steno[60].

The final factor which seems to place Hutton among the earlier geologists is, paradoxically, that thorough-going fluvialism for which he has received so much acclaim. In fact his fluvialism was not the closely reasoned belief expected of a nineteenth-century geomorphologist, but rather an implicit, instinctive faith of exactly the same type as that which had caused so many seventeenth-century scholars to accept the reality of denudation. Had Hutton been a geologist in the modern tradition, he would have been at greater pains to examine the fluvial doctrine in the light of field evidence, and had he done so, he might have returned somewhat less confident of the doctrine's validity. The rejection of fluvialism by many early nineteenth-century geologists has often been remarked upon, but it is insufficiently realised that this rejection was not so much the result of blind stupidity, as the result of the seeming inadequacy of the fluvial doctrine as an explanation of many field phenomena. De Luc's limnological objection to the doctrine was one difficulty, but what was the fluvialist to make of the deep, dry cuts which we now know to be glacial spillways, the dry valleys of chalk and limestone districts, or the Saharan wadis? How was he to explain the fact that many rivers today, far from deepening their valleys, are actually aggrading and building up extensive flood plains? Similarly, one wonders how Playfair would have reconciled his Law of Accordant Junctions with the innumerable hanging valleys he must have seen in the course of his Alpine travels. Hutton was evidently oblivious to the existence of such difficulties until they were pointed out by his critics. His fluvialism was the product not of a profound understanding of Nature's operations, but of a typically superficial and very limited eighteenth-century acquaintance with field phenomena. He was not so much the precursor of the late nineteenth-century school of fluvial geomorphology, as the scholar who blended the simple, uncritical fluvialism of the seventeenth century with the teleology of the eighteenth century. It is one of the strange ironies of history that Hutton, with his intuitive fluvialism, should have been correct, whereas his successors, who sought to test the doctrine in the light of the field evidence, were cruelly misled because of Nature's unsuspected complexity. Unlike Hutton, they knew sufficient to perceive the objections to fluvialism, but at the same time they lacked the knowledge necessary to sweep the objections aside. Their little learning did indeed prove a dangerous thing.

REFERENCES

1. James Hutton, *On Agriculture*, MS. in the Library of the Royal Society of Edinburgh; J. Playfair, 'Biographical Account of the late Dr. James Hutton', *Trans. roy. Soc. Edinb.*, V (3), (1805), pp. 39–99, and *The Works of John Playfair*, IV, pp. 31–118 (Edinburgh 1822); E. B. Bailey, 'James Hutton, Founder of Modern

Geology', *Proc. roy. Soc. Edinb.*, LXIII B (4), (1950), pp. 357–368; Edward B. Bailey, *James Hutton – the Founder of Modern Geology* (Amsterdam 1967).

2. V. A. Eyles and J. M. Eyles, 'Some Geological Correspondence of James Hutton', *Ann. Sci.*, VII (1951), pp. 316–339.

3. S. I. Tomkeieff, 'James Hutton and the Philosophy of Geology', *Trans. Edinb. geol. Soc.*, XIV (1938–51), pp. 253–276, and *Proc. roy. Soc. Edinb.*, LXIII B (4), (1950), pp. 387–400; D. B. McIntyre, 'James Hutton and the Philosophy of Geology', pp. 1–11 in Claude C. Albritton, *The Fabric of Geology* (Stanford, California 1963).

4. *The Antiquity and Duration of the World* (London 1780); *The Antiquity of the World* (London 1783); *The Eternity of the World* (London 1785); *The Eternity of the Universe* (London 1789).

5. James Hutton, *Theory of the Earth, with Proofs and Illustrations*, I, p. 246 (Edinburgh 1795).

6. *Histoire de l'Académie Royale des Sciences* (1771), pp. 705–755; (1773), pp. 599–670.

7. William Ramsay, *The Life and Letters of Joseph Black, M.D.*, p. 124 (London 1918).

8. G. L. Davies, 'George Hoggart Toulmin and the Huttonian Theory of the Earth', *Bull. geol. Soc. Am.*, LXXVIII (1967), pp. 121–123.

9. V. A. Eyles, 'Note on the Original Publication of Hutton's *Theory of the Earth*, and on the Subsequent Forms in which it was Issued', *Proc. roy. Soc. Edinb.*, LXIII B (4), (1950), pp. 377–386; 'A Bibliographical Note on the Earliest Printed Version of James Hutton's *Theory of the Earth*, its Form and Date of Publication', *J. Soc. Biblphy. nat. Hist.*, III (2), (1955), pp. 105–108.

10. J. Cramer and H. K. Swann, *Historiæ Naturalis Classica*, I and II (Weinheim and Codicote 1959).

11. Hutton, *op. cit.* (1795), I, pp. 275 & 276.

12. *Ibid.*, I, p. 375.

13. *Ibid.*, II, p. 547.

14. *Ibid.*, I, p. 273.

15. *Ibid.*, I, p. 15.

16. *Ibid.*, I, p. 223.

17. *Ibid.*, I, pp. 18 & 19.

18. *Ibid.*, II, p. 157.

19. *Ibid.*, II, pp. 116, 379, 401.

20. *Ibid.*, II, pp. 289 & 290.

21. *Ibid.*, II, p. 528.

22. *Ibid.*, II, pp. 308 & 309.

23. James Hutton, *Theory of the Earth*, III, pp. 29 & 30 (London 1899).

24. Hutton, *op. cit.* (1795), II, p. 218.

25. *Ibid.*, II, p. 296.

26. *Trans. roy. Soc. Edinb.* (1788), I (2), p. 216.

27. *Ibid.*, p. 218.

28. Hutton, *op. cit.* (1795), II, p. 550; I, pp. 208, 620.

29. *The Joseph Black Papers*, III, in the Edinburgh University archives; Ramsay, *op. cit.* (1918), pp. 118 & 119.

30. M. J. S. Rudwick, 'Hutton and Werner Compared: George Greenough's Geological Tour of Scotland in 1805', *Br. J. Hist. Sci.*, I (1962), pp. 132 & 133.
31. MS. *Papers of the Natural History Society*, XV (1799 – c. 1807), p. 321. The volumes are in the Edinburgh University archives.
32. R. Kirwan, 'Examination of the Supposed Igneous Origin of Stony Substances', *Trans. R. Ir. Acad.*, V [1793], pp. 51–81; W. Richardson, 'Inquiry into the Consistency of Dr. Hutton's Theory of the Earth with the Arrangement of the Strata, and other Phænomena on the Basaltic Coast of Antrim', *ibid.*, IX (1803), pp. 429–487.
33. Tomkeieff, *op. cit.* (1938–51 and 1950).
34. Hutton, *op. cit.* (1795), II, p. 562.
35. James Bisset, *A Poetic Survey round Birmingham*, p. 12 (Birmingham 1800).
36. *The Life of Samuel Johnson.*
37. *The Botanic Garden: Pt. I Economy of Vegetation*, Canto I, *ll.* 259–262.
38. Elie Halévy, *A History of the English People in 1815*, p. 457 (London 1924).
39. *Dialogues Concerning Natural Religion*, Part II.
40. James P. Muirhead, *The Origin and Progress of the Mechanical Inventions of James Watt*, II, p. 73 (London 1854).
41. Hutton, *op. cit.* (1899), pp. 88 & 89.
42. Hutton, *op. cit.* (1795), I, p. 146.
43. Katharine M. Lyell, *Life, Letters and Journals of Sir Charles Lyell, Bart.*, II, pp. 47 & 48 (London 1881).
44. *Blackwood's Magazine*, I (1817), p. 232.
45. *The British Critic*, VIII (1796), p. 598; William Richardson, *Observations on the Review of Two Memoirs in the Transactions of the Royal Irish Academy* (Dublin 1805).
46. *Telliamed: or, Discourses Between an Indian Philosopher and a French Missionary* (London 1750).
47. John Playfair, *Illustrations of the Huttonian Theory of the Earth*, p. 130 (Edinburgh 1802); *Edinburgh's Place in Scientific Progress*, British Association, pp. 71 & 72, (Edinburgh 1921); R. J. Chorley, A. J. Dunn and R. P. Beckinsale, *The History of the Study of Landforms*, I, p. 34 (London 1964).
48. *The Monthly Review*, LXXIX (1788), pp. 36–38; *The Critical Review*, LXVI (1788), pp. 115–120; *The Analytical Review*, I (1788), pp. 424 & 425.
49. Lyell, *op. cit.* (1881), II, p. 48.
50. *Encyclopédie Méthodique: Géographie-Physique*, I, Paris, year III, pp. 732–763.
51. *An Oration, on the Truth of the Mosaic History of the Creation* (Hartford, 1798).
52. *The Fatal Consequences and the General Sources of Anarchy*, p. 36 (Edinburgh 1793).
53. Edward Nares, *The Bampton Lectures*, pp. 171 & 172 (Oxford 1805).
54. John Williams, *The Natural History of the Mineral Kingdom*, I, pp. lix & lx (Edinburgh 1789).
55. Kirwan, *op. cit.* (1793), p. 64; Williams, *op. cit.* (1789), I, p. lxii.
56. Williams, *op. cit.* (1789). De Luc, four letters to Hutton in *The Monthly Review*, II (1790), pp. 206–227, 582–601, III (1790), pp. 573–586, V (1791), pp. 564–585; a review of the Huttonian theory in *The British Critic*, VIII (1796), pp. 337–352, 466–480, 598–606. Kirwan, *op. cit.* (1793); *Geological Essays* (London 1799). Richardson, *op. cit.* (1803).

57. The *Illustrations* was also included in *The Works of John Playfair*, I (Edinburgh 1822), and a facsimile reprint was published by the University of Illinois Press, Urbana, in 1956.
58. Playfair, *op. cit.* (1802), pp. 102, 113 & 114, 400–404.
59. See, for example, *The Edinburgh Review*, I (1802–3), pp. 201–216, and John Murray, *A Comparative View of the Huttonian and Neptunian Systems of Geology* (Edinburgh 1802).
60. Lyell, *op. cit.* (1881), II, p. 48.

Chapter Seven

Geology's Laggard
1807–1862

It seems, indeed, not a little curious, that notwithstanding the amount of geological knowledge diffused through this country regarding the origin of the various systems and formations which lie beneath the surface, so much ignorance and uncertainty should exist with respect to the origin of the surface itself.

Archibald Geikie
writing in the first edition of his
'Scenery of Scotland' in 1865.

IN 1807 Werner stood supreme as the world's geological oracle, but twenty years later the reputation of the Freiberg Mining Academy lay in ruins, and the palm of world geological leadership had passed from Saxony to Britain. This emergence of a vigorous British school of geology was a remarkably rapid event. During the eighteenth century Britain had produced very few geologists of international repute, but now, in the new century, the situation was dramatically transformed as British scholars threw themselves enthusiastically into the study of Earth-history. Soon the subject was attracting the most able British scientists of the day, and it is hardly an exaggeration to claim that modern geology was brought into being solely as a result of the labours of British geologists between 1807 and 1860.

Geology, however, was not the preserve of a few distinguished savants; it enjoyed an enormous popularity with all classes of society from the Royal Family at Windsor, to Robert Dick the Thurso baker. No science can ever have excited such widespread popular interest; Britain was in fact swept by a geological mania. Soon no gentleman's study was deemed complete without its geological cabinet, and those dilettanti anxious to pose as men of science travelled with hammers, lenses, and collecting-bags on ostentatious display. To judge from Book III of *The Excursion*, Wordsworth's Lake District must have been over-run by the self-styled 'brethren of the hammer' as early as 1814, and ten years later, in Scott's *St. Ronan's Well*, Meg Dods refers to the horde of geologists who

> . . . rin up hill and down dale, knapping the chucky stanes to pieces wi' hammers, like sae mony roadmakers run daft – they say it is to see how the warld was made!

Newspapers carried reports of the latest geological discoveries, while *Punch* repeatedly poked fun at the antics of the geologists. Section C (Geology) regularly drew the largest crowds at the annual meetings of the British Association for the Advancement of Science, and local geological societies sprang up throughout the British Isles in response to the enormous interest in the subject. Even in war geology was not for-gotten. British officers in the Crimea spent their off-duty hours geologising around Balaclava. Ladies too became enthusiastic geologists. The first British *Ichthyosaurus* was found in 1811 by Mary Anning, who for more than thirty years made her living by selling geological specimens at a shop in Lyme Regis, and the Geological Society of London published a paper by a lady (Mrs Maria Graham) as early as 1824. At least one other lady – Mrs Gideon Mantell – must have been less well disposed towards geology; she and her children were forced out of their fossil-filled Brighton home in 1836 when her husband's infatuation with palaeontology led him to throw the house open to the public as a geological museum.

Before 1807, geology and geomorphology had invariably moved forward hand in hand. Men such as Hooke, Ray, Woodward, and de Luc had all understood that any study of Earth-history must take land-forms into account. They had recognised that landforms are the key to the final chapters of Earth-history, just as rocks are the key to the earlier chapters. For the first sixty years of the nineteenth century, however, this dual approach to the problem of Earth-history was largely for-gotten. Geologists became so obsessed with rocks that they forgot to examine the landforms to which the rocks give rise. Their heads were permanently buried underground. It was in 1835 that J. D. Forbes complained that 'geology, of late years, has become little more than a commentary on organized fossils',[1] and a few years later we find geologists of the calibre of Sir Henry De La Beche, Joseph Portlock, and Sir Roderick Murchison, publishing massive tomes on the geology of selected regions of the British Isles, full of geological minutiae, yet offering only the sketchiest accounts of surface morphology. Their disregard of the subject is reminiscent of the similar neglect displayed by the seventeenth-century topographers, but in the works of the nineteenth-century geologists the omission is much less forgivable.

The present chapter, then, is concerned with a period when geomorphology languished in neglect, just as it had a century earlier between 1705 and 1778. There are nevertheless two important distinctions to be drawn between these two periods. Firstly, during the earlier period, the neglect of geomorphology was only one manifestation of a general neglect of the Earth-sciences, whereas during the nineteenth century, geomorphology lay forgotten while other branches of geology surged rapidly forward. Secondly, the earlier period was a time when geomorphology was in almost complete abeyance, whereas the later period was one when the subject suffered merely a relative neglect. There is no dearth of early nineteenth-century geomorphic literature to equal the dearth which so hinders the historian of early eighteenth-century geomorphology, but the proportion of geomorphic material amongst the vast flood of nineteenth-century geological books, pamphlets, and papers is very small. The first five volumes of the *Transactions of the Geological Society of London* (1811–1821), for example, contain a total of 119 papers, but of these only three (two by John Macculloch and one by William Buckland) are deserving of a place in the bibliography of the present work.

Because of its neglect, geomorphology made little progress during the first half of the nineteenth century. While other branches of the Earth-sciences grew rapidly from their infancy, through a sturdy adolescence, to a vigorous adult stage, geomorphology tottered aimlessly along like one of the stunted, misshapen cretins who at once both appalled and fascinated so many early visitors to the Alpine valleys. Had Hutton returned to the geological scene in 1850, he would have found most branches of the subject advanced almost beyond his recognition, but he might well have arrived at the conclusion that geomorphology, far from advancing, had in some respects suffered a regression since his death in 1797. So retarded was the subject, and so little attention did it claim, that it languished nameless until the 1860s when a small group of Geological Survey officers began to interest themselves in landforms. Only then did the subject acquire the title 'Physiographical Geology'.

In order to understand this unfortunate early nineteenth-century eclipse of geomorphology, it is necessary to examine the factors which raised other branches of geology – and especially stratigraphy and palaeontology – to such remarkable heights of popularity. Such an examination makes it clear that geomorphology lacked almost all the attributes which made other branches of geology so attractive. Six factors were chiefly responsible for the geological mania which swept through nineteenth-century Britain.

Firstly, geology gratified the Englishman's delight in the open-air life. As W. H. Fitton wrote in 1817:

> Geology has this great advantage, of which not even Botany partakes more largely, – that it leads continually to healthful and active exertion, amidst the grandest and most animating scenery of Nature.[2]

For the urban dweller, the subject provided escape from the grimy towns of the Industrial Revolution, while among the gentry, Sir Roderick Murchison was not alone in regarding geology as a fascinating alternative to fox-hunting, and J. F. Campbell of Islay, a keen student of glacial relics, certainly considered that the two types of field-sport were complementary.

> Hunting is healthy pastime, and hunting for ice-marks upon hill-tops may be combined with other sport. The spoor leads to the haunts of grouse, deer, and ptarmigan; to grand scenery and to regions of fresh air.[3]

Many certainly combined the hammer with the saddle, and William Buckland of Oxford possessed the doubtful benefit of a horse which refused to pass a geological exposure until its master had plied his hammer to the rock-face.

Secondly, in Britain during the Industrial Revolution the bones of the Earth were for the first time being laid bare on a grand scale. Innumerable sections exposed in new quarries and mines, and during the excavation of cuttings and tunnels for canals and railways, posed a multitude of mute questions about the Earth's history. 'Have I ever told you of the wonderful & surprising curiositys we find in our Navigation?' wrote Josiah Wedgwood to a friend as early as 1767,[4] and it was while working upon the proposed Somersetshire Canal in the 1790s that William Smith first discovered the fascination of geology.

Thirdly, as the Industrial Revolution gathered momentum, a knowledge of geology became a matter of increasing practical importance. The problems that arose during the construction of canals, roads, and railways, during the search for new mineral deposits, and during the provision of augmented urban water supplies, all called into being a small but significant class of professional geologists. John Williams, as we saw earlier, was one of the first of these professionals, and when they were not acting as consultants, such men further stimulated interest in geology by delivering courses of public lectures on the subject – a service for which they were usually able to command a high fee.

Fourthly, geology was then a very simple science, and even the untutored amateur equipped with nothing more than a hammer, a map, and a stout pair of legs, could expect to make a useful contribution to his subject. Also, within the British Isles there was a sufficient diversity of rock types to satisfy almost every geological taste. Yet, despite its simplicity, geology had the fascination of bringing its devotees into contact with one of the most fundamental of problems – the nature and origin of the Earth itself.

Fifthly, geology provided an outlet for one of the strongest of human instincts – the desire to collect. The assembly of rock, mineral, and fossil collections then satisfied the urge which in Britain, at least, is today met by the collection of such items as stamps, coins, or matchbox labels. In the United States, however, the craze for collecting geological specimens continues; few western towns are without their rock shops, and in some places specimens are even available from vending machines at ten cents a time. Nineteenth-century British collectors never knew such a convenience! Even the humblest collector could nevertheless then assemble a large geological collection easily and with a minimum of expense, while the more affluent collectors could purchase anything from a single choice specimen to an entire museum from one or other of the many dealers who entered the business. The demand for complete representative collections was such that in 1841 a Glasgow firm announced a price reduction of up to 25 per cent because their collections were now being mass produced.

Finally, it seems probable that a great stimulus to the study of geology was provided by the gradual extension of the work of the Ordnance

Survey across the British Isles. The earliest Ordnance Survey map was published in 1801, and by the middle of the century the British Isles was for the first time covered by a series of detailed and accurate maps. Now any field-worker knows that there is considerable satisfaction to be found in plotting one's observations upon maps, and in the days when One Inch and Six Inch maps were still novelties, it must have been a fascinating experience to work systematically over a region, map in hand, marking in the geological boundaries. Then, the field-work completed, there came the equally gratifying task of colouring the map to make an attractive wall-piece which the proud amateur could display as the fruit of his own unaided researches. The earliest official Geological Survey map did not appear until 1834, and even in 1879 large areas of England and Ireland, together with almost the whole of Scotland, were still not covered by the official maps. The amateur geologists thus long enjoyed the thrill of mapping virgin country with reasonable hope of making some exciting discovery.

If these were the reasons for the popularity of geology, they were also the reasons for the neglect of geomorphology. True, geomorphology could have offered the same open-air opportunities as geology, but beyond that point everything militated against geomorphology and in favour of other branches of geology. The Industrial Revolution, for example, focused attention not so much upon the form of the Earth's surface, as upon the rocks beneath the surface. Geomorphology seemed to have neither practical nor economic significance. It offered nothing tangible that could be broken with a hammer or examined with a lens, and it certainly provided nothing that could be collected for exhibition in a cabinet. Finally, geomorphology seemed to afford little that could be mapped in the sense that geological boundaries could be mapped. Those infected with the geological mania therefore disregarded geomorphology, and threw themselves wholeheartedly into the pursuit of other branches of geology.

One other reason for the neglect of geomorphology must be mentioned. As a reaction against the wild speculations of the eighteenth-century geologists – and more especially as a reaction against the ill-grounded dogmatism of the Wernerian system – most early nineteenth-century geologists shunned all theorising. Instead, they devoted themselves to the making of careful observations, and to the patient accumulation of the factual basis upon which theories might one day be built. Sir Charles Lyell believed that the object of the Geological

Society of London was to 'multiply and record observations, and patiently to await the result at some future period',[5] and in 1817 the *Edinburgh Review* congratulated the Society upon the fact that its publications were characterised by 'strict experiment or observation, at the expense of all hypothesis, and even of moderate theoretical speculation'.[6] Anything savouring of theory was now suspect, but in geomorphology it was difficult to remain objective and to avoid speculation about the nature of the processes responsible for shaping topography. The flight from theory therefore also became a flight from geomorphology.

In earlier chapters, while tracing the history of geomorphic thought down to 1807, material drawn from many varied sources was used, and the writings of scientists and non-scientists were each allowed equal weight. Such a course seems justified when considering a period before the emergence of geology as a specialist study. The geomorphic views of a bishop are just as deserving of attention as those of a botanist in an age when neither men had made any particular study of the matter at issue. After 1807 the situation changes as result of the emergence of geology as a specialist science, and henceforth our study will be based almost entirely upon the works of those who have some claim to the title 'geologist'. Many non-geologists continued to air their geomorphic theories, but the works of such dabblers need not engage our attention. In particular, no reference will be made to the voluminous writings of what Milton Millhauser[7] has termed 'the Scriptural Geologists' – a large group of authors whose chief concern was to bridge the steadily widening nineteenth-century gulf between *Genesis* and geology. Their works are doubtless of interest to the student of Victorian Britain, but they are hardly relevant to a study of the evolution of the main-stream of geomorphic thought. Although the reconciliation of geology and the Scriptures now became chiefly the concern of the Scriptural Geologists, it is clear that for two decades after 1807, even some of the 'scientific' geologists had not entirely escaped from the religious influences which had so hampered their predecessors. In the present chapter, therefore, it is appropriate to consider first the final emergence of geology from its bibliolatrous stage, and, more especially, the final rejection of the Mosaic chronology. Then we will proceed to an examination of the nineteenth-century conception of Earth-history, and to a discussion of the endogenetic and exogenetic forces as they were understood before 1862.

The Escape from Moses

The influence of religious beliefs upon geology began to diminish soon after 1760. The liberalising tendencies suffered a brief set-back when 'the policeman theory of religion' dominated the scene during the Revolutionary and Napoleonic wars, but the earlier trend was soon restored. As early as 1802, Playfair's *Illustrations* shows a remarkable freedom from all religious inhibitions, but others did not escape from the old constraints quite so easily. As a result, some eighteenth-century attitudes lingered on into the early years of the new century. Bibliolatry survived to some degree, but more surprising is the persistence of teleological considerations, and natural theology, in the works of some of the most eminent of early nineteenth-century geologists. The Rev. Adam Sedgwick, for example, one of the leading lights of the British school of geology, and the Woodwardian Professor at Cambridge (he was also for long the Prebendary of Norwich), informed the Geological Society of London in 1831, during the course of a Presidential Address, that 'Geology lends a great and unexpected aid to the doctrine of final causes'.[8] The same idea pervades *A System of Geology* published in 1831 by Dr John Macculloch, the author of the first detailed geological map of Scotland ever published. It was the Rev. William Buckland, however, the Reader in Geology at Oxford, who carried the old attitudes forward most vigorously into the new century. His famous *Bridgewater Treatise* entitled *Geology and Mineralogy considered with reference to Natural Theology*, first published in 1836, was a work very much in the eighteenth-century tradition. Indeed, some of his arguments can be traced back into the seventeenth century, and his teleological discussion of mountains is clearly derived from Ray's exposition of the same theme dating from 1692. Many of Buckland's contemporaries must have found the work a somewhat embarrassing throw-back, and it is difficult to believe that Buckland really accepted some of the arguments he employed. His efforts to arrive at a teleology of complexly folded and faulted coalfields, for example, reads like a skit upon earlier natural theology. The coal-seams are folded, he claimed, in order to encourage the free drainage of mine galleries, while faults are introduced to prevent the dipping seams from plunging to depths far beyond the miners' reach! How he proposed to account for those faults which, far from raising the coal-seams, throw them down to unworkable depths is not clear, but it comes as no surprise to learn from one of Buckland's letters that he took the precaution of letting the Oxford professors of Divinity and Hebrew vet his manuscript prior to publication![9]

Sedgwick, Macculloch, and Buckland were the exceptions; by the 1830s most geologists had shed those attitudes which had so handicapped their predecessors in the science. Many geologists of course retained a sincere religious faith, but henceforth religious beliefs no longer moulded geological thought. Rather did geological discoveries now begin to influence religious orthodoxy. By the time Lyell wrote to G. J. P. Scrope in 1830 urging him to 'free the science from Moses',[10] the cause for which he pleaded was almost won, and only eight years later the Dean of York was gravely perturbed at the thought that Moses was unlikely to find any champions when the geologists expounded their heresies at meetings of the British Association.[11] When, in the 1840s, a member of the Geological Society of Edinburgh proposed that the society should either affirm its faith in the absolute truth of *Genesis* or else arrange a full discussion of the subject with the clergy represented, his plea fell upon deaf ears.[12] By 1851 the climate had so far changed that Sir Henry De La Beche, the Director of the Geological Survey of Great Britain, relaxing in a Welsh inn after a day in the field, could dismiss the Biblical account of the Flood as 'that funny story'.[13]

In this new age, not only was Ussher's date for the Creation virtually forgotten in geological circles, but the great majority of geologists were no longer even concerned to show that the Earth's early history could be arranged to fall into the six allegorical days of the hexaëmeron. Now, for the first time, the true magnitude of the Earth's age began to be generally appreciated. To some extent this appreciation was part of the slow dawning of truth which was a necessary corollary of the demise of bibliolatry, but two geological discoveries accelerated the development. Firstly, research confirmed that Earth-history, far from being a steady linear progression as the traditional interpretation had suggested, was, rather, a series of complex and seemingly meaningless cycles. Lands had sunk into the sea only to emerge again at some later date; rocks had been formed only to be destroyed by denudation; and innumerable species of plants and animals had come into existence only to be extinguished long before man himself had appeared upon the scene. Clearly, the pre-human period in the history of the Earth could no longer be regarded merely as an overture to the human drama. Man, it now seemed far from being the sole *raison d'être* of the globe, was little more than a recent afterthought in the Creation. The old anthropocentric view of Earth-history had been shattered, and now it began to be generally admitted that an aeon might have elapsed between the creation of the Earth and the creation of man. In his *System of Geology*, published in 1831 but evidently written ten years earlier, John Macculloch wrote of time:

> We are too apt to measure this by our own brief duration, as our vanity dreams that the universe was created for us. Let us contemplate Time as it relates to the Creator, not to ourselves, and we shall no longer be alarmed at that which the history of the Earth demands.[14]

The second development which affected the nineteenth-century appreciation of time, was the discovery of the immense total thickness of the world's sedimentary rocks. In the 1850s A. C. Ramsay, then a young officer of the Geological Survey, calculated this thickness for the benefit of Charles Darwin, and he arrived at the conclusion that the thickness could not be less than 72 584 feet, or $13\frac{3}{4}$ miles.[15] Gone were the days when rocks were supposed to have been precipitated rapidly from some mysterious chaotic fluid, and it was soon generally admitted that the accumulation of such a vast sedimentary pile must have occupied a length of time so long as to be almost beyond human comprehension.

These two developments led geologists to think in terms of tens of millions of years where hitherto a few thousand years had seemed an ample sufficiency. When Lyell published the first volume of his great *Principles of Geology* in 1830, he clearly regarded time as available to the geologists in almost unlimited quantities, and two years later Whewell, reviewing the second volume of the *Principles*, observed:

> We have no occasion to embarrass ourselves for want of thousands, or, if it be necessary, of millions of years.[16]

Even Buckland, the outdated teleologist, accepted the new time-scale. In his *Bridgewater Treatise* of 1836, he was still anxious to preserve some outward conformity with *Genesis*, but he admitted that the interval between the two opening verses of the Bible might represent 'millions and millions of years'. By the middle years of the century, while the fundamentalists still fought for a near-literal interpretation of Moses, the great antiquity of the Earth was generally accepted. In a popular introduction to geology published in 1853, J. B. Jukes, the Local Director of the Irish branch of the Geological Survey, observed:

> Most people are now content to take the authority of Geologists as to
> the age of the globe, on the same footing as they do that of the Astono-
> mers as to the relative size, distance, and motions of the heavenly
> bodies.[17]

Sometimes this faith in the geologists was misplaced. In 1859, for
instance, Darwin, in the *Origin of Species*, went so far as to suggest that
post-Mesozoic time alone might amount to at least 300 million years.
In the following year, in the second edition of the book, he halved this
estimate, but even 150 million years is considerably longer than any
modern geochronologist would allow for the period. Darwin was wise
to delete this particular estimate of time from the third edition of the
book published in 1861. In 1862 William Thomson, the future Lord
Kelvin, chided the geologists for their seemingly extravagant demands
upon the bank of time, and he argued upon physical grounds that the
earth could not possibly be more than 400 million years old.[18] By 1868
he had reduced this figure to a mere 100 million years,[19] but while the
geologists felt some discomfort as these new time shackles were placed
upon them, Thomson's chronology was of a very different order from
that of Moses, and it never seriously inhibited geological thought.

If, as he intimated in a Presidential Address to the Geological Society
of London in 1861, Leonard Horner seriously believed that geologists
were still labouring under the influence of the Mosaic writings and
Ussher's chronology,[20] then he was himself suffering from a strange
delusion. At no time after 1830 was Moses a factor in British geology,
and within little more than thirty years of Hutton's death in 1797, the
Huttonian concept of an extremely ancient Earth was generally accepted.
No longer was there a time problem to retard the development of
uniformitarian geomorphology. Progress in this direction was nevertheless slow, because geologists were tardy in availing themselves of the
opportunities presented by the new time-scale. They were like released
prisoners who find difficulty in adjusting to their new-found freedom.
As Scrope observed as early as 1827:

> It is very remarkable that while the words *Eternal, Eternity, Forever*, are
> constantly in our mouths, and applied without hesitation, we yet
> experience considerable difficulty in contemplating any definite term
> which bears a very large proportion to the brief cycles of our petty
> chronicles.[21]

Similarly, but forty years later, Darwin, who had long been writing airily of periods of hundreds of millions of years, confessed to James Croll:

> I now first began to see what a million means, and I feel quite ashamed of myself at the silly way in which I have spoken of millions of years.[22]

Not until after 1862 did geologists really begin to understand the geomorphic possibilities inherent in the new time-scale, and Lyell was entirely justified when he wrote to his father-in-law in 1857 saying:

> A generation must die off before geologists will know how to make use of an ample allowance of time.[23]

The Cyclic Theory Accepted

Hutton's writings were never widely studied among nineteenth-century geologists. Even as late as 1839 the libraries of the Royal Society, the Linnæan Society, and the Geological Society possessed no copies of Hutton's *Theory* of 1795.[24] Hutton's cyclic interpretation of Earth-history – the cardinal principle upon which the entire theory rested – was nevertheless by then firmly established as the basic tenet in the creed of modern geology. As early as February 1828, W. H. Fitton, the President of the Geological Society of London and one of Jameson's former students, announced that the Neptunian theory, with its linear view of Earth-history, was dead, and that it had been replaced by what he termed 'a modified Volcanic theory'. He continued:

> Whatever, therefore, be the fate of the Huttonian theory in general, it must be admitted, that many of its leading propositions have been confirmed in a manner which the inventor could not have foreseen.[25]

Perhaps Fitton was a trifle hasty in pronouncing the obsequies of Neptunism, because only the previous year Jameson had reaffirmed his faith in many of the theory's leading assumptions,[26] and we have Darwin's assurance that Jameson was at that time still teaching the Neptunian origin of basalts, dykes, and veins. Seemingly, however, even Jameson's faith in the Freiberg system was soon to be shaken because on 17 April 1831 we find him writing to De La Beche to say that he proposes to recommend the latter's *Geological Manual* – an orthodox 'Huttonian' work published that year – as a text for his students.[27]

The death of the Neptunian system, and the triumph of its Huttonian rival, were the results of a steadily increasing knowledge of rocks in the field. Four discoveries were here of particular importance. Firstly, field-mapping elsewhere soon revealed that a knowledge of the geology of Saxony was not the microcosmic key to an understanding of the geology of the whole world. Secondly, familiarity with the contorted rocks of regions such as the Alps demonstrated the absurdity of Werner's claim that all dipping strata had been formed in their present tilted positions. Thirdly, the lessons which Hutton had learned from the unconformities in Arran and the Southern Uplands, were re-learned around the world; periods of sedimentation and denudation had clearly alternated with each other throughout geological time. Finally, there was the realisation that granite was indeed of igneous origin, and this admission finally put paid to the old belief in the existence of primitive topography.

The cyclic Huttonian theory earned its acceptance because of its ability to account for observed field phenomena, but, interestingly, there were still a few early nineteenth-century geologists to whom the theory appealed for another reason – because it offered a neat resolution of the denudation dilemma. With the early nineteenth-century divorce of geology from natural theology, the old dilemma ceased to be of any real significance, but as late as the 1820s there were still a few scholars who chose to view the Huttonian theory in the light of the dilemma. Thus W. T. Brande in his lectures at the Royal Institution, where he succeeded Sir Humphry Davy in the chair of chemistry, explained the theory to his class in the following terms:

> But as the Author of Nature has not given laws to the universe, which, like human institutions, carry with them the elements of their own destruction; as he has not permitted in his works any symptom of infancy or of old age; it was necessary in this theory to provide for reproduction, and such a provision Dr. Hutton has derived from the agency of submarine fire, by which the substances now accumulating in the depths of the sea are to be again agglutinated and elevated.[28]

Once the cyclic theory of Earth-history had found general acceptance, geology ceased to be concerned with the origin of the Earth, and instead focused its attention upon the decipherment of the complex cyclic patterns revealed in the continental rocks. An important corollary of this development was the recognition that the Earth's present topography, far from being a final end-product doomed to endure a perpetual *status quo*, really represents nothing more than one ephemeral stage in a cyclic landscape evolution. This appreciation found expression in two well-known stanzas from Tennyson's *In Memoriam*:

> There rolls the deep where grew the tree.
> O earth, what changes hast thou seen!
> There where the long street roars, hath been
> The stillness of the central sea.

> The hills are shadows, and they flow
> From form to form, and nothing stands;
> They melt like mist, the solid lands,
> Like clouds they shape themselves and go.

Thus, at long last, did the fundamental philosophy of modern geomorphology become firmly established.

The Nature of Earth-movements

Implicit in the cyclic interpretation of Earth-history is the notion that fresh land masses are periodically raised from the ocean floor, and that such land masses must owe something of their initial form to the action of the endogenetic forces. By the 1830s such ideas were commonplace in Britain, but there remained strong disagreement as to the precise nature of the earth-movements responsible for continental uplift. There were two schools of thought. On the one hand there were the catastrophists – the lineal descendants of Hooke, Burnet, and de Luc. Theirs was a dramatic theory of Earth-history. Like their predecessors, they were impressed by the gigantic forces unleashed during an event such as the Lisbon earthquake. They believed that similar, but even greater convulsions, had shaken the crust at intervals throughout the history of the globe, suddenly heaving up new land masses and submerging the old. On the other hand, at the opposite end of the geological forum, there stood the uniformitarians – the lineal descendants of Hutton and Playfair. They argued that cataclysmic earth-storms were a myth, and that the endogenetic forces have never operated with any greater intensity than they display in the world today. The uniformitarians recognised that small vertical continental movements are currently taking place around us, and they argued that even the highest mountain or the deepest sea is merely the cumulative result of such slow movements operating through vast periods of time. They of course never believed that *all* points on the Earth's surface were undergoing constant endogenetic modification, but they did maintain that at any given moment *some* portions of the crust are always being either uplifted or depressed. The catastrophists thus envisaged a cyclic history of the Earth in which prolonged stillstands are regularly interrupted by sudden, seismic convulsions, whereas the uniformitarians believed that the cyclic events were carried through as a result of the Earth's crust being in a state of constant agitation. These two schools of thought must be examined in turn.

The catastrophists of the seventeenth and eighteenth centuries had been able to adduce little real evidence to confirm their views, but their nineteenth-century successors could find some phenomena which seemed to support a cataclysmic interpretation of Earth-history. It was argued, for example, that the tightly folded structures displayed in the strata of a region such as the Alps, could only have been formed when the rocks were still in a plastic state. The convulsion responsible must therefore have been a sudden, but brief event, occurring after the sedimentation was complete, but before the strata had had time to assume any rigidity. Murchison, the leading British catastrophist, writing towards the end of his career, summarised his experience as follows:

> I may say that I never examined any extensive area without recognizing evidences of fracture, displacement, and occasionally inversion of the strata, which no amount of gradual, continuous action could possibly explain.[29]

Similarly, unconformities displaying horizontal strata resting on contorted beds, were interpreted as evidence of a sudden earth-storm followed by a period of tranquillity when sedimentation was renewed. Conglomerates, such as those of the Old Red Sandstone, lying above the plane of an unconformity, were regarded as debris torn off the undermass during the cataclysm. Again, if, as was widely believed during the early decades of the nineteenth century, all erratic blocks owed their dispersal to great tidal waves, which had swept across the continents, then the most obvious cause of such waves was violent seismic upheavals. The catastrophists could even adduce palaeontological evidence in support of their cause. Study had demonstrated the existence of numerous sharp breaks in the fossil record where many species, and even entire genera, die out at the close of a geological period. Such mass extinctions, it was claimed, could only be the result of world-wide catastrophes, and bone-beds, such as that in the British Rhaetic, were regarded as the charnel-houses of the cataclysm's victims.

Such an interpretation of the fossil record led in the biological field to the doctrine of Progressionism – the idea that after each catastrophe life-forms are largely re-created, but at a new and slightly higher level of complexity. Amongst the geologists, one of the most influential catastrophists and progressionists was Élie de Beaumont, professor of geology at the *École nationale supérieure des mines* in Paris. Beaumont, conforming to the orthodox teaching of the catastrophic school, viewed Earth-history as a series of stillstands punctuated by brief periods of great violence. His own particular contribution to this interpretation was the claim that the mountains heaved up during any particular earth-storm display a general parallelism wherever they occur in the world. He first propounded this theory in 1829, and by 1831 he had identified twelve different mountain systems in Europe alone, each representing the work of a different catastrophe.[30] In Britain his many disciples included such eminent figures as Sedgwick, De La Beche, and Murchison.

The leader of the opposing, uniformitarian school of thought, was Sir Charles Lyell. True to his belief in the present as the key to the past, Lyell claimed all continental movement as the work of precisely the same forces as those operative today. He divided these forces into two types. On the one hand he recognised extremely slow movements operating continuously throughout long periods of time, and on the other hand he recognised sudden movements caused by earthquakes which raise or lower an area by several feet, but which, by their repeated occurrence, may have enormous cumulative effects. Here, it must be emphasised, Lyell was not reverting to catastrophism; whereas the catastrophists employed a single gigantic heave to form a mountain range, Lyell and his disciples attributed the same topographical feature to the action of innumerable seismic shocks spread through a period of perhaps millions of years. Clearly, not all of the Earth's crust is today in a condition of instability, and Lyell explained the uniformitarian theory of shifting belts of earth-movement in the following passage from his *Principles* of 1830:

215

... the energy of the subterranean movements has been always uniform as regards the *whole earth*. The force of earthquakes may for a cycle of years have been invariably confined, as it is now, to large but determined spaces, and may then have gradually shifted its position, so that another region, which had for ages been at rest, became in its turn the grand theatre of action.[31]

Two other points about Lyell's views on the causes of relative changes in the level of land and sea deserve mention. Firstly he, in common with other geologists of his day, attributed such changes to the uplift or depression of the land, rather than to eustatic changes of sea-level. This was doubtless because it seemed easier to invoke local movements of the land, than to invoke world-wide changes of sea-level, and because field-studies showed there to be no uniformity in the direction of movement; some coasts were emerging from the sea while others were being submerged. There was nevertheless one other factor which militated against any form of eustatic theory; such a theory would have seemed dangerously close to a reversion to the Neptunian theory which had freely employed the idea of a fluctuating sea-level without ever succeeding in explaining how such fluctuations might be caused. Secondly – a minor point, but one of historical interest – the views to which Lyell subscribed on the subject of earth-movements are precisely the views which Hutton had advocated. Lyell, however, defeated by Hutton's prolixity and obscurity of style, had so far misunderstood Hutton as to believe that his system employed 'violent and paroxysmal convulsions' to elevate new continents.[32]

The uniformitarians could find ample evidence to support their belief in slow and continuing earth-movements. As early as 1802 Playfair had adduced numerous examples of secular changes in the relative levels of land and sea as proof of the reality of movements of this type. One of his examples was the slow emergence currently taking place in Southern Sweden, and this phenomenon later attracted much attention. Lyell himself made a tour of southern Scandinavia in order to examine the evidence of the movement, and he outlined his conclusions in his Bakerian Lecture to the Royal Society in 1834.[33] Three years later Darwin added further substance to the uniformitarian cause when, in May 1837, he outlined his theory of coral-reef formation to the Geological Society of London.[34] In order to explain the presence of coral in waters far deeper than those in which the coral polyps can survive, Darwin postulated a steady and prolonged subsidence of the entire

Pacific floor – precisely the type of earth-movement which the uniformitarians demanded. There was no need, however, to go to the Pacific, or even to Scandinavia, in search of evidence to support the uniformitarian doctrine; such evidence was available within the British Isles. If, for example, the famous Parallel Roads of Glen Roy in the Scottish Highlands were, as many believed, uplifted marine strand-lines, then clearly their present horizontality seemed striking proof that the uplift must have been gentle without any hint of a catastrophe. Similarly, if, as was widely accepted after 1840, there had been a glacial submergence of Britain, then the restoration of the land to its present level must have been non-violent, otherwise the glacial moraines would have been thrown into disarray.

The favourite proof adduced by the uniformitarians did nevertheless come from outside the British Isles – from the so-called Temple of Jupiter Serapis near Pozzuoli on the Bay of Naples. Playfair had referred to this intriguing ruin, and Lyell discussed it in some detail in 1830.[35] The platform of the temple then lay one foot below high-water mark, but the three marble columns rising above the platform were all pitted with marine lamellibranch borings up to a height of 23 feet above high-water mark. Now the temple is known to have stood above sea-level during the reign of Marcus Aurelius in the second century A.D., but since then it had clearly been submerged to a depth of 23 feet, kept there sufficiently long for the lamellibranchs to do their work, and then re-elevated – all this happening without any violent disturbance to topple the slender columns. The ruin seemed such a splendid proof of his doctrine, that Lyell used an illustration of the three marble columns as a frontispiece to the first volume of his *Principles* (Plate VI). Later, the scene was featured still more prominently as a cover illustration in numerous subsequent editions of the *Principles*, and shortly before his death in 1875, Lyell selected the temple as the subject for the reverse of the Lyell Medal, which is presented annually by the Geological Society of London.

The uniformitarians thus had no difficulty in finding evidence of slow-acting earth-movements; their belief that other major continental movements might result from innumerable small but locally violent shocks was obviously less easy of proof. In this context evidence from Chile attracted much attention. In 1822, and again in 1835, large areas of Chile were shaken by earthquakes which in some places left the land standing many feet above its former level. Lyell regarded these earthquakes merely as two in a long series of similar shocks, and he observed:

> We know that one earthquake may raise the coast of Chili for a hundred miles to the average height of about five feet. A repetition of two thousand shocks of equal violence might produce a mountain chain one hundred miles long, and ten thousand feet high.[36]

Darwin was visiting Chile in the *Beagle* at the time of the 1835 tremor, and he was profoundly impressed by the mountain-building potential of a succession of such events. In his report on the geology of South America he suggested that such a succession of shocks had indeed lifted the area around Valparaiso through 1300 feet within comparatively recent times.[37]

The geomorphic implications of the catastrophic, as opposed to the uniformitarian interpretation of earth-movements, will emerge in later sections, and only one further point on this subject needs to be mentioned here. It is often implied, if not expressly stated, that the uniformitarians held a monopoly of the truth in this controversy over the true nature of earth-movements, but such is hardly the case, and in this context the catastrophists certainly deserve much more credit than they have been accorded by modern historians. Indeed, it might be argued that it was the catastrophists who approached most closely to the truth. Lyell and his disciples were mistaken in their belief that earth-movements have acted incessantly and with the same intensity throughout geological time, and their opponents, with their theory of catastrophes alternating with periods of calm, came closer to the modern conception of Earth-history as a series of orogenies separated from each other by prolonged periods of relative quiescence. As was noted earlier, the sole mistake of the catastrophists was to regard the earth-storms as sudden cataclysms occupying a period to be measured in days rather than in the millions of years demanded by modern geology.

The Discovery of Denudations

Nineteenth-century geologists – whether catastrophists or uniformitarians – were well aware that the features produced on the Earth's surface as a result of earth-movements had all been modified to some degree by the action of the exogenetic forces. Now, for the first time, it began to be admitted generally that the Earth's topography is in large measure a residue left as a result of the long-continued action of weathering and erosion, although the precise forms of weathering and

218

erosion invoked were not necessarily those employed by modern geomorphology. Some continued to emphasise the significance of earth-movements in the shaping of topography, but henceforth landforms were increasingly viewed not as constructional features produced by forces working inside the Earth, but rather as destructional features produced by forces acting upon the Earth's skin from the outside. Thus the view which Catcott and Hutton had advocated with so little success during the previous century, now found widespread acceptance.

This development was a direct result of geological research, and, more especially, of the careful mapping of rocks in the field. The discovery that many once extensive formations today survive only as a few scattered outcrops, led to an appreciation of the immense quantities of rock that have been removed since such formations were first exposed to the attack of Nature's destructive forces. Similarly, as Hutton had pointed out long before, the existence of innumerable faults displaying a large throw, but yet devoid of any topographical expression, was further striking proof of the denudation of the continents. One fault system of this type in the vicinity of Newcastle-upon-Tyne – the very region from which Hutton had drawn his own example of this particular phenomena – was brought to the notice of the British Association in 1823 by W. D. Conybeare. The faults in question possess throws of about 1000 feet and should, Conybeare reasoned, give rise to scarps of that height at the surface,

. . . yet is the actual level of the surface found absolutely uniform, and affording no trace whatever of the vast subterraneous disturbances; – a most striking proof of the vast mass of materials which must have been removed subsequently to their occurrence.[38]

Again, once it was established that granite was an igneous rock formed deep in the crust, then it followed that every granite body presently visible owes its exposure to the removal of great thicknesses of cover rocks. Folded strata too, with the fold-structures truncated abruptly at the Earth's surface, were further proof of rock destruction. In some areas it proved possible to make a tentative reconstruction of the former topography by projecting the dips of the surviving strata skywards above present ground-level. One of the earliest such reconstructions came from Thomas Webster, who later became the first professor of geology in University College, London. In 1816 he published a geological cross-section of the Isle of Wight and the Isle of Purbeck, reconstructing the geology as it must have been at a time when the surface of the region stood more than 3000 feet above its present level.[39] Such diagrams were a graphic reminder of the scale of Nature's depredations. Even the thickness of the sedimentary column itself, as Hutton had wisely observed, stood as a silent testimony to the scale of former rock destruction, because sedimentation and wastage must obviously proceed in parallel. As Macculloch wrote in his *System of Geology* of 1831:

> The immense deposits of materials which now form the alluvial tracts of the globe, the enormous masses of secondary strata which have been produced by antient materials of the same nature, all prove the magnitude of the destruction which mountains have formerly experienced, which they are now daily undergoing. Let imagination replace the plains of Hindostan on the Himálaya, or rebuild the mountains which furnished the secondary strata of England, and it needs not be asked what is the extent of ruin, modern or antient.[40]

Geological mapping thus exerted a profound influence upon geomorphic thought. As Darwin remarked in 1881 in his last book:

> It was long ago seen that there must have been an immense amount of denudation; but until the successive formations were carefully mapped and measured, no one fully realised how great was the amount.[41]

Geological mapping brought to light one other extremely important fact illustrative of the potency of Nature's destructive forces; it was found that over wide areas of Britain great thicknesses of the younger strata had been peeled away to reveal the older rocks beneath. Such areas were described as 'denudations'. In the 1819 edition of Abraham Rees's *New Cyclopædia*, this term was defined as the one used

> ... to express those disappearances of the upper strata of the earth in particular districts, by which the lower strata are partially exposed to view.

Lest there be any misunderstanding, it must be emphasised that the term 'denudation' was then used simply as an abstract noun carrying none of the word's present connotation as a verbal noun, and it certainly carried no implications as to the mode of the denudation's formation. Numerous British denudations were described, and four examples – one taken from each constituent part of the United Kingdom – will serve to illustrate their general nature.

Firstly, in England, there was the Derbyshire Denudation described by John Farey in 1811.[42] Farey was a friend and pupil of William Smith, and between 1807 and 1810, at the instance of Sir Joseph Banks, he carried out a geological survey of Derbyshire. Following Smith's work in the English scarplands, Farey expected to find that western and north-western England consisted of 'an uninterrupted series of basset-edges of strata, dipping to the SE. and ranging in continuity from SW. to NE. in certain undulating lines'. In fact he found no such simplicity of structure. Instead, he was forced to the conclusion that the whole of western and north-western England had originally been mantled in 'red strata' (the Trias), but that subsequently in Derbyshire what would now be termed block faulting had raised older rocks towards the surface. He outlined his views in the following passage:

> Instead, however, of seeing the middle and all the western and northern parts of Britain covered by the same red strata, we find now, in this space, numerous local and many very large tracts of strata, surrounded by vertical and connected faults, and greatly lifted and tilted; from the surface of which lifted tracts, the upper red earth, and vast and very unequal thicknesses of strata, that lay in regular succession below this red earth, have been denuded, 'abrupted', or carried off, leaving thus, a great variety of what have been called coal-fields, or mineral-basins. . . .[43]

The faults which Farey invoked are of course almost entirely fictitious, but the Derbyshire Denudation is a real enough feature of English geology.

Secondly, in Ireland, William Richardson, the cleric-geologist whom we encountered earlier in the context of the denudation dilemma and as one of Hutton's opponents, identified an Antrim Denudation which he described in 1808 and 1812.[44] In Antrim, Tertiary basalts rest widely upon Cretaceous chalk, and an examination of the lava flows outcropping on either side of valleys, and of the basalt-capped outliers, convinced Richardson that the present landscape had been formed as a result of the dissection of a former extensive sheet of chalk and basalt. He likened Antrim

> . . . to a vast tablet or block of stratified marble, upon which a mighty operator has been set at work to form in bass-relief our present surface. . . . According to this idea, our prominences, of whatever size, are undisturbed parts of the original block, while the materials, that once

filled the hollows and cavities, have all vanished (as in our diminutive bass-reliefs) under the hand of the operator. . . . Our mountains and hills, I have said, were not formed, but left behind. . . .[45]

Thirdly, in Scotland, Hugh Miller, the former Cromarty stone-mason and bank clerk, traced his beloved Old Red Sandstone southwards along much of the eastern Scottish coast, and thence followed it westwards through the Midland Valley. Then, having mistaken the Torridonian Sandstone of the Western Highlands for the Old Red Sandstone, he jumped to the conclusion that the Scottish Highlands constitute a denudation set within a Devonian frame. In his *Old Red Sandstone*, first published in book form in 1841, Miller wrote:

> I entertain little doubt, that when this loftier portion of Scotland, including the entire Highlands, first presented its broad back over the waves, the upper surface consisted exclusively, from the one extremity to the other, – from Ben Lomond to the Maidenpaps of Caithness, – of a continuous tract of Old Red Sandstone.[46]

Finally, from Wales, there is Ramsay's well-known paper 'On the Denudation of South Wales and the adjacent Counties of England' which was presented to the British Association at Cambridge in 1845, and then published the following year in the first volume of the *Memoirs of the Geological Survey*. The title of the paper, incidentally, might today be misunderstood; Ramsay was employing the term 'denudation' in its then commonly accepted sense as an abstract noun, and not, as a modern reader might suppose, as a synonym for the gerundive term 'denuding'. In the paper, Ramsay showed how the fold-structures mapped in detail by the officers of the Geological Survey could be projected above the present land-surface to give some indication of the thickness of the rocks removed from the denudation. This thickness, he claimed, varies from a mere 3 500 feet of strata stripped from the Vale of Woolhope, to almost 11 000 feet removed from over the site of the Vale of Towy, and he reinforced his argument by presenting a series of dramatic cross-sections illustrating the immense volume of rock removed as compared with that still remaining above sea-level.

Geologists of the first half of the nineteenth century thus had a very sound grasp both of the scale on which rock destruction has occurred and of the relict nature of the Earth's present topography. These four discussions of denudations are thoroughly typical of their age, but it must be emphasised that a belief in the existence of denudations did not lead to a fluvialistic interpretation of landscapes. The geologists were conversant enough with the work of rain and rivers, but a fluvialistic explanation of the denudations was in most cases firmly dismissed. This is well illustrated by reverting to the four examples of denudations just considered. A modern reader, misunderstanding the early nineteenth-century use of the term 'denudation' will be sadly disappointed if he turns to Farey's paper of 1811 in expectation of a discussion of Derby-shire's denudational history. The paper is really an essay on the supposed tectonic geology of the region, and Farey made no attempt to explain how great thicknesses of rocks had been stripped from the area. His sole effort in this direction was to suggest that after the rocks had been destroyed in some way, the debris was drawn off into space as a result of the close approach of some heavenly body which temporarily caused a reversal of gravity.[47] Richardson too pleaded his inability to discover the mode of formation of the Antrim Denudation, but he agreed with Farey that once formed, the debris had probably been lost into space. Hugh Miller was bolder; he suggested that marine currents had stripped the Old Red Sandstone off the Scottish Highlands during their final emergence. Ramsay took a similar view. He belittled the role of fluvial processes in the formation of the South Wales Denudation, and instead he attributed its creation to marine agencies operating during a submergence which converted Ireland into an archipelago and allowed the Atlantic waves to beat directly upon the Welsh mountains. Thus while Hutton's belief in the circumdenudational origin of topography had now found general acceptance, his fluvialism was by many authors not even deemed worthy of serious consider-ation.

The Fate of Fluvialism

The history of the fluvial doctrine during the years immediately after Hutton's death has been strangely misunderstood. Historians have viewed the doctrine as one of the points at issue in the controversy between the Neptunists and the Plutonists, and one history of geo-morphology refers to a Wernerian counter-offensive against Huttonian geomorphology. In reality no such event occurred. A vigorous con-troversy between the Neptunists and the Plutonists certainly did take place, and it reached a remarkable pitch of intensity in Edinburgh where Hutton's disciples found themselves confronted by Jameson's troupe of Wernerians. The dispute permeated into every aspect of the city's life and was even carried into the theatre; a play by an Huttonian, although graced with a prologue by Walter Scott, is said to have been condemned on its first night by a house deliberately packed with Neptunians.[48] Geomorphology, however, had no place in this controversy; the disputants were concerned almost entirely with one problem – were granite and basalt of chemical or igneous origin?

The early nineteenth-century position so far as fluvialism is concerned is really very simple. Even if we include such essentially eighteenth-century figures as de Luc, Kirwan, and Richardson, all of whom lingered on into the second decade of the nineteenth century, it is still difficult to muster more than half a dozen British geologists of the nineteenth century who seriously doubted the destructive potential of rain and rivers. The overwhelming weight of geological opinion – Wernerians and Huttonians alike – accepted the reality of fluvial denudation, and many, following in the tradition established long before by Carpenter, Ray, and Burnet, freely admitted that the long continued action of rain and rivers must result in the eventual re-modelling of the Earth's topography. On the other hand, almost all geologists – again Wernerians and Huttonians together – equally firmly rejected the fluvialistic interpretation of the Earth's present landscapes. Thus the early nineteenth-century geologists preserved the same dichotomous approach to denudation which we have already observed among their seventeenth- and eighteenth-century predecessors; they drew a clear distinction between the immense potential of future denudation as compared with the very limited amount of work that rain and rivers were believed to have performed in the past. This distinction must now engage our attention more closely.

The aged de Luc, in a series of volumes published between 1809 and 1813, was the last British geologist to make a serious attack upon the concept of denudation. All later British writers, without exception, willingly admitted the reality and potential of the forces of weathering and erosion. The Wernerians certainly never doubted the reality of denudation, and those historians who have imagined that Neptunism and a belief in denudation were incompatible have overlooked the fact that the Neptunists regarded the Transition, Floetz, and Alluvial classes of rock as consisting largely or entirely of clastic sediments. Jameson himself, while belittling the erosive power of rivers, certainly never doubted the reality of denudation. Examples of his writing on the subject were offered in an earlier chapter, and his continuing belief in denudation emerges clearly from his *Manual of Mineralogy* of 1821. Therein he discusses the rocks according to the standard Wernerian classification, but he adds comments upon the resistance of the various rocks to denudation, and upon the kind of topography into which they are commonly moulded.

Down to the 1830s there were a surprising number of geologists who still viewed denudation from a teleological angle, and who emphasised its role in the conversion of barren mountains into fertile soils. In an introductory geological text published by Robert Bakewell in 1828 for example, it was explained to students that the strata yielding the most fertile soils had been expressly designed to weather more rapidly than the rocks which provide only a poor tilth.[49]

At this point it is hardly necessary to offer quotations to illustrate the early nineteenth-century understanding of the various processes involved in denudation, and it suffices to note that the subject was then being discussed in thoroughly modern terms. Numerous texts of the period contain full accounts of such denudational minutiae as spheroidal and honeycomb weathering, the part played by plant-roots in mechanical weathering, and the role of stream-bed pot-holes in river degradation. Even quantitative estimates of the rate of denudation were now beginning to be made. In 1832 the Rev. Robert Everest (brother of the more famous Sir George Everest) informed the Asiatic Society of Bengal that observations showed the Ganges to be carrying 6,368,077,440 cubic feet of debris to the sea each year.[50] Similarly, in 1853, Alfred Tylor calculated that the same river was lowering its basin at the rate of one foot in 1751 years, as compared with a figure of one foot every 9000 years in the case of the basin of the Mississippi.[51] Thomas Oldham, the Irish Superintendent of the Geological Survey of India from 1850 until 1876, certainly had no doubts about the energy of the rivers in the Ganges basin.[52] He claimed that during the monsoon season he had himself seen a boulder weighing 350 tons being swept along by a stream in the Khasi Hills in Assam, and while he regarded fluvialism as a totally inadequate general theory of landscape evolution, he did admit that the gorges in the Khasi Hills could only be fluvial in origin:

> ... they appear to me to offer a magnificent instance of the almost incredible power of degradation and removal, which atmospheric forces may exert under peculiar and favourable circumstances.

In 1831 Macculloch affirmed that the end-product of all such denudation must be a plain, and he outlined his concept of base-levelling in the following passage which is strongly reminiscent of the much later writings of J. W. Powell and the members of the American school of geomorphology:

> As the general declivity diminishes, so does the ratio of destruction; and thus has it continued to diminish from the moment when the rivers first began to flow. But as the inequalities of the earth waste away, so will ages be hereafter required to do that for which years once sufficed. Yet we may still imagine the ultimate ratio; that point at which the forces of destruction and those of resistance shall be balanced. This is the speculative period at which the earth will become a plain, when the altitude of the land will scarcely exceed that of the sea, and the channels of rivers shall refuse to conduct their waters. Offensive marshes and arid sands will then take the place of this fair variety of hill and dale, and the last chaos will be worse than the first.[53]

Macculloch recognised the existence of a relationship between climate and rates of denudation, and he suggested that in theory this base-levelled plain should form more rapidly in humid Scotland than in the drier climate of southern England, and more rapidly in southern England than in arid Egypt. In practice, however, he doubted whether any such dank, base-levelled plain would ever really come into existence. He believed, with the earlier teleologists, that mountains are essential to the entire terrestrial economy, and he held that long before the ultimate base-levelled condition was attained the Deity would intervene and raise a new generation of mountains through the agency of an earth-storm.[54]

In view of this general acceptance of the reality of denudation, it is strange that the fluvialistic interpretation of topography should have been so firmly rejected. The fate of fluvialism is nevertheless only too clear; down to 1862 British geologists were almost unanimous in their belief that fluvial processes could at best have played only a trivial part in shaping the Earth's present landforms. Sir James Hall, for example, one of Hutton's closest friends and a leading Plutonist, was at pains to disassociate himself from the fluvial doctrine. He fully endorsed Hutton's views on the origin of igneous rocks – what Hall termed the essence of Hutton's theory – and by his experimental researches he did much to establish their truth, but a study of Lake

Geneva convinced him of the force of the limnological objection to
fluvialism and of Hutton's error in the geomorphic field. In March and
June 1812, Hall presented two papers to the Royal Society of Edinburgh
drawing attention to his master's supposed mistake and seeking to
rectify it by introducing cataclysmic tsunami into the Huttonian
theory, just as Whiston had introduced comets into Burnet's theory.[55]

In 1822, in a work which is today one of the classics of geological
literature, W. D. Conybeare and William Phillips expressed amaze-
ment that any serious scholar could ever have believed in the fluvial
origin of valleys:

> . . . this hypothesis must be abandoned at once by any one who will
> take the trouble of subjecting it to a rigorous application to the vallies of
> any extensive district. . . .[56]

In the following year, W. H. Fitton, the earliest historian of British
geology and one of Hutton's most ardent champions, wrote:

> . . . it is now almost universally admitted, that valleys have been
> excavated by causes no longer in action, – contrary to the opinion of
> Dr. Hutton and Mr. Playfair. . . .[57]

In 1828 Robert Bakewell observed:

> Geologists seem now generally agreed, that the action of rivers is not
> sufficient to explain all the phenomena of vallies.[58]

By 1835 John Phillips, the nephew of William Smith, was more
emphatic:

> Dr. Hutton's opinions on the origin of valleys have been rejected as a
> general theory.[59]

Professor Ramsay, perhaps the outstanding field-geologist of his day,
was a little less dogmatic in his inaugural address at University College,
London, in 1848. He nevertheless rejected fluvialism, and he observed
of Hutton himself:

> Still he was not warranted in attributing the formation of all valleys
> to the erosive influence of streams and rivers; for it cannot be doubted
> that many of the principal features of our continents are partly due to the
> effects of ancient marine denudations.[60]

227

None of the eminent geologists just cited had any doubts as to the future ability of rain and rivers to modify landscapes, but they one and all refused to allow these agents a role of any significance in the shaping of the Earth's present topography. Two figures from the first half of the nineteenth century are nevertheless sometimes thought of as thorough-going fluvialists in the Hutton–Playfair tradition. These two are G. J. P. Scrope and Colonel George Greenwood, and we must pause here to consider how far they deserve their reputations as fluvialists.

George Julius Poulett Scrope's real, paternal name was Thomson, but he assumed his wife's name in 1821 when he married the daughter and heiress of William Scrope, the artist and sportsman. Scrope was for long prominent in the affairs of the Geological Society of London, and from 1833 until 1868 he was the Member of Parliament for Stroud in Gloucestershire. Numerous pamphlets advocating free-trade and various social reforms came from his pen, but his reputation with posterity rests chiefly upon his *Memoir on the Geology of Central France* which he completed in 1822. Surprisingly, in view of the geological mania then sweeping Britain, he had some difficulty in finding a publisher for the book, and it was 1827 before it finally appeared in print. It will be suggested later that this time lag of five years between the book's completion and its publication, may be a matter of some significance in assessing the true nature of Scrope's fluvialism.

Scrope's book – the outcome of six months' field-work in 1821 – is concerned chiefly with the geology of Auvergne – that fascinating region of central France where recent vulcanism has left dozens of beautifully preserved cinder cones overlooking acre upon acre of black basalt. His field studies showed that the great basaltic pile has been deeply dissected, and that formerly expansive lava flows are to-day interrupted by numerous deeply cut valleys. In many places the exposures on the opposing sides of the valleys reveal identical sequences of flows, thus leaving no room for doubt as to the post-basaltic age of the incisions. Still more impressive are the areas where the basalts have been reduced to a series of small outliers, and nowhere is this better seen than in the Limagne, to the south-east of Clermont Ferrand. There the basalts survive only as the resistant cap-rocks to a few striking, but scattered mesas rising above the Eocene and Oligocene clays which floor the lowland drained by the Allier river. Scrope in fact demonstrated that the region afforded a dramatic example of a denudation. In a general way, the denudation is not unlike that which Richardson had already described in the basalts of Antrim, but whereas the Irishman had been unable to suggest a cause for the Antrim Denudation, Scrope could adduce significant evidence bearing upon the origin of the French example. He pointed out that the denudation could not be the result of catastrophic causes, because any such violence must have destroyed the delicately poised cinder cones of the local volcanoes. The denudation must, rather, be the result of gentle, slow-acting agents, and only the fluvial processes, operating with nothing more than their present intensity, seemed to offer an adequate explanation of the phenomenon. He was confirmed in this conclusion by the discovery that in certain of the region's valleys, successively younger lava flows lie at steadily lower levels, thus preserving a visible record of the valley's slow excavation at a time when the local volcanoes were still active.

Lyell reviewed Scrope's book for *The Quarterly Review* and hailed it as the most convincing proof of the fluvial doctrine to appear since Playfair's *Illustrations* a quarter of a century earlier.[61] In 1828, when Lyell, accompanied by Mr and Mrs Murchison, visited Auvergne to see the evidence for himself, he was entirely convinced of the validity of Scrope's reasoning. Even Murchison, the die-hard catastrophist, returned persuaded that the Auvergne landscapes at least were fluvial in origin.[62] At a first glance, Scrope's sound discussion of the Auvergne Denudation might lead us to conclude that in him we have a unique phenomenon – a thorough-going fluvialist among the ranks of the early nineteenth-century geologists. Such an assessment, however, is false. He was in reality nothing more than a semi-fluvialist. Even in his *Memoir* he invoked subterranean forces in explanation of certain topographical features in a manner that would certainly never have met with Hutton's approval. After writing about fluvial activity, Scrope continued:

> In reality, it cannot be doubted that vast irregularities of level, elevations and depressions of the earth's surface, on every scale of magnitude, have been occasioned by other causes, chiefly subterranean expansion. The basins of our seas, lakes, and rivers, no doubt owe to circumstances of this nature their *primary* forms.[63]

Here, of course, we again encounter a familiar problem; what did Scrope imagine to be the relationship between his primary landforms and those extant today? In other words, did he view the Earth's present topography as essentially primary or essentially denudational in origin? To find an answer to this question we must turn from Scrope's French *Memoir* to another of his works.

In 1825 he published a volume entitled *Considerations on Volcanos*. Now although this book was published two years before the *Memoir*, it appeared more than three years after Scrope had completed both his field-work in France and the *Memoir* in manuscript. The book on volcanoes may therefore stand alongside the *Memoir* as embodying his mature geomorphic judgments. It is true that in the book he does suggest that many valleys – perhaps even a majority of valleys – are river-cut, but at the same time he repeatedly emphasises the role of earth-movements in forming the Earth's present topography. He for example claimed the basins of lakes Lugano, Como, Iseo, and Garda as tectonic features. He regarded all the strike valleys of the English Weald, together with such of the region's dip valleys as those of the rivers Wey, Mole, Medway, Arun, Adur, and Ouse, as what he termed 'fissures of elevation', and in the following passage he specifically rejects a fluvialistic interpretation of the Wealden topography.

> That these districts have suffered much subsequent denudation, may be, and is no doubt, true; but the cause of their peculiar form lies in the elevation, fracture, and subsidence of the strata which once covered this space.[64]

Other valleys he regarded as cut by torrents draining off the continents after their emergence. Even many of the features which are fluvial in origin, he claimed, were shaped not under the present rainfall regime, but during some former pluvial period.

Scrope thus seems to have inclined to the view that denudation had merely added the detail to a topography which was essentially tectonic in origin, and he was certainly no Huttonian geomorphologist. In a paper delivered to the Geological Society of London in 1830 he did argue cogently that the valleys of rivers flowing in deeply incised meanders had most probably been excavated by the rivers themselves,[65] but thereafter his geomorphic thinking stagnated. When in 1858 he published a second, enlarged edition of his French *Memoir*, he retained his original discussion of topography almost intact. The only changes introduced were retrograde; he now added mountains and valleys to his list of primary landforms, and included submarine currents and marine erosion among his valley-forming processes. A comparison of the two editions of the book serves as a sharp reminder not merely of Scrope's personal failure to advance his geomorphic thought, but also of the more general lassitude which afflicted geomorphology during the first sixty years of the last century.

If any further proof be needed that Scrope was nothing more than a semi-fluvialist, it lies in the pages of *The Geological Magazine* for 1866.[66] There we find a correspondence between Scrope and Jukes on the subject of fluvialism. Jukes was then one of a rapidly growing band of converts to the thorough-going fluvialism of Hutton and Playfair, but in the correspondence, far from welcoming Jukes into the fluvialist camp, Scrope actually cautioned him against exaggerating the efficacy of rain and rivers. Scrope, now near the end of his life, summarised his own thinking upon the subject in the following passage:

> So far, then, from agreeing with what seems to be the opinion of Professor Jukes, I would maintain that all the grander features of the earth's surface have been fashioned by internal rather than by external forces – the influence of the latter being confined to what may be called the minor details, the planing and chiselling, rather than the moulding, of the subject matter. And in respect even to some of the minor details, such as the transverse valleys, that act as tributaries to these grander depressions of the surface, there seems good ground for believing many of them to owe their *origin*, and consequently the course of the superficial waters or ice-streams that have, since their emergence from the sea, widened and deepened them by erosion, to the transverse cracks and fissures which could not fail to accompany the violent elevation of more or less solid strata, even though effected by gradual throes.

Thus even after the fluvialistic revival of the 1860s had begun, Scrope was still not prepared to accept an unqualified fluvialistic interpretation of topography.

The second figure deserving of special examination in the present context is Colonel George Greenwood of the Life Guards. He was born at Alresford in Hampshire in 1799, and after an education at Eton, he was gazetted to the Second Life Guards in 1817.[67] He soon acquired a great reputation as a dashing cavalry officer, and King William IV is said once to have toasted him as the most accomplished horseman in the British army. Appropriately enough, Greenwood's earliest publications were a book of cavalry sword exercises and a very successful treatise on horsemanship. Had he stayed in the army he

231

would doubtless have risen to high rank, and he might well have earned fame for himself in the Crimea. Such, however, was not to be. In 1840 he learned that he was suffering from a heart condition, and he immediately resigned his commission and retired to his Hampshire estates where he died in 1875. He devoted his long retirement to arboriculture, and he invented a device – the tree-lifter – for transporting entire trees from one site to another. It was in 1853, in the second edition of the book in which he explained the use of this device,[68] that he first revealed his interest in geomorphology and, more especially, his belief in the efficiency of the fluvial processes. He expanded his geomorphic ideas further in his book *Rain and Rivers*, first published in 1857, and he later became a regular advocate of fluvialism through the correspondence columns of *The Geological Magazine* and other periodicals.

Greenwood was a uniformitarian. Those of his geomorphic writings published before 1862 unfortunately lack precision, but he evidently believed that current processes, operating with no more than their present intensity, are entirely adequate to have formed the Earth's modern topography. He certainly accepted the fluvial origin of valleys, and he had no doubts about the erosive potential of a river in flood.

> A torrent swollen by rain to perhaps twenty times the volume of its usual spring water, and hurling fragments of rocks along of all sizes, is in point of excavating and destructive power as much more formidable than its usual self, as a shotted gun is more formidable than an unshotted gun . . . when the torrent is turbid with the wash of rain, we can *hear* its huge cannon balls rattling down, and grinding each other and their rocky bed and banks till what has started from the mountain's brow as a huge rock arrives at the sea in the form of pebbles, or of sand.[69]

Important though rivers might be, it was to rainwash that he attached the greatest significance in the moulding of a landscape. This process, he observed is 'the destroyer, the dissolver, of continents'. Indeed much of his writing is concerned, in a very general sort of way, with what we now know as soil erosion. He observed, for example, in a passage which typifies his none-too-happy attempt to be witty and entertaining:

> . . . no drop of rain *runs* an inch on the surface of the earth without, as far as it goes, setting some soil forward on its road to the sea. And it won't run back again. No return tickets are given. It will wait there, and go on by the nex-*t-rain*.[70]

He recognised the soil piled up behind transverse walls on hill-slopes as clear evidence of a down hill movement of material, and he advised:

> If your neighbour's land lies below you on a steep hill side, unless you wish to make him a present of your best soil, pound it back on to your own land by a fence, and when it accumulates against your fence cart it up the hill again.[71]

One final passage from his works is worth citing, partly because it introduces the interesting question of relative rates of uplift and denudation, and partly because it provides some slight insight into Greenwood's conception of the scale on which denudation occurs.

> It does not follow that while mountains are rising they are increasing in height. They may be decreasing in height. Suppose the Alps to have been rising six inches in a century for myriads of years, if their denudation has been seven inches in a century, they have been decreasing in height.[72]

In recent years, Greenwood has been accorded a position of some eminence in the history of the fluvial doctrine, and he has even been hailed as 'the father of modern subaerialism'.[73] It is true that he hedges his fluvialism about with none of the reservations which so mar the work of his contemporaries, but the claims made on his behalf do nevertheless seem somewhat extravagant. Three points need to be emphasised. Firstly, Greenwood merely dabbled in geology. Unlike the writings of so many other amateur brethren of the hammer, Greenwood's works published before 1862 display no real grasp of modern geology and its techniques. His knowledge of the subject is at best described as rudimentary. He was so ignorant of geological literature that, as he later confessed, when he first went into print on the subject of fluvialism, he had never even heard of Hutton. As late as 1857, Greenwood was still so misinformed as to believe that Lyell was personally responsible for the subversion of a long-established fluvial doctrine, hence the full title of Greenwood's book: *Rain and Rivers; or Hutton and Playfair against Lyell and all Comers*. In view of this ignorance of his subject, and in the absence of any evidence to the contrary, one must conclude that Greenwood was blissfully oblivious of the many solid objections which had been raised against the fluvial doctrine.

Secondly, Greenwood's writing on the subject of fluvialism is

233

diffuse and ineffective. He did point out some contradictions in Lyell's theory of valley formation by marine erosion, but his own works are almost entirely devoid of any reasoned presentation of the fluvial doctrine. We must sympathise with Scrope when he dismissed Greenwood as an 'eccentric author' who was never likely to win support for the fluvial cause.[74] Apart from some brief comments on the denudation of the Weald, he has little to say about the history of any landscapes. Much of the time he is concerned with nothing more than the down slope movement of the surface mantle, and his interest in geomorphology is that of a gentleman-farmer rather than that of a geologist.

Finally, Greenwood exerted no influence upon the thought of his contemporaries. So far as can be judged from the literature, only Scrope and Jukes seem to have taken any immediate notice of his writings. Later, in the 1860s, when Greenwood's name did become more familiar among geologists, his advocacy of fluvialism was of little consequence, because by then the doctrine had been embraced by a group of young Geological Survey officers who, unlike Greenwood, were able to support their theories with closely reasoned arguments based upon field evidence. Their writings carried the weight and conviction which is so sadly lacking from Greenwood's own work.

In the light of these three points it does seem that Greenwood's significance has been rather overrated. He may have been a fluvialist in the Huttonian tradition – although his writings are so nebulous that judgment on this question is difficult – but if he was, the present writer believes that his fluvialism was not that of a rational and praiseworthy mind clinging to its conviction in a full awareness of all the objections the doctrine must face. It was, rather, a naïve fluvialism born of an ignorance of those counter-arguments which had led to the unanimous rejection of the doctrine among his more experienced geological contemporaries. If Greenwood has to be categorised, he is perhaps best placed with that group of eighteenth-century authors who had accepted the reality of fluvial denudation as a common-sense belief, but who had never paused to consider the implications and difficulties inherent in the belief.

One other geologist who leaned towards a fluvialistic interpretation of topography during the first half of the nineteenth century deserves mention alongside Scrope and Greenwood – Dr John Macculloch. Macculloch's name has already cropped up several times in the present work. He was one of the outstanding geologists of his day, but a man who has received only faint praise from the historians of geology, and scarcely even a mention from the historians of geomorphology. He was born in 1773,[75] and he doubtless first encountered the Huttonian theory when, as an Edinburgh medical student, he attended Black's chemistry lectures early in the 1790s. In 1803, after a brief spell as a practising physician, he became Chemist to the Board of Ordnance, and in this capacity, and later as geologist to the Trigonometrical Survey, he undertook a series of geological surveys in Scotland which gave him an intimacy with Scottish geology such as only Jameson could rival. Macculloch was undoubtedly a man of talent, but at the same time it must be admitted that he was a somewhat unpleasant character. He was dishonest in his financial dealings with the government, he resented the appearance of any other geologists in what he regarded as his own Scottish preserve, and he was outspokenly contemptuous of all amateur geologists. He aroused deep resentment in Scotland by his scathing comments upon the character of the Highlanders, he minimised the significance of all advances in geology (he once referred to stratigraphers in general as 'namby pamby cockleologists and formation men'[76]), and above all he displayed a vain egotism reminiscent of the character of John Woodward. Eventually he ostracised himself from all geological society, and there were few friends to mourn him when in 1835 he died of injuries received in a carriage accident while on his honeymoon in Cornwall.

Macculloch had a keen interest in landforms. Among his many papers in the *Transactions of the Geological Society of London* are studies of the Cornish tors, of the Parallel Roads of Glen Roy (which he correctly interpreted as lake strand-lines), and of the coastline of Heligoland. In his books on Scotland he repeatedly turns aside to consider the origin of topographical features, and in his *System of Geology*, published in 1831, he presents a detailed discussion of both the weathering processes, and the relationship between rock-type and topography. During his Scottish travels he encountered innumerable features demonstrating the potency of Nature's destructive forces. In Skye he found deeply weathered basalts that could be dug with a spade, while he noted that the flanks of the island's Red Hills were mantled 'with long torrents of red rubbish'. Of the hills of Raasay he observed:

> ... huge piles of ruin cover their slopes with fragments, advancing far into the sea and strewing the shore with rocks.[77]

In the Scottish Highlands he recognised a denudation in Sutherland and Ross where residual masses of almost horizontal Torridonian Sandstone rest upon the Lewisian Gneiss and form such striking mountains as Suilven, Cul Mòr and Stac Pollaidh. Of this area he wrote:

> It is impossible to avoid inferring that a continuous bed of sandstone once covered the whole: the divisions of the strata in these separate mountains corresponding accurately in level, while no instance of a position more angular occurs; and the summits, every where, bearing the usual marks of waste. They have all therefore been shaped out of a continuous mass; or the intermediate country has been denudated, and that to a great extent.[78]

Further to the south, after ascending the highest peak in the British Isles and noting the evidence of denudation on the mountain's flanks, he observed:

> Nature did not intend mountains to last for ever; when she is so fertile in expedients as to lay plans for destroying a mountain so apparently unsusceptible of ruin as Ben Nevis.[79]

Macculloch believed, as we saw earlier, that such denudation, if uninterrupted, could result only in the base-levelling of the continents, and his wide experience of Scottish topography undoubtedly brought him to the verge of a fluvialistic interpretation of landforms. Repeatedly he seems about to throw in his lot with Hutton, but on every occasion he draws back at the last moment. He offered no real objections to the fluvial doctrine, but his difficulty – implicit rather than expressed – was a reluctance to admit that fluvial processes, operating with no more than their present intensity could have produced the observed results within the life-span of the Earth. He had no doubts as to the antiquity of the Earth – repeatedly he emphasises that the Earth's age is to be measured in terms of countless millions of years – and he certainly never doubted the future potential of the fluvial processes, yet strangely he could not bring himself to believe that those same processes were responsible for moulding the Earth's existing topography. He was guilty of that early nineteenth-century failing already noted; he could write gaily of the Earth as being millions of years old without having any conception of what such aeons really meant in terms of Earth processes.

236

Of all the British geologists who entered into print on the subject between 1800 and 1862, only John Playfair displayed a firm and un-questioned allegiance to the thorough-going fluvialism embodied in the Huttonian theory, but even he wrote nothing in support of the fluvial doctrine apart from his *Illustrations* of 1802. Indeed, one cannot help but wonder whether his faith in the doctrine may not have been shaken as a result of his travels in Europe subsequent to 1815, and whether his failure to produce the promised second edition of the *Illustrations* may not be a fact of some significance. It cannot be empha-sised too strongly that the rejection of fluvialism during the first half of the nineteenth century was not, as is often implied, the result of crassi-tude and ignorance. It was a studied and reasoned rejection, and it is now time to examine its rationale.

Reasons for the Rejection of Fluvialism

Almost all the objections raised against the fluvial doctrine during the early years of the nineteenth century were based upon sound field observations, but we may conveniently open our review of the ob-jections by reverting to that old problem the denudation dilemma. The dilemma ceased to be a major factor in British geomorphology before 1800, but the last vestiges of the dilemma lingered on into the early decades of the nineteenth century. We noted above that a few early nineteenth-century geologists still regarded the Huttonian theory as a convenient resolution of the dilemma, and similarly there were still a few geologists who sought escape from the dilemma by refusing to admit that the God-given continents are today mouldering and on their way to becoming dank, base-levelled plains. John Kidd in 1815,[80] William Knight in 1819,[81] and Robert Bakewell in 1828,[82] all protested that prodigious denudation was impossible because rain and rivers become impotent once a landscape is clothed with soil and vegetation, while at the coast, they claimed, erosion and deposition exactly balance each other so that a state of equipoise is maintained.

The last British geologist to seek escape from the denudation dilemma through a denial of the possibility of prodigious denudation, was Adam Sedgwick on no lesser occasion than his retirement address from the chair of the Geological Society of London in 1831. In the course of his address he reviewed the first volume of Lyell's recently published *Principles of Geology*, wherein Lyell discusses some aspects of the work of denudation. Sedgwick, after affirming his own belief in the reality of limited denudation, continued by casting doubts upon the possibility of denudation ever occurring on the grand scale.

But are there no antagonist powers in nature to oppose these mighty ravages – no conservative principle to meet this vast destructive agency? The forces of degradation very often of themselves produce their own limitation. The mountain torrent may tear up the solid rock, and bear its fragments to the plain below: but there its power is at an end, and the rolled fragments are left behind to a new action of material elements. And what is true of a single rock is true of a mountain chain; and vast regions on the surface of the earth, now only the monuments of spoliation and waste, may hereafter rest secure under the defence of a thick vegetable covering, and become a new scene of life and animation.

It well deserves remark, that the destructive powers of nature act only upon lines, while some of the grand principles of conservation act upon the whole surface of the land. By the processes of vegetable life, an incalculable mass of solid matter is absorbed, year after year, from the elastic and non-elastic fluids circulating round the earth, and is then thrown down upon its surface. In this single operation, there is a vast counterpoise to all the agents of destruction. And the deltas of the Ganges and the Mississippi are not solely formed at the expense of the solid materials of our globe, but in part, and I believe also in a considerable part, by one of the great conservative operations by which the elements are made to return into themselves.[83]

Thus did the old ideas of a protective mantle and a mysterious renovating force find their final advocate in Britain. This, however, was still not quite the end of the denudation dilemma because Lyell felt impelled to reply to Sedgwick in the second volume of the *Principles* published in 1832. In the volume Lyell devoted an entire chapter (Chapter XII) to showing that vegetation could not possibly protect a landscape from the ravages of time. His discussion of this topic survived down to the ninth edition of the *Principles* published in June 1853, but it disappeared during the major revision which resulted in the tenth edition of 1867. Lyell, in June 1853, was thus the last British geologist to publish a work displaying concern for the denudation dilemma, although by that date the true nature of the dilemma must have been completely forgotten.

Only a handful of nineteenth-century geologists ever saw the denudation dilemma as an obstacle to their acceptance of the fluvial doctrine; gone were the days when such religio-scientific problems were major factors shaping geological thought. What did turn geologists against the doctrine was not the metaphysical problem represented by the dilemma, but, rather, a number of solid observational facts gleaned during the course of patient field investigation. Such investigation amply demonstrated the impossibility of explaining all features in terms of fluvial processes. This point was first argued in detail by de Luc in 1810,[84] and his arguments against fluvialism seemed so telling that even as late as 1857 so balanced and impartial a scholar as William Whewell could hail de Luc's critique as the complete antidote to Hutton's fluvial extravagance.[85] De Luc's arguments were used and augmented by many later authors, and the following ten objections are the chief reasons for the rejection of fluvialism in Britain in the years before 1862.

1. Pride of place here must be given to de Luc's old, but still very telling limnological objection. As we have seen, it was an examination of Lake Geneva with this particular objection in mind which led that resolute Plutonist, Sir James Hall, to make a public renunciation of his master's fluvialism. In 1842 Sedgwick used precisely the same argument, based upon the continued existence of lakes, to prove that the topography of North Wales and the English Lake District could not possibly be fluvial in origin.[86]

2. As de Luc pointed out, Playfair's Law of Accordant Junctions has no application in most of the mountain regions of Europe. There, where hanging valleys abound, discordance of level at a junction is more common than accordance. Playfair himself must have become painfully aware of this fact as he journeyed through the Alps in 1816.

3. The fluvial doctrine was held to be incapable of explaining the courses of those rivers which, after flowing for some distance over broad lowlands, then plunge into deep transverse gorges lying across the axis of some hill or mountain upland. The water gaps through the North and South Downs were favourite examples of this phenomenon, but in 1825 Sedgwick drew attention to two similar valleys in the Isle of Wight. There both the Medina and the Yar rise on the lowlands on the southern side of the island but flow northwards through a ridge of chalk hills and empty themselves into the Solent and Spithead.[87] Similarly, in 1840, J. A. de Luc junior drew attention to numerous other transverse gorges ranging from that of the Danube at the Iron Gate, to the gorges of the Breede and Hex rivers in South Africa, and those of the Potomac and Susquehanna in the New World. Such

anomalous features, he claimed, could only be fissures opened as a result of seismic convulsions.[88]

4. Many valleys, from the dry valleys of the English Downs to the wadis of Egypt, contain no rivers. How could the fluvialist hope to explain such features? As the young Lyell wrote in his Journal in 1817, after visiting the uplands of Yorkshire's North Riding:

> In ascending the Black Hambleton Hills, we passed up a very deep narrow glen, without a river (unfortunately for the Huttonians). . . .[89]

Similarly, what was the fluvialist to make of the deep, dry cuts which are now known to have been excavated by glacial meltwaters? It was in 1822 that George Young and John Bird asked the fluvialists to explain what manner of river could have formed The Fen which notches the watershed between the Vale of Goathland and Newton Dale in the North Yorkshire Moors.[90] Exactly eighty years were to elapse before P. F. Kendall answered the question in his classic paper on the glacial lakes of the Cleveland Hills.[91]

5. If valleys have been excavated by rivers, then clearly there should be some relationship between the size of a river and the size of its valley. Even a cursory examination of a few examples, however, was sufficient to show that in many cases no such relationship exists. In the mountain regions of Europe, in particular, many small rivers are misfits occupying major troughs.

6. In a river, it was argued, erosion takes place chiefly on the river-bed. As a result of this vertical degradation, therefore, genuine river-excavated valleys must be deep, narrow canyons quite unlike the broad, open valleys which are so widespread in the world today. As Lyell observed in a letter written at Madeira in 1854:

> In general this island confirms his [Charles Darwin's] doctrine, that if all valleys were cut by rivers alone, they would be very narrow, though they might be of any depth, and that the sea is the great widening power.[92]

Similarly, it was argued that rivers eroding vertically could never have formed those denudations where the reduction of former expansive sheets of rock to a few scattered outliers seemed to imply that the erosive agent had operated chiefly in a lateral direction.

7. Today innumerable rivers, far from eroding, are quite clearly filling their valleys in with alluvium and other debris. Again, this

process is well displayed in the mountain districts of Europe, and the amount of infill present in such valleys makes it obvious that the rivers have for long been aggradational, rather than degradational agents.

8. A river-excavated valley should have a smooth, gently sloping rock floor, but in fact many valleys in highland areas have rough floors; long gentle reaches alternate with steep rocky steps which give rise to waterfalls and rapids.

9. Careful geological mapping demonstrated that some valleys are aligned along faults, and where such a relationship exists, it was easy to dismiss the possibility of fluvial excavation and to regard the valley as a fissure formed during the same earth-movement as produced the fault itself. Some earlier writers had claimed that any valley displaying identical sequences of strata on its opposing walls could only be erosional in origin, but nineteenth-century geologists, much better informed than their predecessors in the field of tectonic geology, were well aware that faulting might take the form of horizontal movement with little or no vertical displacement of the strata. If some valleys are fissures opened as a result of crustal tension, why should all valleys not be so explained? Perhaps a fault would be found in the rocks beneath almost every valley if only the alluvial infill could be removed.

10. The sea-lochs of western Scotland, and the rias of Devon, Cornwall, and western Ireland, are each merely the seaward extension of a subaerial valley. The two sections of such a valley – the subaerial and the submarine – seemingly have a common origin, but that origin must, it was argued, be quite independent of fluvial erosion because rivers clearly cannot continue their work below sea-level.

Apart from the denudation dilemma, and these objections to fluvialism arising from field observations, there was one other reason for the doctrine's failure to find acceptance. The rejection of Moses had left the geologists with aeons at their disposal, but, as we noted earlier, there was a general failure to appreciate the true magnitude of the new time-scale. The transition from the old chronology to the new required a long period of mental readjustment, and although geologists might now write of the Earth as being millions of years old, such a time-span was really too much for their comprehension. As a result they were slow to appreciate the possibilities inherent in the new time-scale, and in particular they were slow to recognise that the time-scale was of such dimensions that only a fraction of Earth-history affords ample time for even the most feeble of Nature's forces to effect enor-

241

mous changes. Nineteenth-century geologists, fully acquainted with Nature in the field, were perhaps more aware than their predecessors of the slowness with which rain and rivers do their work, but this fact of itself was no obstacle to the acceptance of fluvialism; the real obstacle here, as we saw in the case of Macculloch, was a general inability to grasp the true extent of the time-span throughout which such processes have been in action.

Strangely, there were even some nineteenth-century geologists who, while proclaiming the Earth's great antiquity, at the same time still saw a period of a few thousand years as representing a considerable proportion of Earth-history. So deluded, they sought for secular evidence of changes effected by the fluvial processes within the historic period, and finding none, they rejected the entire fluvial doctrine. In a paper read to the Geological Society of London in 1829, for example, Conybeare pointed out that the prehistoric earthworks and Roman sites in the Thames basin – even those in the most exposed of positions – are still recognisable after the passage of almost two thousand years. In view of this, he reasoned, the valley of the Thames, many hundreds of feet deep, could never have been excavated by the river itself, no matter how long we allow it for the task.[93] Interestingly, this paper was commissioned by Buckland as an antidote to the fluvialistic interpretation of the Auvergne Denudation presented to the Society a few months earlier by Lyell and Murchison, and there is no doubt that Conybeare and Buckland had the majority of British geological opinion on their side. Thus, even in the nineteenth century, there was still a seeming time deficiency to inhibit the acceptance of the fluvial doctrine.

The passage of time, bringing with it a greater intellectual maturity, eventually gave geologists a more sound understanding of the true scale of Earth-history, and as such a perception developed, one objection to the fluvial doctrine slipped from the scene. The objection to fluvialism which arose from the denudation dilemma was never a serious matter during the nineteenth century, and it was in any case virtually forgotten during the 1830s. The objections which arose out of field observations, on the other hand, were much less easy of disposal. A few of these latter objections could perhaps have been removed immediately had more attention been devoted to landforms, but most of the objections were of a much more fundamental nature. They were incapable of removal so long as geologists remained blissfully unaware of the complexity of the recent history of the Earth's surface. The solution of the problem presented by sea-lochs and rias, for example, had to await the development of the isostatic and eustatic theories, and similarly the problem of dry valleys was difficult of solution in the absence of some understanding of the climatic changes which have affected the Earth's surface since the close of the Tertiary era. Above all, however, the most telling objections to fluvialism could not be removed until the glacial theory was firmly established, and, more especially, until there was some understanding of the erosive power of glacier ice. In the absence of such an understanding, the limnological objection, the existence of hanging valleys, the problem of stepped valley floors, and the aggradational work of so many modern rivers, remained four damning arguments against the fluvial doctrine.

Since the fluvial doctrine had been rejected, other forces had to be invoked in explanation of the Earth's surface phenomena. Five such forces – or sets of forces – were employed: seismic convulsions; erosion by the swollen rivers of some former pluvial period; erosion by de-bacles released when the barriers impounding lakes were removed; erosion during some diluvial catastrophe when water surged over the continents following earthquakes; and, finally, erosion by waves and currents during some bygone marine submergence. Several of these forces will already be very familiar to readers of earlier chapters, and it must be admitted that early nineteenth-century writers showed little advance in this area upon the works of their seventeenth- and eighteenth-century predecessors. As in the earlier periods, of course, no nineteenth-century author believed that any one of these processes had alone been responsible for shaping all of the earth's topography; landscapes were viewed as basically the result of the combined action of two or more of these forces, while the fluvial processes were nor-mally admitted to have added a final flourish of topographical detail. Although they were never mutually exclusive, it is convenient to consider each of these five nineteenth-century substitutes for fluvialism separately in turn.

Earthquakes still in Vogue

The reasons for the nineteenth-century persistence of a belief in convulsive earth-movements have already been examined, and those catastrophists who subscribed to this belief very naturally claimed that such revolutions must have diversified the continents with a wide variety of landforms. In 1813 de Luc was still arguing that most topo-

243

graphical features are the result of crustal collapse, and the following passage was written not in the seventeenth century by Robert Hooke, but in 1828 by Robert Bakewell, one of the leading 'practical' geologists of his day:

> ... we can scarcely conceive the possibility of a whole district being covered with new mountains and another soil, in the space of a single night; yet such changes have been produced, by the united agency of earthquakes and volcanoes.[94]

Bakewell's eminent contemporary, William Buckland, held similar views. In 1825 he drew the attention of the Geological Society of London to such English denudations (breached anticlines) as those of the Weald, the Vale of Kingsclere in Hampshire, and the Vale of Pewsey to the east of Devizes in Wiltshire, and he claimed the strike valleys within the denudations as crustal fissures opened during the arching of the rocks. For such features he proposed the term 'valleys of elevation' in contradistinction to the 'valleys of denudation' which he believed to have been excavated by diluvial torrents.[95]

Ten years later such views were given the high respectability of a mathematical imprimatur when the Cambridge mathematician William Hopkins turned his attention to the examination of the supposed causal relationship between earth-movements on the one hand, and such geological structures as mineral veins, faults, fissures, dykes, and valleys on the other.[96] He coined the term 'Physical Geology' to describe these studies, and after a careful mathematical analysis, he concluded that all such structures conform to general mechanical and physical laws, and that all of them – including valleys – had been formed as a direct result of crustal uplift. In 1841, with this theory in mind, he examined the Weald, and in a paper to the Geological Society of London he employed his mathematics to prove that during the uplift of the Wealden pericline, transverse and longitudinal fissures must have opened at precisely those points where dip and strike valleys exist today.[97] Few geologists can have possessed a sufficient mastery of mathematics to allow them to follow Hopkins's reasoning, but they were nevertheless impressed by the evident ease with which mathematics was able to resolve a whole series of perplexing geological problems. Emboldened by the reception accorded his work, Hopkins now pressed his studies still further. In 1842, using the same mathematical techniques, he arrived at the conclusion firstly that most of the valleys of the English Lake District are aligned along a series of radial faults formed during the region's uplift, and secondly that all the region's major lakes are not merely fault-controlled, but are areas of actual and recent subsidence.[98] Had he been alive, de Luc would have been happy to see mathematical confirmation of the collapse theory of lake basins which he had advocated for so long.

Nobody championed the seismic origin of topography more strenuously than Murchison. In his massive *Geology of Russia* published in 1845 he observed:

> We conceive, indeed, that nearly all transverse gorges by which rivers escape across ridges from one water basin to another, are nothing more than ancient apertures in the crust of the earth.[99]

He believed that all the mammoths which formerly inhabited Russia in Europe had been killed off by the same recent catastrophe as heaved up the Ural Mountains and left the crust riven with the fissures which today form the region's valleys. Even in those places where the strata were clearly once continuous across the site of a modern valley, he argued:

> . . . we believe that even in such exceptions, the transverse chasm has been mainly produced by a great vibratory movement, giving rise to a fissure, the depth and size of which has been augmented by powerful denudation when the land and waters were changing their relations.[100]

Murchison had thus completely forgotten the lessons which he had learned in Auvergne seventeen years earlier. His cataclysmic creed is repeatedly reiterated in his papers and books, and in the many presidential addresses which he delivered over the decades to the Geological Society of London, the British Association, and the Royal Geographical Society. Late in his life, in the 1860s, he did concede that members of the new school of fluvial geomorphology were presenting their case well, but he nevertheless went to his grave in 1871 an unrepentant catastrophist.

Topography from a former Pluvial Period

The theory that some topography had been shaped by the greatly swollen rivers of a bygone pluvial period was an effort to reconcile a predilection for fluvialism with a belief that the Earth's present rivers could never have achieved the observed results within the time avail-

able. Hooykaas has suggested that those who supposed present processes
to have operated with greater power in the past, should be termed
'actualists' to distinguish them from the strict uniformitarians who
believed that natural forces never vary in their intensity.[101] This
distinction is doubtless of value, but such actualism was never a factor
of any importance in geomorphology. In the 1790s Richard Kirwan did
toy with the idea of explaining modern topography in terms of erosion
performed during a former pluvial period, and twenty years later the
same idea found a place in the writings of John Macculloch. Another
to whom the theory appealed was one John Carr who wrote to the
Philosophical Magazine in 1809 from an address in Manchester.[102] Carr
disagreed with Farey's suggestion that the debris from the Derbyshire
Denudation had been lost into space, and he claimed instead that the
denudation was entirely the work of the fluvial processes. To support
his case, Carr, without any acknowledgment to Playfair, presented
Playfair's proofs of the fluvial origin of valleys. Carr, however, was no
Huttonian geomorphologist; he evidently believed that the denudation
had been produced under some former fluvial regimen and he certainly
regarded the present rivers as impotent.

The fullest and most realistic application of the actualist principle in
the field of denudational geomorphology is contained in a paper on the
valleys of southern England and northern France which was read to the
Royal Society early in 1862 by Joseph Prestwich.[103] He argued that
the present geology and morphology of such valleys as the Great Ouse,
the Seine, and the Somme are explicable only in terms of the existence
of rivers

> . . . of power and volume far greater than the present rivers, and
> dependent upon climatal causes distinct from those now prevailing in
> these latitudes. The size, power, and width of the old rivers is clearly
> evinced by the breadth of their channel, and the coarseness and mass of
> their shingle beds.

He related the onetime existence of these powerful rivers to floods
caused by the seasonal melting of the local Pleistocene ice sheets, and
to recent climatic changes generally.

> In all valleys connected with mountain-chains the result of these
> climatal changes must have greatly increased the power of the annual
> floods – whence the greater excavation of the valleys connected with
> such regions.

246

Such views, however, were never very influential, and this particular theory of landscape development need detain us no longer.

Topography from Debacles

Many early nineteenth-century geologists believed that the Earth's surface had formerly been dotted with innumerable lakes, and that at intervals catastrophic earthquakes had destroyed the barriers impounding the lakes, thus releasing violent debacles. Those subscribing to this theory maintained that these debacles had played a major part in the shaping of the Earth's topography.

This principle of erosion by debacles had long been known to mineral prospectors through the exploratory technique known as 'hushing'.[104] This involved the construction of a small dam across a stream so as to form a pond or even a miniature lake. Then, once sufficient water had gathered, it was allowed to sweep the dam away, and the resultant debacle stripped away the surface mantle on the slopes below to reveal any mineral veins outcropping in the so-called 'hush-gutter'. Precisely the same process was exemplified on the grand scale in Switzerland in 1818. In April of that year, ice from the Getroz glacier avalanched into the Val de Bagnes above Sembrancher, blocked the course of the river Dranse, and impounded a lake estimated to have held 800,000,000 cubic feet of water. The district engineer, one Ignace Venetz of whom more later, ordered the construction of a tunnel through the upper part of the dam, and by this means some 300,000,000 cubic feet of water were drawn off. Then, in mid-June the dam suddenly gave way, and the remaining lake-water surged towards Lake Geneva, devastating the valley and causing heavy loss of life.[105] The geological world took note of the tragedy, and many viewed it as an example of a type of event which they believed to have been common in the Earth's earlier history. J. D. Forbes described the disaster as 'an awful but a grand lesson for the geologist',[106] and Lyell was profoundly impressed by the scene of desolation when he visited the Val de Bagnes only two months after the debacle. Years later the scene was still fresh in his mind when he wrote an account of the catastrophe for inclusion in the first volume of his *Principles*.

Sir Henry De La Beche was the foremost advocate of debacles as an important geomorphic process,[107] and in his *Geological Observer* of 1851[108] he discussed the various types of dam which might be thrown across the course of a river to impound a lake and result in an eventual debacle. Apart from an ice dam such as had existed in the Val de Bagnes, he suggested that the dam might take the form of a lava flow, a landslide, or a large alluvial fan built into the main valley by a tributary stream. By 1851, however, there were probably few geologists who still attached much geomorphic significance to debacles; other, seemingly more credible geomorphic processes had come into favour.

The Last Days of Diluvialism

The Mosaic deluge had always been a favourite geomorphic process, and belief in a catastrophic flood persisted long after British scientists had shed their former bibliolatry. Although most early nineteenth-century geologists no longer accepted the literal truth of the Genesis account of the Deluge, the flood which they invoked in explanation of so many natural phenomena was essentially the same diluvial event as had featured so prominently in the geological writings of the seventeenth and eighteenth centuries. It was still pictured as a sudden and comparatively recent event when waters had swept over the continents, drowning a multitude of creatures, and effecting major topographical changes within a very short space of time. There was, nevertheless, one significant difference between the early diluvialism and that of the nineteenth century. Hitherto belief in a deluge had rested largely upon the Mosaic records backed up by a minimum of field evidence, but in the nineteenth century, as Moses slipped from the scene, field investigation brought to light a considerable weight of facts all seeming to confirm the reality of a recent deluge. By far the most important element in this evidence was the thick drift mantle which covers so much of the northern hemisphere. Today this deposit is known to be a legacy left by the Pleistocene glaciers, but a century and a half ago, in the absence of the glacial theory, the heterogeneous and unsorted character of the drift seemed ample proof that it had been laid down in the turbulent waters of a universal flood. Buckland termed the deposit 'diluvium' to distinguish it from the 'alluvium' formed by rivers, and these two terms were soon in general use.

The idea that the diluvium was a flood deposit received valuable support in 1831 as the result of a discovery which was destined to confuse and mislead British geologists for almost half a century. In that year Joshua Trimmer found fragments of marine shells at a height of 1392 feet in the diluvium on Moel-tryfan to the south of Caernarvon in North Wales.[109] Five years later John Scouler recorded the existence of similar shells in the diluvium around Dublin,[110] and soon other deposits of shelly diluvium had been located throughout the British Isles. The wide distribution of the deposits seemed to confirm that the inundation had indeed been on a grand scale, and the Moel-tryfan deposit in particular was immediately accepted as affording a minimum depth for the submergence. Here was another conclusion which seemed entirely reasonable in the light of the evidence as it was then understood, and for the error to be corrected there was need not merely for the glacial theory, but also for some knowledge of the flow-mechanics of glacier-ice.

Some geologists were a little puzzled to find the diluvium commonly so completely devoid of bedding and sorting as to allow sand, pebbles and huge boulders to appear in a confused juxtaposition. Surely, they observed, even a diluvial deposit should show some evidence of its formation in water. Here icebergs came to their aid. It was suggested that during the diluvial catastrophe, icebergs had been swept from polar regions into warmer latitudes, and that as they travelled, they melted, scattering their morainic cargoes at random over the surface of the drowned continents. Such a mechanism explained not only the heterogeneous confusion of the diluvium itself, but also the dispersal of the large individual erratic blocks which had been perplexing geologists since the closing decades of the previous century.

The earliest notable nineteenth-century diluvialist was that leading Plutonist, Sir James Hall, although, for reasons that will emerge later, his diluvialism was not entirely typical of his age. As we have seen already, Hall rejected Hutton's fluvialism, and in his two papers to the Royal Society of Edinburgh in 1812 he argued that it was diluvial water which had really played the major part in shaping the Earth's topography.[111] He claimed that continental emergence occurs not slowly, as Hutton had believed, but in a series of sudden starts, each start so violent that it sends huge tsunami sweeping over the continents. It was to the action of such tsunami that Hall attributed not only the formation of all the world's erratics and unsorted superficial deposits, but also the excavation of all those valleys which were not seismic fissures. As a keen experimentalist, Hall, like Hooke and Catcott in earlier centuries, sought to test his geomorphic theories in the laboratory. He exploded small charges of gunpowder under water in an effort to produce tsunami in miniature, but he tells us little about the results obtained from these somewhat dangerous investigations.

The manuscript journals of Hall's continental travels reveal him to have been a perceptive observer of Nature.[112] He certainly had a splendid eye for geomorphic detail, and in his papers of 1812 he backed up his particular version of the diluvial theory with evidence drawn from the Edinburgh region. This evidence was of four types. Firstly, he drew attention to the region's widespread superficial mantle of unsorted debris. Secondly, he pointed out the existence of smooth polished rock pavements which he termed 'dressed surfaces' and which he attributed to the abrasive action of the debris-laden diluvial waters. Thirdly, he recorded the discovery of fifteen sites around Corstorphine Hill, to the west of Edinburgh, where the dressed surfaces display striations. These he likened to the slickensides produced by movement along a fault plane, but he believed that on Corstorphine Hill, as elsewhere in Scotland, the striations had been etched into the rocks by angular material being swept along by the deluge. Where this phenomenon occurs, he wrote:

> . . . the surface is found to resemble that of a wet road, along which a number of heavy and irregular bodies have been recently dragged; indicating that every block that passed, and every one of its corners, had left its trace behind it; and these are rendered very distinctly visible, when the surface is drenched with water.

Finally, he adduced the evidence provided by the present shape of Edinburgh Castle Rock, Calton Hill, and Corstorphine Hill. Each of these hills is markedly asymmetrical, with steep slopes on the western side, and very gentle slopes to the east. Hall explained the asymmetry in terms of diluvial waves sweeping over the region from the west and eroding the western, northern, and southern slopes of the hills, while on the protected leeward sides the gentle pre-diluvial slopes survived virtually intact. To describe such asymmetrical features he coined the term by which they are still known – crag and tail topography. By measuring the orientation of both the tails and the striations, Hall satisfied himself not merely that the diluvial currents had swept eastwards over the site of Edinburgh, but that their precise bearing had been 10° north of east.

Hall's two papers are memorable as the first attempt to discuss the geomorphology of a small region in some detail, and especially interesting is his inclusion of a map to show the precise location of the sites at which he had observed striations. His observations are beyond reproach, and he must be forgiven for his diluvialism. Surely a deluge of water was a much more plausible occurrence than the deluge of ice which we now know to be the true explanation of Hall's Edinburgh phenomena. Again, Nature's unexpected complexity was leading even the most astute field observers into false theories.

Although Hall attributed the topography around Edinburgh to the work of a deluge, he evidently believed that this particular deluge was but one in a whole series of similar events caused by the jerky uplift of the continents. It is in this respect that his diluvialism was unusual for its day. Most of his contemporaries, like their predecessors in earlier centuries, thought not in terms of multiple and perhaps local inundations, but, rather in terms of one single, recent, and universal diluvial catastrophe. This simple diluvialism found its fullest expression in the early writings of William Buckland – the eminent Oxford geologist whose adherence to diluvialism gave the doctrine its seal of authority. His best-known work on the subject is his *Reliquiæ Diluvianæ* published in 1823, and therein he voiced the conviction that the Earth's continents had recently been completely submerged beneath the waters of a universal deluge which had acted as an important geological and geomorphic process. Unlike so many of the earlier diluvialists, Buckland made no attempt to explain the cause of the deluge, but in 1823 he clearly believed that the flood which he was discussing was one and the same event as the Noachian Deluge described in *Genesis*. This belief entitles Buckland to the doubtful honour of being the last British geologist of note to relate the discoveries of modern geology to the Mosaic writings. Surprisingly, in discussing the geomorphic significance of the flood, Buckland repeatedly refers to Catcott's book of 1761 as authoritative upon the subject, and so far as this particular aspect of the catastrophe is concerned, the views of the two men were very much in accord. Apart from leaving innumerable erratics scattered over the continents, and a widespread mantle of diluvium, Buckland believed that the inundation had re-shaped most of the Earth's topography, and he claimed features such as the Straits of Dover, most outliers, and all valleys of denudation (as distinct from his valleys of elevation) as diluvial in origin.

Faith in this simple brand of diluvialism was seriously shaken during the 1820s and 1830s as a result of two geological discoveries. Firstly, it was found that in middle and low latitudes the continents have no mantle of diluvium. Secondly, study of the diluvium itself had revealed

251

something of the complexity of drift stratigraphy. In Europe, for example, it was found that two distinct types of diluvium were present; in the north there was the so-called Northern Drift consisting chiefly of debris of Scandinavian origin, while further south there was another type of diluvium consisting largely of material from Alpine sources. Clearly it was difficult to reconcile such discoveries with a belief in a single, universal inundation, and in place of the old, simple diluvialism, there arose a new diluvialism which employed not one diluvial catastrophe, but a whole series of such events. This neo-diluvialism was a faith such as Hall had propounded in 1812, and it accorded well with Élie de Beaumont's now popular theory that the Earth had experienced a large number of catastrophic mountain-building revolutions. Each such revolution, it seemed, must have sent gigantic tsunami surging around the world, inundating and remodelling the continents.

In Buckland's case the change from the old diluvialism to the new took place in the late 1820s. In April 1829 Lyell wrote to Gideon Mantell after hearing Conybeare's paper on the Thames valley at the Geological Society of London:

> He admits three deluges before the Noachian! and Buckland adds God knows how many *catastrophes* besides, so we have driven them out of the Mosaic record fairly.[113]

Fifteen months later Lyell again wrote – this time to his sister – saying that Buckland had indeed renounced his original brand of diluvialism,[114] and in his *Bridgewater Treatise* of 1836 Buckland himself freely admitted that the deluge of such significance to geology was not the comparatively recent Noachian Deluge, but a very much earlier event which occurred long before man arrived upon the Earth. The Biblical Flood, he now argued, was merely a gentle rise and fall of the waters without any of the turbulence necessary to effect topographical changes. It was 211 years since Nathanael Carpenter had arrived at precisely the same conclusion. Adam Sedgwick, Buckland's opposite number at Cambridge, publically renounced his simple diluvialism in 1829 and again, more fully, in his presidential address to the Geological Society of London in 1831.[115] By 1850 the simple traditional diluvialism was virtually dead.

The neo-diluvial theory was never formulated in any very precise terms. Maybe this was part of its appeal because the theory allowed its adherents freedom to invoke as many catastrophic inundations as they thought necessary to explain any given set of phenomena. The theoretical basis of the new diluvialism did nevertheless receive some examination. In 1842, for example, William Hopkins, the mathematician, discussed the dynamics of what he termed 'currents of elevation' occurring within 'waves of translation' caused by the sudden uplift of portions of the sea-bed.[116] Similarly, but five years later, William Whewell calculated that the deluge which swept the Northern Drift over Europe must have been generated by the sudden uplift of 45,000 square miles of sea-bed through a vertical distance of 500 feet.[117] So far as the shaping of topography was concerned, the deluges were supposed to have worked upon the continents in three ways. Firstly, there was the damage inflicted by the violent initial impact of the waters. Secondly, there was the erosive work performed by currents within the diluvial waters during the submergence. Finally, there was the combing action of torrents of water draining off the continents as the flood abated. It was to this latter process that most of the neo-diluvialists attached the greatest significance, and in this respect their views differed little from those expressed by Catcott and Williams half a century earlier.

Neo-diluvialism found its most ardent British disciple in Murchison; even in 1867, in the fourth edition of *Siluria*, he was still proclaiming the importance of diluvial torrents as geomorphic agents. Despite the allegiance of so influential a geologist, however, neo-diluvialism was of little significance in the evolution of geomorphic thought. Diluvialism, in whatever its form, was merely one facet of the catastrophic conception of Earth-history, and by the time neo-diluvialism appeared upon the scene, catastrophism itself was fast dying. It was being replaced by the rival uniformitarian interpretation of Earth-history, and this interpretation in fact cut at the very roots of diluvialism. If, as the uniformitarians claimed, continental uplift was normally a relatively gentle affair, and if they were correct in their belief that Nature never unleashes greater forces than those displayed at Lisbon in 1755, or in Chile in 1822 and 1835, then where was the force capable of sending gigantic tsunami sweeping across the continents? Thus, as the catastrophic interpretation of Earth-history gave way to the uniformitarian, diluvialism became obsolete, and geomorphology relinquished a process which from the earliest days had always featured prominently in explanations of topography.

Although diluvialism whether in its old or new forms, was no longer a factor of importance in British geomorphology after the middle years of the nineteenth century, the basic belief in a recent submergence of the continents proved remarkably resilient. Indeed, no sooner had diluvialism entered upon its decline, than the idea of a recent submergence arose phoenix-like in a fresh guise. In a sense, this new submergence theory was closer to the old diluvialism than to neo-diluvialism. It held simply that at some time in the recent past the continents had been submerged, and that the normal processes of marine erosion had then worked over the continental surfaces and given them their present configuration. Such marine erosion was the last of the five topography-forming processes invoked during the first half of the nineteenth century in lieu of a fluvialistic interpretation of landscapes.

Topography from Marine Erosion

Apart from those few eighteenth-century geologists who had sought a resolution of the denudation dilemma through a complete denial of the reality of all forms of denudation, no British writers had ever doubted the potency of the sea as an agent of erosion. It is therefore strange that Gabriel Plattes's notion that the Earth's topography had been shaped by marine erosion – a notion dating from 1639 – should have long remained unique. The idea that topography owed much to a diluvial catastrophe had always been popular, but the idea that topography might have been formed by normal wave and current action during a time of high sea-levels, was virtually unknown. Two centuries elapsed before the theory which Plattes had propounded came into favour, and when it did achieve popularity in the 1830s, it had no lesser authority than Lyell as its leading champion. His ideas relative to the date and duration of the submergence were never very precisely expressed, but his theory attributing topography to marine erosion was really nothing more than a uniformitarian version of the diluvial theory of landscape development. Lyell merely took the diluvial theory, stripped it of its catastrophism, replaced the catastrophic agencies by such acceptable uniformitarian processes as wave and current action, and then adopted the refurbished theory as his own. It was thus no accident that the marine erosion theory came into vogue at the very moment when support for catastrophism and the diluvial theory was rapidly fading.

In the 1840s British geologists received a forceful reminder of the power of the sea when Thomas Stevenson, the father of Robert Louis

Stevenson, made his famous observations at Skerryvore Lighthouse off Tiree in the Inner Hebrides, and showed that Atlantic storm-waves sometimes break with a force in excess of 6000 lbs per square foot.[118] Granted such power in the sea, what evidence was available to support the view that waves and currents had really worked over the entire surface of the present continents? Here Lyell and his followers could draw upon much the same evidence as the diluvialists had adduced in support of their own particular conception of a recent continental submergence. During the 1830s, for example, in the absence of the glacial theory, the diluvium – and particularly the shelly diluvium – was just as satisfactory a proof of recent submergence for the uniformitarians as it had been earlier for the diluvialists. Indeed, where the Earth's superficial deposits were concerned, Lyell borrowed various ideas from the diluvialists. He attributed the unsorted character of so many of the deposits to their having been formed from debris dropped by melting icebergs, and similarly he invoked ice-rafting to explain the dispersal of the world's erratic blocks. For obvious reasons, however, the adherents of the new school of thought found the old term 'diluvium' unacceptable. In 1840 Lyell himself, with the notion of drifting icebergs in mind, suggested that the diluvium should be rechristened 'drift',[119] and this new term remains in use to the present day.

When, in the 1840s, some of the drifts were shown to be glacial in origin, rather than marine, the case for believing in a recent submergence was somewhat weakened. The shelly drifts, and the bedded drifts which are now known to be of glaciofluvial origin, nevertheless continued to be regarded as marine formations, and the advent of the glacial theory certainly did not stifle the earlier submergence theory. Rather did the arrival of the glacial theory lead to a refinement of thinking concerning the recent submergence. Before 1840 it was believed that there had been only one modern submergence, and that during this inundation the continental topography had been moulded and the drifts deposited. After the advent of the glacial theory, however, it was generally admitted that there must have been two submergences separated by the main glacial episode. The first of these submergences was pre-glacial, and being the longer of the two, it was usually held to have been the most important so far as the shaping of topography was concerned. The second submergence, occurring within the glacial period, was of shorter duration and therefore of lesser geomorphic significance, although it was supposed to have seen the deposition of the widespread shelly and bedded drifts. This second, glacial submergence will engage our attention further in the next chapter.

One other type of deposit which seemed to support the submergence theory also deserves a mention. In 1857 Prestwich discussed the now famous Lenham Beds of Kent before the Geological Society of London and offered them as evidence of a Pliocene sea-level standing some 600 feet higher than the present sea-level.[120]

The importance of the drift deposits to Lyell's marine submergence theory must nevertheless not be exaggerated. They afforded the theory nothing more than strong corroborative evidence, and the theory would have flourished even had the continents been devoid of their drift mantle. As we have seen, the theory was to some extent a uniformitarian version of catastrophic diluvialism, but the theory possessed a much more fundamental relationship to the uniformitarian doctrine. The stratigraphical record showed clearly that marine transgressions had occurred regularly throughout geological time, and once the uniformitarian concept of slow continental uplift was established, then it followed as a natural corollary that during each transgression and regression, marine waves and currents must have had the opportunity to work over vast areas of the continental surfaces, cutting cliffs, excavating valleys, and re-shaping the topography generally.

Once the concept of a recent, non-violent submergence of the continents was firmly established, then field phenomena began to be interpreted in the light of the concept, and innumerable features were found, all seeming to confirm that topography had indeed been shaped by marine processes. The relationship of topography to former wave and current action is a constantly recurring theme in Lyell's writings over a period of more than forty years, and his application of the marine erosion theory to the Wealden Denudation obtained wide currency. The bold scarp faces of the cuestas within the denudation, he explained, are nothing less than wave-cut cliffs from which the sea has only recently retreated. Lyell, however, was only one of many eminent British geologists who claimed to see the work of waves and currents in the world's landscapes. For some thirty years down to 1862, the marine erosion theory was the most generally accepted theory of landscape development, and reference to a few works of the period will serve to illustrate the general nature of the ideas then current.

In 1839 Charles Darwin interpreted the Parallel Roads of Glen Roy as the strand lines of a former marine inundation,[121] and seven years later, in his *Geological Observations on South America*, he observed:

> Finally, the conclusion at which I have arrived, with respect to the relative powers of rain and sea water on the land, is, that the latter is far the most efficient agent, and that its chief tendency is to widen the valleys; whilst torrents and rivers tend to deepen them, and to remove the wreck of the sea's destroying action. As the waves have more power, the more open and exposed the space may be, so will they always tend to widen more and more the mouths of valleys compared with their upper parts: hence, doubtless, it is, that most valleys expand at their mouths, – that part, at which the rivers flowing in them, generally have the least wearing power.[122]

In his 1846 memoir, Ramsay attributed the formation of the South Wales Denudation to the work of a transgressive sea, and – like many a twentieth-century geomorphologist – he claimed the gently sloping upland plains around Cardigan Bay as the result of marine planation. In 1848 Robert Chambers of Edinburgh published his book *Ancient Sea-Margins* wherein he discussed a variety of supposed shore-line features dating from the submergence, and he offered the conclusion that in the British Isles there is evidence of no less than twenty-seven strand-lines lying between present sea-level and a height of 545 feet. In 1851 De La Beche thought that he had detected evidence on the western flanks of the Leinster Mountains of 'the destructive influence of Atlantic breakers rolling in from the westward'.[123] Two years later John Phillips explained how the topography of Yorkshire had been moulded by waves and currents, and of the region's limestone scars he wrote:

> The great inland cliffs, which are among the most striking phænomena of Yorkshire, only differ from sea cliffs, because the water no longer beats against them.[124]

In the following year Joshua Trimmer claimed even the solution pipes in the English chalk as the work of waves,[125] while further afield, Daniel Sharpe, the Treasurer of the Geological Society of London, returned from the Alps convinced that he had found evidence of former Alpine sea-levels at heights of 9000 to 9300 feet, 7500 feet, and 4800 feet.[126]

Two final quotations relevant to the marine erosion theory of topography deserve inclusion. They are taken from two works published by J. B. Jukes in 1853 and 1857 respectively, and they are of interest on two counts. Firstly, they epitomise the views of those many geologists who subscribed to this particular school of geomorphic thought. Secondly, the two passages afford an interesting contrast to the opinions which Jukes expressed after 1862 because in that year he became the first British geologist to renounce the marine erosion theory of landscapes in favour of a full-blown Huttonian fluvialism.

> Every valley and hollow, every slope of a hill, every cliff, every ravine, has been formed mainly by the action of the sea, when that portion of the land was at or but a little below its surface. It may have subsequently been modified by the action of the atmosphere, the frost, or the river, but its principal feature has been formed by the sea. Unless a previously formed valley and system of valleys had existed, the river and its tributaries could not have commenced.[127]

> Just as actual sea cliffs are proofs of the erosive action of the sea now in operation, so, in almost all cases, inland cliffs, crags, scars, and precipices, as well as all valleys and ravines, gorges and mountain passes, are proofs of the former erosive action of the sea, in times when the land stood at a lower level with respect to it. . . . Rivers are not the producers of their own valleys; they are the results of those valleys, but they are their immediate results. The river could not be formed till after the valley, with all its tributary branches, had been marked out.[128]

During the first sixty years of the nineteenth century, geomorphology trailed far behind as other branches of geology swept triumphantly forward. Fluvialism was rejected as being an inadequate explanation of topography, and for sixty years a variety of other sets of forces were allowed to occupy the position which rightfully belonged to rain and rivers. It needs to be re-emphasised that no geologists of the period ever believed that any one of these other sets of forces had alone been responsible for shaping all of the Earth's topography; attempts to categorise nineteenth-century writers on landforms as diluvialists or marine erosionists can only result in a grossly over-simplified picture. Hall and Buckland, for example, explained topography in terms of seismic shocks and diluvial currents. Macculloch made use of seismic shocks, torrents draining off the continents as they emerged, and erosion by the swollen rivers of a pluvial period. Sedgwick employed a combination of seismic shocks, diluvial torrents and submarine currents.

Scrope favoured seismic shocks, marine action, and fluvial erosion during a pluvial period. Murchison invoked gigantic convulsions and tsunami, while Lyell advocated debacles and marine erosion, and in the fourth edition of the *Principles*, published in 1835, we actually find him criticising Hutton for his refusal to admit that some valleys are the result of crustal subsidence. In addition to these various sets of forces, all geologists of the first half of the last century did admit that the fluvial processes must have played some part in shaping the Earth's topography. With the very doubtful exception of George Greenwood, however, not one writer of the period can be hailed as a fluvialist in the Huttonian tradition. Thanks to Lyell's influence, it was marine waves and currents which by the middle years of the century were generally accepted to have been the most important of Nature's geomorphic tools, and even today this predilection for marine erosion is still evident in British geomorphology. The widespread planation surfaces of the British Isles lying below the 650-foot contour are still regarded as the work of marine abrasion during a Plio-Pleistocene submergence, despite the fact that in Highland Britain, at least, there is far less corroborative evidence of such a submergence than there seemed to be a century ago when the glacial drifts were generally believed to be marine in origin. Not until after 1862 did fluvialism come into its own, and, paradoxically, the key to the acceptance of the fluvial doctrine was the glacial theory.

REFERENCES

1. *Edinb. New Philos. Journ.*, XXI (1836), p. 3.
2. *Edinb. Rev.*, XXIX (1817), p. 74.
3. *Frost and Fire*, II, pp. 2 & 3 (Edinburgh 1865).
4. Eliza Meteyard, *The Life of Josiah Wedgwood*, I, p. 500 (London 1865).
5. Charles Lyell, *Principles of Geology*, I, p. 71 (London 1830).
6. *Edinb. Rev.*, XXVIII (1817), p. 175.
7. 'The Scriptural Geologists: An Episode in the History of Opinion', *Osiris*, XI (1954), pp. 65–86.
8. *Proc. geol. Soc.*, I (1831), No. 20, p. 315.
9. Katharine M. Lyell, *Life, Letters and Journals of Sir Charles Lyell, Bart.*, I, p. 446 (London 1881).
10. *Ibid.*, I, p. 268.
11. William Cockburn, *A Remonstrance . . . upon the Dangers of Peripatetic Philosophy* (London 1838).
12. *Trans. Edinb. geol. Soc.*, XIII (1931–38), p. 233.
13. Archibald Geikie, *Memoir of Sir Andrew Crombie Ramsay*, p. 191 (London 1895).

14. John Macculloch, *A System of Geology*, I, p. 506 (London 1831).
15. Charles Darwin, *Origin of Species*, Chap. IX (London 1859).
16. *The Quarterly Review*, XLVII (1832), p. 108.
17. Joseph B. Jukes, *Popular Physical Geology*, pp. 325 & 326 (London 1853).
18. *Trans. roy. Soc. Edinb.*, XXIII (1864), pp. 157–169.
19. *Trans. geol. Soc. Glasg.*, III (1), (1868), pp. 1–28.
20. *Quart. J. geol. Soc. Lond.*, XVII (1861), pp. xxvii–lxxii.
21. George J. P. Scrope, *Memoir on the Geology of Central France*, p. 165 f (London 1827).
22. James C. Irons, *Autobiographical Sketch of James Croll*, p. 200 (London 1896).
23. Lyell, *op. cit.* (1881), II, p. 253.
24. *Edinb. Rev.*, LXIX (1839), p. 455 f.
25. *Proc. geol. Soc.*, I (1827–28), No. 6, pp. 55 & 56.
26. Georges Cuvier, *Essay on the Theory of the Earth with Geological Illustrations by Robert Jameson* (Edinburgh 1827).
27. *The Sir Henry De La Beche Papers* in the National Museum of Wales, Cardiff.
28. *Outlines of Geology*, p. 209 (London 1829).
29. *Siluria*, fourth edition, p. 489 (London 1867).
30. *Annls. Sci. nat.*, XVIII (1829), pp. 1–25; *Phil. Mag.*, N.S. X (1831), pp. 241–264; Henry T. De La Beche, *A Geological Manual*, pp. 496–499 (London 1831).
31. Lyell, *op. cit.* (1830), I, p. 64.
32. *Ibid.*, I, p. 64.
33. *Philos. Trans.* (1835), pp. 1–38.
34. *Proc. geol. Soc.*, II (1837), No. 51, pp. 552–554.
35. Lyell, *op. cit.* (1830), I, pp. 449–459.
36. *Ibid.*, I, p. 80.
37. Charles Darwin, *Geological Observations on South America*, Chap. II (London 1846).
38. *Rep. Brit. Ass.*, I–II (1831–32), p. 381.
39. Henry C. Englefield, *The Isle of Wight* (London 1816).
40. Macculloch, *op. cit.* (1831), I, p. 154.
41. *The Formation of Vegetable Mould, through the Action of Worms*, p. 231 (London 1881).
42. *General View of the Agriculture and Minerals of Derbyshire* (London 1811–17); 'An Account of the Great Derbyshire Denudation', *Philos. Trans.* (1811), Pt. II, pp. 242–256; 'Geological Observations . . . on the Use and Abuse of Geological Theories', *Phil. Mag.*, XXXVII (1811), pp. 440–443.
43. *Philos. Trans.* (1811), Pt. II, p. 245.
44. *Philos. Trans.* (1808), pp. 187–222; John Dubourdieu, *Statistical Survey of the County of Antrim*, Appendix II (Dublin 1812).
45. Dubourdieu, *op. cit.* (1812), Appendix II, pp. 41 & 43.
46. *The Old Red Sandstone*, Chap. II.
47. *Phil. Mag.*, XXXIII (1809), pp. 257–263.
48. Henry Holland, *Recollections of Past Life*, p. 81 (London 1872).
49. Robert Bakewell, *An Introduction to Geology*, p. 437 (London 1828).
50. *J. Asiat. Soc., Calcutta*, I (1832), pp. 238–242.

51. *Phil. Mag.*, Ser. 4, V (1853), pp. 258–281.
52. *Mem. geol. Surv. India*, I (1859), pp. 173 & 174.
53. Macculloch, *op. cit.* (1831), II, p. 53.
54. John Macculloch, *Proofs and Illustrations of the Attributes of God*, I, Chap. VI (London 1837).
55. *Trans. roy. Soc. Edinb.*, VII (1815), pp. 138–211. See also *ibid.*, V (1805), p. 68f; VI (1812), pp. 171 & 172.
56. *Outlines of the Geology of England and Wales*, p. xxiii (London 1822).
57. *Edinb. Rev.*, XXXIX (1823), p. 227.
58. Bakewell, *op. cit.* (1828), p. 472.
59. *Illustrations of the Geology of Yorkshire*, Pt. I, p. 43 (London 1835).
60. *Passages in the History of Geology*, pp. 33 & 34 (London 1848).
61. *The Quarterly Review*, XXXVI (1827), p. 477.
62. *Proc. geol. Soc.*, I (1828–29), No. 9, pp. 89–91; *Edinb. New Philos. Journ.*, VII (1829), pp. 15–48.
63. Scrope, *op. cit.* (1827), p. 164 f.
64. *Considerations on Volcanos*, p. 213 f (London 1825).
65. *Proc. geol. Soc.*, I (1829–30), No. 14, pp. 170 & 171.
66. *Geol. Mag. Lond.*, III (1866), pp. 193–199; 241–243; 331–333; 379–380.
67. See George Greenwood, *River Terraces*, pp. ix–xv (London 1877).
68. *The Tree-Lifter* (London 1853).
69. *Ibid.*, pp. 184 & 185.
70. *Ibid.*, p. 182.
71. *Rain and Rivers*, p. 121 (London 1857).
72. *Ibid.*, p. 102.
73. *Scot. geogr. Mag.*, LXXVI (1960), pp. 108–110.
74. *The Geology and Extinct Volcanos of Central France*, p. 208 f (London 1858).
75. V. A. Eyles, 'John Macculloch, F.R.S., and his Geological Map', *Ann. Sci.*, II (1937), pp. 114–129.
76. Katharine M. Lyell, *Memoir of Leonard Horner*, I, p. 174 (London 1890).
77. *A Description of the Western Islands of Scotland, including the Isle of Man*, I, p. 242 (London 1819).
78. Macculloch, *op. cit.* (1831), II, p. 36.
79. *The Highlands and Western Isles of Scotland*, I, p. 326 (London 1824).
80. *A Geological Essay on the Imperfect Evidence in Support of a Theory of the Earth*, p. 187 (Oxford 1815).
81. *Facts and Observations towards forming a new Theory of the Earth* (Edinburgh 1819).
82. Bakewell, *op. cit.* (1828), p. 421.
83. *Proc. geol. Soc.*, I (1831), No. 20, pp. 303 & 304.
84. *Geological Travels*, I, pp. 10–54 (London 1810).
85. *History of the Inductive Sciences*, III, p. 505 (London 1857).
86. *Geology of the Lake District in Letters addressed to W. Wordsworth Esq.*, Letter 1.
87. *Ann. Philos.*, N.S. X (1825), pp. 18–37.
88. *Edinb. New Philos. Journ.*, XXVIII (1840), pp. 32–42.
89. Lyell, *op. cit.* (1881), I, pp. 45 & 46.
90. *A Geological Survey of the Yorkshire Coast*, p. 285 (Whitby 1822).

91. *Quart. J. geol. Soc. Lond.*, LVIII (1902), pp. 471–571.

92. Lyell, *op. cit.*, (1881) II, p. 192.

93. *Proc. geol. Soc.* I (1829), No. 12, pp. 145–149.

94. Bakewell, *op. cit.* (1828), p. 330.

95. *Trans. geol. Soc. Lond.*, N.S. II (1829), pp. 119–130.

96. *Trans. Camb. phil. Soc.*, VI (1838), pp. 1–84.

97. *Trans. geol. Soc. Lond.*, N.S. VII (1845–56), pp. 1–51.

98. *Quart. J. geol. Soc. Lond.*, IV (1848), pp. 70–98.

99. *The Geology of Russia in Europe*, I, p. 345 (London 1845).

100. *Ibid.*, I, p. 345.

101. *The Principle of Uniformity in Geology, Biology and Theology* (Leiden 1963).

102. *Phil. Mag.*, XXXIII (1809), pp. 385–389 and 452–459; XXXIV (1809), pp. 19–26 and 190–200.

103. *Philos. Trans.*, CLIV (1864), pp. 247–309.

104. Westgarth Forster, *A Treatise on a Section of the Strata, commencing near Newcastle upon Tyne*, pp. 139–142 (Newcastle 1809).

105. One of the earliest English accounts of the disaster is in *Edinb. Philos. Journ.*, I (1819), pp. 187–191.

106. *Travels through the Alps of Savoy*, p. 263 (Edinburgh 1845).

107. *Researches in Theoretical Geology*, p. 172 (London 1834).

108. Henry T. De La Beche, *The Geological Observer*, pp. 46–49 (London 1851).

109. *Proc. geol. Soc.*, I (1831), No. 22, pp. 331 & 332.

110. *Ibid.*, II, 1836–37, No. 48, pp. 435–437; *Journ. geol. Soc. Dublin*, I (1838), pp. 266–276.

111. *Trans. roy. Soc. Edinb.*, VII (1815), pp. 138–211.

112. National Library of Scotland MSS. 6329–6332.

113. Lyell, *op. cit.* (1881), I, p. 253.

114. *Ibid.*, I, p. 276.

115. *Ibid.*, I, p. 253; *Proc. geol. Soc.*, I (1831), No. 20, pp. 313 & 314.

116. *Quart. J. geol. Soc. Lond.*, IV (1848), pp. 70–98.

117. *Ibid.*, III (1847), pp. 227–232.

118. *Trans. roy. Soc. Edinb.*, XVI (1849), pp. 23–32.

119. *Proc. geol. Soc.*, III (1840), No. 67, p. 171.

120. *Quart. J. geol. Soc. Lond.*, XIV (1858), pp. 322–335.

121. *Philos. Trans.* (1839), Pt. I, pp. 39–81.

122. Darwin, *op. cit.* (1846), pp. 68 & 69.

123. De La Beche, *op. cit.* (1851), p. 660 f.

124. *The Rivers, Mountains, and Sea-Coast of Yorkshire*, p. 11 (London 1853).

125. *Quart. J. geol. Soc. Lond.*, X (1854), pp. 231–240.

126. *Ibid.*, XII (1856), pp. 102–123.

127. Jukes, *op. cit.* (1853), p. 152.

128. *The Student's Manual of Geology*, pp. 89 & 92 (Edinburgh 1857).

Chapter Eight

The Glacial Key
1826-1878

Notice to Geologists. – At Pont-aber-glass-llyn, 100 yards below the bridge, on the right bank of the river, and 20 feet above the road, see a good example of the furrows, flutings, and striæ on rounded and polished surfaces of the rock, which Agassiz refers to the action of glaciers. See many similar effects on the left, or south-west, side of the pass of Llanberris.

> *William Buckland*
> *An entry made in the visitor's book at the Goat Hotel, Beddgelert, North Wales, on 16th October, 1841.*

THE arrival of the glacial theory was an event of enormous significance in the history of geomorphology. The discovery that the landscapes of northern Europe owed much of their present form to the activities of glacier ice was itself clearly a matter of the utmost importance, but the glacial theory brought in its train another development of scarcely less moment. Once the erosional work of glaciers had begun to be recognised, it was soon appreciated that many of the landforms which formerly had been adduced in evidence against the fluvial doctrine, were in reality the work of glacier ice. It was now seen that such landforms were in fact entirely irrelevant to the debate over the doctrine's validity. It slowly dawned that the doctrine had been condemned largely through a misunderstanding of the true nature of field phenomena. Above all, once the power of glaciers to excavate rock-basins was admitted, then that most effective of all the objections to fluvialism – the limnological objection – was at long last removed. This

263

relationship of the glacial theory to the fluvial doctrine has never been adequately appreciated by historians, and the glacial theory was indeed the key which in the 1860s opened the way to a fluvialistic interpretation of most of the world's landscapes.

De Luc, de Saussure, and many other eighteenth-century naturalists, had all recognised the ability of modern glaciers to transport debris, and de Luc had actually used the morainic accumulations around the snouts of the Swiss glaciers as one of his natural chronometers proving the Earth's novity. From such a recognition it was a comparatively simple step to the suggestion that debris lying far beyond the present glacial limits had been deposited by the swollen glaciers of some bygone era. A Swiss cleric, Bernard Kuhn, made this suggestion as early as 1787. He was followed by Hutton in 1795, by Playfair in 1802, and by the mountaineer Jean-Pierre Perraudin of Lourtier in the Val de Bagnes in 1815,[1] but it was not until the 1820s and 1830s that the glacial theory really began to make progress. Its development then took place in the minds of three men: Ignace Venetz, Jean de Charpentier, and Louis Agassiz.

Ignace Venetz was an engineer, and it was he who had the difficult task of trying to avert the disastrous debacle in the Val de Bagnes in 1818. In 1821 he prepared a memoir in which he timidly suggested that the snouts of the Alpine glaciers had once lain several miles downstream of their present positions. Unfortunately the memoir lay gathering dust for more than ten years and it was not published until 1833. In 1829, however, at a meeting of the *Société Helvétique des Sciences Naturelles* at the Hospice du Grand St. Bernard, Venetz read a paper in which he advanced a very much bolder thesis than that embodied in his original memoir. He now claimed that the former glaciation had been on such a grand scale that the Swiss glaciers had then extended from the Alps and across the Swiss plain, to impinge upon the Jura, where they deposited the innumerable Alpine erratics which in Venetz's day were the subject of so much puzzled attention. Further, he claimed that other erratics throughout northern Europe had been similarly transported by glaciers which have now completely disappeared. Surprisingly, Venetz's ideas evoked little response in the world at large, but he did succeed in capturing the interest of his friend Jean de Charpentier. De Charpentier was a graduate of the Freiberg Mining Academy, and a very successful director of the salt mines at Bex. It was he who first placed the glacial theory upon a secure scientific basis when in July

1834 he read a paper to the *Société Helvétique* at Lucerne on the subject of erratics as evidence of a former Alpine glaciation.

It was now that Louis Agassiz came upon the scene. He was only a young man aged twenty-nine, but his ichthyological researches had already given him an international scientific reputation. At first he was very sceptical about the glacial theory, but he was converted as a result of a field excursion with Venetz and de Charpentier around Bex and in the Valais during the summer of 1836, and he returned from the field a glacial enthusiast. The theory doubtless appealed to Agassiz because it allowed his mind to roam freely over the face of the globe, and because it permitted him to make those sweeping generalisations in which he indulged all too frequently. His emergence as a glacialist was neverthe-less an event of great significance. He brought to the aid of the theory his powers of quick perception, and his remarkable ability to interpret and classify facts, and in addition, his great scientific reputation ensured that the theory would henceforth receive the serious attention which had been denied it hitherto. Agassiz, collaborating with his brilliant but dissolute friend Karl Schimper, rapidly developed the ideas of Venetz and de Charpentier. Indeed, in his unbounded enthusiasm, he elaborated the theory far too rapidly for the liking of the theory's two pioneers. Understandably they resented what they regarded as Agassiz's forceful intrusion into glacial affairs, and the cavalier manner in which he seemed to have taken the glacial theory over as his own. As a result, Agassiz and his two former friends became estranged.

The outcome of the studies of Agassiz and Schimper was the concept of *die Eiszeit* – the Ice Age. The concept was first outlined in detail in Agassiz's famous *Discours de Neuchâtel*, written in one night in July 1837 and presented on the following day to a meeting of the *Société Helvé-tique* at Neuchâtel. In the *Discours* Agassiz discussed the distribution of Swiss erratics, moraines, and polished and striated rock surfaces. After demonstrating the inadequacy of the diluvial theory as an explanation of such phenomena, he went on to show that the presence of these features proves the one-time existence of Swiss glaciers far longer and thicker than the glaciers of the present day. Agassiz, however, like de Charpentier, believed that there was evidence of glaciers having former-ly existed far beyond the narrow confines of Switzerland. He threw scientific caution to the winds and boldly claimed that the entire northern hemisphere from the North Pole down to the latitude of the Mediterranean and Caspian seas had until recently been shrouded beneath a thick mass of glacier ice. This was his *Eiszeit* – not merely a local extension of the Swiss glaciers, but a full-scale ice-sheet glaciation of continental proportions.

Naturally, Agassiz was tempted to account for this recent refrigeration of the northern hemisphere, and here he has the doubtful distinction of having outlined what must be the most fantastic explanation of the glacial period ever offered. Almost with the doctrine of the macrocosm and the microcosm in mind, he claimed that just as the human body turns cold at death, so too does the earth periodically grow frigid as the various episodes in its life draw to their conclusion. Perhaps we should not be too severe on him for this flight of fancy, however, because even after almost a century and a half of scientific speculation, the cause of the glacial period is still scarcely less baffling than it was in 1837. The *Discours* was certainly premature in many respects, and Agassiz, like Hutton in 1785, was guilty of publishing a theory – albeit a pregnant theory – without first making a detailed study of the evidence upon which the theory was supposed to rest. Not until July 1838 did he go to the Bernese Oberland to begin his glacial researches in earnest, and it was 1840 before he proved that his intuition had been correct in leading him to postulate a massive glaciation of northern Europe. The results of his early Swiss studies were published in September 1840 as his beautifully illustrated *Études sur les glaciers*, printed in French and German editions, and dedicated, despite their quarrel, to the two men upon whose shoulders Agassiz had climbed – to Venetz and de Charpentier. The world at large, however, scarcely knew even the names of these two pioneers of the glacial theory. In the minds of geologists everywhere the theory was inseparably linked with the name of Agassiz himself, and by the close of 1840 his forceful championing of the theory had sparked off a major scientific controversy.

Early British Interest in the Glacial Theory

James Hutton, in the expanded version of his theory published in 1795, was the first British author to suggest that the Alpine glaciers had once extended far beyond their present limits. It will be remembered that he believed these extended glaciers to have existed at a period before denudation had reduced the Alps to their present level, and that he regarded the glaciers as the agents responsible for the dispersal of the Alpine erratics. In his *Illustrations* of 1802 Playfair adopted the same reasoning as his preceptor, and when he was able to visit the Alps at the

266

conclusion of the Napoleonic War, he seems to have paid particular attention to the erratic phenomena. One of the erratics he examined was a monster block of granite weighing 2520 tons lying near Neuchâtel, and he was evidently satisfied that the ponderous movement of a former glacier was the only force adequate for the transportation of such a block to its present site.[2] As we noted earlier, Hutton's theory concerning the cause of his postulated Alpine glaciation might equally well have been applied to other mountain ranges which have undergone great denudation. Hutton believed, for example, that the Southern Uplands of Scotland were merely the denuded stumps of very much higher mountains, and had he wished, he could easily have argued that the region's ancient peaks had been sufficiently lofty to nourish extensive snowfields and glaciers. Neither Hutton nor Playfair, however, ever carried their thinking this far. They never invoked glacial transport to explain the dispersal of erratics within the British Isles, and neither of them ever discussed the former glaciation of any region other than the Alps.

Not until the 1840s was it generally recognised that glaciers had once existed within the British Isles, and the introduction of the glacial theory to Britain is commonly attributed to Agassiz himself. It was in the autumn of 1840 that he toured the British Isles and made the earliest positive identification of the relics of former British glaciers. This tour was a momentous occasion in the history of British geomorphology, but in reality the priority of introducing the glacial theory into Britain does not rest entirely with Agassiz. Apart from Hutton and Playfair, there are three other British geologists whose pre-1840 writings deserve mention in the context of the glacial theory. These are Robert Jameson the Edinburgh Neptunist, William Buckland the Oxford diluvialist, and James Smith of Jordanhill, a very successful Glasgow merchant who combined a passion for geology with a taste for yachting.

Jameson has received little attention in the context of the glacial theory. There is nevertheless abundant evidence showing that he was keenly interested in the theory and in ice phenomena generally, and he was probably the first geologist to appreciate that glaciers had once existed within the British Isles. He never published anything bearing upon this subject himself, but in 1845 J. D. Forbes, one of the leading early glaciologists, recorded that as a student in Edinburgh in 1827 he had heard Jameson refer to the erratic phenomena of Scotland 'as perhaps requiring to be explained by the former presence of glaciers'.[3]

Evidence to corroborate Forbes's statement is to be found among the *Jameson Papers* preserved in Edinburgh University. The papers include a manuscript syllabus for one of Jameson's lecture courses, and it reveals that he lectured upon the following subject:

> Proofs of former glaciers in countries where they are no longer met with. Norway, Scotland, etc.[4]

The syllabus is only a rough note, and even the most pedantic reader will surely forgive Jameson for including Norway among the countries which are today devoid of glaciers; it was clearly his intention to focus upon those regions of Norway which today contain no glaciers, but which abound with the relics of former glaciation. Unfortunately, the syllabus is undated, but for reasons to be explained below, the prominence given to the Norwegian glacial relics suggests that the document was prepared after 1825. No firm terminal date for the compilation of the syllabus can be offered but, it seems likely that it was written no later than 1836. Thus at some period between 1825 and 1836 Jameson was evidently teaching his classes that Scotland, and other northerly regions, had recently been held in the grip of glacier ice. In view of the novelty of his views in this field, it is with eager anticipation that we turn to the edition of Cuvier's theory of the Earth which Jameson published in 1827. Only disappointment awaits us. Jameson in print was evidently much more circumspect than Jameson in the lecture-theatre, and in the notes which he wrote to accompany Cuvier's theory he invoked a debacle to explain the transport of the Alpine erratics to the Jura, while he attributed the dispersal of the more northerly European erratics to the ice-rafting of debris during a general submergence.[5]

Although Jameson chose to express thoroughly orthodox views upon the subject of erratics in 1827, the pages of *The Edinburgh New Philosophical Journal*, which he edited for almost thirty years, make it abundantly clear that his interest in glacial phenomena never waned. He published a large number of papers and notes on snow, ice, glaciers, frozen ground, and other related topics, and in 1843 he admitted that it was his continuing policy to give his readers as much material as possible relating to past and present glacial regions. To this end he was clearly at some pains to persuade continental glaciologists to write for the *Journal*, and he was equally punctilious in obtaining and publishing translations of important relevant papers which had originally appeared in foreign periodicals. The first important fruit of this latter policy appeared as early as 1826 when Jameson published a paper by the Danish mineralogist Jens Esmark, the paper having originally been printed in a Norwegian journal.[6] In the paper Esmark used the evidence provided by erratics, perched blocks, and moraines to prove that Norway had recently undergone a widespread glaciation, and he even claimed to see the handiwork of former glaciers in the subdued landscapes of his native Denmark. This paper was evidently the first work published in Britain to expound the theory that large areas of northern Europe had until recently been submerged beneath glacier-ice, and to Jameson must go the credit for giving the paper a British airing. It seems likely that it was the existence of this paper which caused Jameson to give prominence to the glaciation of Norway when he came to draw up the lecture syllabus mentioned earlier.

Almost ten years elapsed before Jameson was able to offer his readers another really important paper on a glacial topic, but between October 1836 and October 1839 he made up for this earlier dearth of material by publishing no less than five major papers dealing with past or present glaciers. The first of these papers appeared in October 1836 and was a translated and amplified version of the paper which de Charpentier had presented to the *Société Helvétique* at Lucerne in July 1834.[7] In its English version the paper is a carefully reasoned exposition of the thesis that the Swiss tills and erratics could only have been distributed by former extensive glaciers, and de Charpentier effectively demonstrated the absurdity of all theories attributing such deposits to the work of deluges or debacles. The paper is concerned chiefly with the glaciation of Switzerland, but while de Charpentier had no conception of an ice-sheet glaciation of Europe as a whole, he did go so far as to admit that erratics occurring in mountainous regions other than the Alps might also have been transported by former valley glaciers. He regarded his glacial transportation theory as applicable to erratics

> . . . generally in the valleys and at the base of all *high* mountain chains, with the exception of those which lie in equatorial districts, and on which the masses of perpetual snow cannot be converted into glaciers.

The second important paper appeared in *The Edinburgh New Philosophical Journal* in January 1837 and also came from the pen of de Charpentier.[8] It was a discussion of the evidence proving that the Alpine glaciers had once extended across the Swiss plain to rest upon the flanks of the Jura. To the paper Jameson himself added a footnote which clearly harks back to Esmark's paper of 1826:

> The great moraines in valleys in Norway, where now no glaciers occur, are here deserving of notice.

Next, in January 1838, Jameson published a paper by Agassiz himself on the subject of the glacial transportation of erratic material from the Alps to the Jura,[9] and in April of the same year Jameson offered his readers his pièce de résistance – an English translation of Agassiz's *Discours de Neuchâtel*.[10] The *Actes de la Société Helvétique réunie à Neuchâtel* in which the *Discours* originally appeared was printed in only limited numbers, and historians have therefore commonly supposed that Agassiz's paper was not widely read. In fact, however, thanks to Jameson's interest and enterprise, the *Discours* must have reached a very wide readership through the pages of the Edinburgh *Journal*. The last of the five important papers was also from Agassiz's pen, and it appeared in Jameson's *Journal* in October 1839.[11] The paper was a translation of a memoir which Agassiz had delivered to the *Société géologique de France* on the subject of the former extent of glaciers, and in it Agassiz confidently reaffirmed his belief that the whole of Europe had recently lain beneath a gigantic ice-sheet.

Thus between 1836 and 1839 Jameson gave British geologists the opportunity to read accounts of the glacial theory written by two of its leading exponents. Many British geologists had used *The Edinburgh New Philosophical Journal* as a vehicle for the announcement of their own discoveries, and we must therefore presume that the journal was widely read in geological circles. For some reason, however, Jameson's efforts to arouse his readers' interest in the glacial theory met with no success. This fact is difficult to understand because British geologists were already keenly aware of many of the problems presented by the Earth's drift mantle. It has been suggested that their ice-blindness was a result of there being no glaciers available for study within the British Isles, but this is hardly a valid excuse. Most of the leading British geologists had made pilgrimages to the Alps, and in consequence they were thoroughly familiar with glaciers and glaciated landscapes. Perhaps a more realistic reason for the early British neglect of the glacial theory was the conceit of the British geologists. It was generally admitted that they led the world in almost every branch of the subject, and it may be that they were slow to appreciate that important new geological theories were just as likely to emerge from remote Swiss valleys as from the assembly rooms of the Geological Society of London. Whatever the reason for the neglect of the glacial theory in Britain between 1836 and 1840 may have been, there is one charge which cannot justly be levelled against the British geologists of the period; thanks to Jameson's initiative, they cannot be accused of having been ignorant of the work of the Swiss glacial geologists.

Apart from Jameson, William Buckland seems to have been the only other British geologist who devoted much attention to the glacial theory during the 1830s. Buckland and Agassiz entered into correspondence early in the decade, and they became close friends during Agassiz's first visit to Britain in 1834. Their friendship was renewed when Agassiz returned to Britain in 1835, and in the autumn of 1838 Buckland, accompanied by his wife, visited Agassiz in Switzerland in order to examine some of the evidence upon which the glacial theory rested. Buckland, it will be remembered, had renounced his old style diluvialism about 1829. Although he was now nominally a neo-diluvialist, he was evidently not entirely happy with his new faith, and he was anxious to see whether the glacial theory offered a more acceptable explanation of the drift deposits in which he was so keenly interested. At first the Swiss evidence left him unmoved. An excursion with Agassiz into the Jura near Neuchâtel to see the erratics and polished and striated rocks did nothing to convince the Oxford geologist of the former presence of glaciers. Later, during a tour of the Bernese Oberland – the very district which Agassiz himself had explored a few weeks earlier – Mrs Buckland wrote to Agassiz:

> We have made a good tour of the Oberland, and have seen glaciers, etc., but Dr. Buckland is as far as ever from agreeing with you.[12]

Soon after this letter was written, Buckland began to waver, and a few days later he became a convert to the glacial theory.[13] He was now convinced that Agassiz was correct in believing that the Alpine glaciers had once extended far downstream of their present snouts.

As he journeyed from Rosenlaui and Grindelwald back to Neuchâtel, Buckland began to draw mental parallels between the glacial relics which surrounded him and those similar British features which he had for so long attributed to diluvial action. At Neuchâtel, he described some of these British features for the benefit of Agassiz. Buckland recollected that in 1811 he had seen striking examples of moulded and

271

polished rocks at Dunkeld in Perthshire, and that in 1824, when in company with Lyell, he had found some beautiful sets of striations on the eastern flanks of Ben Nevis. Similarly, he remembered the striations which he had found by following Sir James Hall's footsteps around the Edinburgh region in 1824 and 1834, and the puzzling feature which he had seen about 1824 on the slopes of Ben Wyvis overlooking the Cromarty Firth – a feature which he now suspected of being a glacial moraine.[14]

As he admitted two years later, Buckland returned from his Alpine excursion of 1838 convinced that the British Isles had undergone a recent glaciation, but, strangely, it was evidently not until 1840 that he began to re-examine the British landscape in the light of his Alpine experience. In that year, on his way north to attend the Glasgow meeting of the British Association, he decided to pause in Dumfries to seek for the relics of former British glaciers. His search was soon rewarded. Just to the east of Thornhill, in Crichope Linn, he found a ridge 20 to 30 feet in height looking like the vallum of some ancient encampment, and consisting of unsorted material from the nearby slates of the Lowther Hills, together with a few granite erratics from Loch Doon lying 25 miles away to the west. This, Buckland tentatively concluded, must be a glacial moraine, but he knew that only a few days later he was to join forces with Agassiz himself, and Buckland was content to leave the first positive identification of a British glacial landform to Agassiz's more experienced eye.

The third member of our trio of geologists of interest in connection with the early history of the glacial theory in Britain is James Smith of Jordanhill. As an enthusiastic yachtsman and an equally enthusiastic geologist, Smith combined the helm with the hammer by making a detailed study of the coastlands of the Firth of Clyde, and his discoveries in the region's estuarine clays have led to his being hailed as 'the father of the post-Tertiary geology of Scotland'. His relationship to the glacial theory is less intimate than that of either Jameson or Buckland, but his researches did provide the theory with significant corroborative evidence. Smith enters the story in 1835 when his friend Lord John Campbell, later the seventh Duke of Argyll, was building a new sawmill on a marine foreland lying at Ardencaple near Helensburgh, at the entrance to the Gare Loch.[15] While excavating for the mill's foundations, Lord Campbell's labourers exposed a stiff blue clay containing marine shells. He invited his geological friend along to examine the deposit, and

Smith soon showed that the clay contained an arctic fauna. Smith reported his findings to the Geological Society of London in April 1839,[16] and he thus provided sound palaeontological backing for the theory that the British Isles had recently experienced conditions very much colder than those prevailing today. The scene was now set for the appearance in Britain of the glacial theory's leading proponent – Louis Agassiz himself.

Agassiz comes to Britain

During his discussions with Agassiz in Switzerland in 1838 Buckland evidently suggested that Agassiz should visit Britain in order to examine some of the many features which the Oxford geologist now suspected of being glacial relics. Further, it was agreed that if Agassiz did visit Britain, Buckland himself would serve as his travelling companion and guide. Agassiz spent the summer of 1839 completing his Swiss investigations, but by 1840 he felt ready to accept Buckland's invitation. He explained the motives underlying his visit to Britain in the following passage:

> After having obtained in Switzerland the most conclusive proofs, that at a former period the glaciers were of much greater extent than at present, nay, that they had covered the whole country, and had transported the erratic blocks to the places where these are now found, it was my wish to examine a country where glaciers are no longer met with, but in which they might formerly have existed.[17]

In fact, by visiting Britain he hoped to find some traces of the great European ice-sheet which he had invoked three years earlier in his *Discours de Neuchâtel*, but of which he as yet had little real evidence.

As if to herald his forthcoming British visit Agassiz prepared a paper 'On the polished and striated surfaces of the rocks which form the beds of Glaciers in the Alps' which was one of eleven communications presented to the Geological Society of London at its meeting on 10th June 1840.[18] On that date Agassiz was still in Switzerland, and although the Geological Society's records shed no light upon the subject, it was presumably Buckland, the president, who communicated the paper to the society. In the paper Agassiz described the striations and polished surfaces which his investigations had shown to be so widespread in Switzerland. He argued very effectively that such phenomena could not possibly have resulted from the action of avalanches, debacles or a

273

deluge, and he presented his own conclusion that both the striations and the polish were the result of abrasion effected by quartz grains trapped beneath the swollen glaciers of the glacial period.

Agassiz, still continuing his glacial researches, remained in Switzerland until later in the summer. On 20th August he wrote the Preface to his *Études sur les Glaciers* at the Hospice de Grimsel, but a few days later he set off for England. There his quest for former British glaciers was at first not taken very seriously, and he later wrote of his reception:

> Even when I arrived in England, many of my friends would fain have dissuaded me from my expedition, urging me to devote myself to special zoölogical studies, and not to meddle with general geological problems of so speculative a character.[19]

Soon after his arrival in Britain he set off for Scotland in order to attend the meeting of the British Association which opened in Glasgow on Thursday 17th September. On his way north he visited Sir Philip Egerton at Oulton Park in Cheshire, and from there Agassiz wrote to Roderick Murchison on 8th September:

> Je regrette infiniment de ne vous avoir pas vu à la réunion de la Soc. Géol. J'espère avoir encore une discussion avec vous sur le sujet de glaciers, car l'affaire est réellement trop importante pour être traitée legèrement.[20]

Further to the north Agassiz probably travelled over the same roads that Buckland had traversed a few days earlier, but there is no record of Agassiz having made any glacial observations while en route to Glasgow. He must nevertheless have dallied somewhere on the way because unlike Buckland, who is known to have been in Glasgow at least as early as 18th September, Agassiz evidently missed the opening days of the Association's meeting.[21]

Agassiz reached Glasgow on or before 21st September, because on that date he presented the first of the four communications which he made to the Association that year. Only one of these communications needs claim our attention – that which he read to Section C(Geology) on Tuesday 22nd September. This was a lengthy paper, which he delivered in French, entitled 'On Glaciers and Boulders in Switzerland'. In the paper Agassiz discussed the ability of glaciers to polish and striate rocks, and he reminded his audience that such markings, together with moraines, are widespread far beyond the limits of modern ice action. This proves, he observed,

. . . that at a certain epoch all the north of Europe, and also the north of Asia and America, were covered with a mass of ice. . . .

Strangely, as if to placate Murchison and the other diluvialists and neo-diluvialists in his audience, Agassiz sought to show that belief in the glacial theory was by no means incompatible with belief in a recent catastrophic submergence. He explained that after the melting of the glaciers, the newly released waters had inundated the continents, and that during this submergence marine currents had imported boulders and other debris and deposited them widely over the continental surfaces. This idea was evidently an attempt to account for the bedded and well sorted gravels – now known to be glaciofluvial in origin – which are commonly associated with the true morainic deposits, and the concept of a glacial submergence was soon to find widespread acceptance in Britain.

A summary of Agassiz's paper was published in the official report of the Glasgow meeting,[22] but this abstract credits Agassiz with no specific reference to the former existence of glaciers within the British Isles. It is therefore interesting to find the following prescient passage in the report of the meeting published in *The Athenæum*:

> Prof. Agassiz is also inclined to suppose that glaciers have been spread over Scotland. . . . If we understood him rightly, he means to follow up his valuable researches in the Highlands of Scotland during his stay in this country, and that he confidently expects to find evidence of such glaciers having existed, particularly around Ben Nevis.[23]

Thus on 22nd September 1840 Agassiz became the first geologist to make a confident assertion that the British Isles had indeed experienced a recent glaciation.

After reading his paper Agassiz displayed and explained the fine plates which were to accompany his forthcoming *Études sur les Glaciers* (Plate VII), and there followed a lively discussion in which Charles Lyell, Henry De La Beche, and Murchison participated. Murchison briefly described the occasion for the benefit of the absent Adam Sedgwick in a letter dated 26th September 1840:

> Agassiz gave us a great field-day on Glaciers, and I think we shall end in having a compromise between himself and us of the floating icebergs. I spoke against the general application of his theory.[24]

Interestingly, but very appropriately, Agassiz's chairman at this meeting of Section C was James Smith of Jordanhill. On the first day of the Glasgow meeting Smith had given a second reading of the paper on the arctic fauna of the Clyde estuarine clays which he had originally presented to the Geological Society of London in April of the previous year. Because of his late arrival in Glasgow, Agassiz must have missed hearing the paper delivered, but he was doubtless happy to know that the Glasgow region afforded clear palaeontological evidence supporting his belief in a British Ice Age.

Active though Agassiz was in the proceedings of the Glasgow meeting, not all his time was spent in the sectional assembly-rooms. He devoted many hours to combing the city and its environs for evidence of glacial action, and his search was soon crowned with success. He recorded that

> [I] had scarcely arrived in Glasgow, when I found remote traces of the action of glaciers. . . .[25]

Two years later he wrote of his days in Glasgow:

> There is no locality in which I have been able to study the till more completely than at Glasgow, where the numerous works carried on in 1840 for the embellishment of the town had exposed it at many points; but everywhere it presents the same characters; the rounded, polished, and scratched blocks of very various dimensions, are every where indiscriminately mixed together in a marly or clayey paste.[26]

It was thus in and around Glasgow that Agassiz first identified the work of former British glaciers.

The British Association adjourned on 23rd September, and Agassiz, now accompanied by Buckland, headed northwards into the Scottish Highlands (Figure 2). Buckland's tentative identification of a moraine near Dumfries, coupled with Agassiz's discoveries in Glasgow, must have given the two men high hopes of further exciting glacial finds amidst the Scottish mountains. The search for glacial relics was nevertheless not the sole objective of the tour; Agassiz was anxious to visit the Moray Firth to see the Old Red Sandstone from which Hugh Miller, the former Cromarty stone-mason, had extracted such a wonderful wealth of fossil creatures. From Glasgow the two friends proceeded north-westwards to Loch Lomond, and they followed the shores of the loch as far as Tarbet. There they turned westwards to Arrochar, passed around the head of Loch Long, and followed the road through Glen Croe across to Loch Fyne. Strangely, they seem to have identified

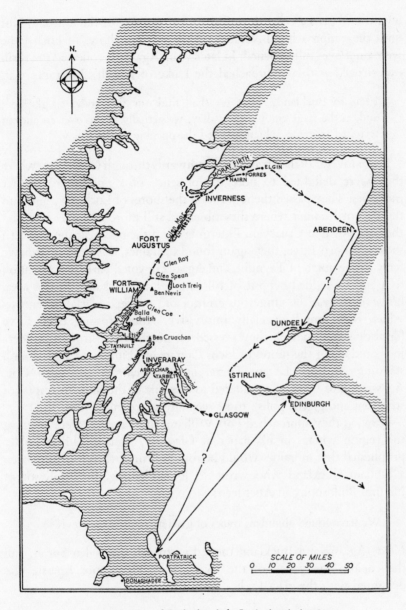

Fig. 2 *The Tour of Scotland made by Louis Agassiz in 1840*

277

few, if any glacial relics during this early portion of the journey. Not until they approached Inveraray on the western shores of Loch Fyne were their eyes fully opened. In later years Agassiz recollected the thrill experienced as they approached the Duke of Argyll's Inveraray seat.

> I said to Buckland: 'Here we shall find our first traces of glaciers'; and, as the stage entered the valley, we actually drove over an ancient terminal moraine, which spanned the opening of the valley.[27]

From Inveraray they travelled northwards through Glen Aray, where they were delighted to note the presence of a number of further moraines. They passed thence around the shores of Loch Awe and into the Pass of Brander where they observed still more moraines lying on the flanks of Ben Cruachan. Emerging from the pass near the southern shores of Loch Etive, they again found moraines, this time at Bonawe ferry to the north of Taynuilt, and Agassiz devoted particular attention to a rudely bedded deposit of till exposed at Muckairn. In this area, however, it was the numerous erratics and polished glacial pavements which they found especially impressive. After crossing Loch Etive (doubtless at Bonawe), they pressed on northwards through Benderloch, and along the shores of lochs Creran and Linnhe, to Ballachulish standing on Loch Leven. There, near the entrance to Glen Coe, Agassiz found what he considered to be some of the finest examples of striations and glacial polish to be seen anywhere in Scotland.

Now, as they approached Fort William, the travellers were entering the region where, in his paper at Glasgow, Agassiz had so boldly prophesied that moraines would be found. It must therefore have been a jubilant Buckland who wrote to John Fleming, the Professor of Natural Philosophy in Aberdeen:

> We have found abundant traces of glaciers round Ben Nevis.[28]

Next, Agassiz and Buckland moved eastwards into Glen Spean, and thence into the valley of the river Treig. Here, too, Agassiz was delighted with the glacial relics.

> I shall never forget the impression I experienced at the sight of the terraced mounds of blocks which occur at the mouth of the valley of Loch Treig, where it joins Glen Spean; it seemed to me as if I were looking at the numerous moraines of the neighbourhood of Tines, in the valley of Chamounix.[29]

It was at the entrance to the Treig valley that Agassiz found his most choice examples of Scottish roches moutonnées.

In this area, one natural phenomenon above all others demanded their attention – the Parallel Roads of Glen Roy. The Roads were then regarded as one of the chief geological curiosities of the British Isles, and since their first description by Thomas Pennant in 1776,[30] no other morphological feature in Britain had been the object of so much puzzled attention. Many varied explanations seeking to account for the phenomenon had appeared in print, and when Agassiz and Buckland visited the glen, the most recent attempt at explanation belonged to Charles Darwin. Only the previous year, in a memoir presented to the Royal Society – a memoir which he later ruefully regretted as being 'one long gigantic blunder from beginning to end'[31] – Darwin had claimed the Roads as former marine strand-lines. Familiar as he was with the Märjelensee, Agassiz immediately recognised the true explanation of the Parallel Roads; they are the successive shore-lines of a former ice-dammed lake. It was in a letter written on 3rd October at Fort Augustus, and addressed to Jameson in Edinburgh, that Agassiz first announced his brilliant solution to the problem. The letter was intended for inclusion in *The Edinburgh New Philosophical Journal*, but the next number of the *Journal* had already gone to press, and Jameson therefore passed the letter on to his friend Charles Maclaren, the editor of *The Scotsman*. Maclaren was himself a keen geologist, and his publication of Agassiz's letter on 7th October gave *The Scotsman* something of a journalistic scoop;[32] the paper was the first publication to break the startling news that relics of former Scottish glaciers had now been positively identified. The following is an extract from the published letter:

> . . . at the foot of Ben Nevis, and in the principal valleys, I discovered the most distinct *moraines* and polished rocky surfaces, just as in the valleys of the Swiss Alps, in the region of existing glaciers; so that the existence of glaciers in Scotland at early periods can no longer be doubted. The parallel roads of Glen Roy are intimately connected with this former occurrence of glaciers, and have been caused by a glacier from Ben Nevis. The phenomenon must have been precisely analogous to the glacier-lakes of the Tyrol, and to the event that took place in the valley of Bagne.

From Fort Augustus, Agassiz and Buckland travelled along Glen Mor towards the Moray Firth, where Murchison joined them in an examination of the Old Red Sandstone. Hugh Miller himself was evidently not present; perhaps his newly undertaken duties as editor of the *Witness* prevented his leaving Edinburgh.

The two travellers now began their return journey. They passed through Inverness and along the southern shores of the Moray Firth via Nairn, Forres, and Elgin, to Aberdeen. There they examined a cluster of moraines lying to the north of the city, and there Agassiz and Buckland evidently parted company. Buckland continued his journey via Stonehaven, Fordoun and Brechin into Strathmore where he visited Charles Lyell at Kinnordy House, the Lyells' Scottish home near Kirriemuir. There he stayed for a few days before continuing on through Blairgowrie, Dunkeld, Logierait, and Aberfeldy to join a house party at Taymouth Castle, the seat of the Marquis of Breadalbane, who had been President of the British Association for the Glasgow meeting. Agassiz's route southward from Aberdeen is less easy to discover. He evidently avoided Strathmore and travelled to Dundee either by sea or along the coastal road through Montrose. He certainly visited Stirling, and of this region he later wrote:

> Among the localities of Scotland where the indications of glacial action are most marked is the region about Stirling. Near Stirling Castle the polished surfaces of the rocks with their distinct grooves and scratches show us the path followed by the ice as it moved down in a northeasterly [*sic. recte* south-easterly] direction toward the Frith of Forth from the mountains on the north-west.[33]

His Highland tour completed, Agassiz now crossed to Ireland on the steam-packet which then plied between Portpatrick and Donaghadee, and he went to stay at Florence Court near Enniskillen, the seat of the Earl of Enniskillen. There he received news of an important convert to the glacial theory in a letter from Buckland dated at Taymouth Castle on 15th October.

> ... Lyell has adopted your theory *in toto* ! ! ! On my showing him a beautiful cluster of moraines, within two miles of his father's house, he instantly accepted it, as solving a host of difficulties that have all his life embarrassed him.[34]

The conversion had taken place during the course of an excursion into nearby Glen Prosen made while Buckland was staying at Kinnordy House. In the same letter Buckland advised Agassiz to examine one of the Irish eskers.

There are great reefs of gravel in the limestone valleys of the central
bog district of Ireland. They have a distinct name, which I forget. No
doubt they are moraines; if you have not, ere you get this, seen one of
them, pray do so. But it will not be worth while to go out of your way to
see more than one; all the rest must follow as a corollary.[35]

One of the attractions which drew Agassiz to Ireland was the large
collection of fossil fish maintained at Florence Court by his friend
Lord Cole, the third Earl, but Agassiz was also eager to search for relics
of Irish glaciers. These he found in abundance. In the immediate vicinity
of Florence Court he discovered what he regarded as his best Irish
examples of polished and striated limestones, while nearby, around the
flanks of Cuilcagh Mountain, he saw a series of moraines 'more
distinct than any that I have seen in the United Kingdom'. His Irish
observations were by no means confined to Ulster, although his Irish
itinerary remains obscure. In his writings he claimed to have searched
for glacial features in 'the north, centre, west and south-west of Ireland',
and he suggested, evidently as a result of field observations, that the
mountains of Antrim, Wicklow, and the west of Ireland had all been
ice-centres from which erratic material had been dispersed onto the
surrounding lowlands. He certainly visited Dublin, because he identified what he believed to be moraines lying to the south-east of the city,
and he evidently travelled along the road from Dublin to Florence
Court because he noted the presence of striated rocks near Virginia, Co.
Cavan. Later in the year, after his return to Switzerland, and in a letter
to Alexander von Humboldt, Agassiz wrote of his observations in the
Irish midlands:

> . . . j'ai observé mes surfaces polies et stries *jusqu' au niveau de la mer*
> sur toute la plaine qui s'abaisse d'Enniskillen vers Dublin; là les stries
> sont dirigées du N.O. au S.E. . . .[36]

In view of the thick blanket of drift which obscures the limestone over
most of the Irish midlands, one cannot help but feel that he was here
presenting Humboldt with a rather misleading impression of the extent
of the Irish glacial pavements.

The visit to Ireland lasted less than two weeks, and on 24th October
Agassiz arrived in Edinburgh.[37] There he devoted his time to the
examination of fossil collections, to searching for traces of ice-action,
and to social calls, although he evidently refused all dinner invitations
'owing to his late fatigues &c.' On 27th October a small group of

Edinburgh geologists, including Maclaren of *The Scotsman,* took Agassiz southward from the city on a search for glacial markings. Firstly they went to Blackford Hill, where initially Agassiz was sceptical about some of the features which were shown to him as possible striations. When they arrived on the southern side of the hill, however, near the quarry overlooking the Braid Burn, they discovered clear glacial markings on the concave surface of the rocks forming a shallow grotto. There it was a confident and delighted Agassiz who exclaimed 'That is the work of ice'.[38] The section displayed was of some significance because in such a sheltered position the markings could not possibly have been cut by the stranded icebergs which many earlier writers had invoked in explanation of all such phenomena. The site of this discovery has for long been famous and it is today marked by a commemorative plaque bearing the following inscription:

Agassiz Rock
In 1840
Louis Agassiz Swiss Geologist
stated that this rock was
polished and grooved by
ice during the great ice age.

From Blackford Hill the party went north-westwards to Corstorphine Hill in order to inspect the scratches which Sir James Hall had described in 1812, and here too Agassiz was satisfied that he was seeing true glacial striations. On the following day, the search for glacial relics continued when Jameson himself conducted Agassiz to Calton Hill to see what Jameson suspected of being glacial markings. Afterwards Jameson noted in his diary that Agassiz was

. . . much gratified by the display which he considered as very characteristic of the former existence of Glaciers there.

On 29th October Agassiz was preparing to leave Edinburgh for his return to England. On his way south he passed through Wallington in Northumberland, and he then cut across the Pennines via Haydon Bridge and Alston to Penrith. This was the route which Buckland had followed on his southward journey a few days earlier, and as he travelled along the eastern margin of the Lake District, Agassiz was able to confirm his friend's glacial observations and inferences. In this area he observed moraines at Penrith and Kendal, and striations and roches moutonnées between Shap and Kendal. He also noted that the Shap granite had been dispersed radially by the region's former glaciers. He continued on southwards through Lancashire where, without giving any precise locations, he claimed to have seen some fine examples of glacially polished limestones.

Agassiz evidently arrived back in London on or before 4th November because on that day his paper entitled 'Glaciers, and the Evidence of their having once existed in Scotland, Ireland, and England' was presented to the Geological Society of London.[39] In the paper he again discussed the features to be seen in the vicinity of the modern Swiss glaciers – erratics, polished and striated rocks, roches moutonnées, moraines, and spreads of till – and he now declared himself completely satisfied that precisely the same suite of features was to be found in Britain. Such features, he claimed, could never have been formed by diluvial activity; they are, rather, conclusive proof

> . . . that great masses of ice, and subsequently glaciers, existed in these portions of the United Kingdom at a period immediately preceding the present condition of the globe. . . .

Here, with some insight, Agassiz was recognising that the glaciation had consisted of more than one stage. He believed that during the earliest glacial phase the British Isles had been submerged under an ice-sheet (he termed it a 'nappe') such as today covers Greenland, and that at some later stage the British lowlands had been ice-free while the last relics of the former ice-sheet persisted in the mountain regions in the form of valley glaciers. With this in mind he was careful to emphasise that in continental Europe

> . . . the glaciers did not advance from the Alps into the plains, but that they gradually withdrew towards the mountains from the plains which they once covered.

In the British Isles he specifically mentioned the mountains of the Grampian highlands, Ayrshire, the Lake District, Northumberland, Wales, Antrim, the west of Ireland, and Wicklow as the last redoubts from which the vestigial valley glaciers had sallied forth onto the adjacent lowlands.

Agassiz must have written his paper either while still experiencing all the distractions of his journey, or during the twenty-four hours or so (it can hardly have been a much longer period) that elapsed between his return to London and the assembly of the Geological Society on the evening of 4th November. Had he been able to give the matter more attention he would doubtless have produced a better paper, but as it stands his communication can hardly have had much impact upon his audience. Strangely, Agassiz failed to make effective use of the immense wealth of field observations which he had acquired during the previous two months, and he seems not to have realised that British geologists, used to concerning themselves with geological minutiae, would only be swayed by a closely reasoned argument securely grounded upon carefully presented field data. Much more effective were the contributions which now came from Agassiz's two chief converts – Buckland and Lyell.

As soon as Agassiz had completed the presentation of his paper, Buckland left the chair to begin the reading of his own discourse entitled 'Memoir on the Evidence of Glaciers in Scotland and the North of England'.[40] Buckland knew exactly what the Fellows of the Geological Society expected in the way of scientific argument, and in his paper he offered a detailed account of the discovery which he had made near Thornhill on his way north to Glasgow, together with a full discussion of the glacial relics which he had seen in Scotland and northern England after parting company with Agassiz in Aberdeen. Like Agassiz Buckland adduced the widespread occurrence of till, moraines, polished and striated surfaces, and what he termed 'mammillated rocks', to prove the former existence of British glaciers, and he described an interesting test to which he had subjected the glacial theory. He explained that he had taken a map of the region around Comrie in Strath Earn, and that he had then marked in those places where, assuming the glacial theory to be valid, traces of former glaciers might be expected to occur. He then visited the sites in question, and he informed his audience that 'the results coincided with the anticipations'. In particular, he emphasised that the radial dispersal of erratics from mountain ranges could only be explained satisfactorily by invoking glacial transport. Diluvial currents, he now admitted, could never account for the conveyance of Shap granite boulders northwards to Penrith, eastwards over Stainmore Forest into the Tees valley, and southwards to Morecambe Bay. Buckland, nevertheless, had not entirely renounced his diluvialism, and he adopted the theory which Agassiz had propounded at Glasgow. He claimed that the melting of the glaciers had caused a flood which drowned the continents and allowed icebergs to deposit their morainic cargoes widely over the submerged continental surfaces.

It is interesting to note that at one time Buckland seriously considered making this glacial submergence the subject of a companion volume to his *Reliquiæ Diluvianæ* of 1823.[41]

Buckland's was a long paper, and its presentation was spread over three successive meetings of the Geological Society. Not until 2nd December was it completed, but on 18th November, and again on 2nd December, sufficient time was found in the programme to allow Lyell to make his contribution to the glacial debate. This contribution took the form of an important paper 'On the Geological Evidence of the former existence of Glaciers in Forfarshire'.[42] Lyell in 1840 had clearly been ripe for conversion to the glacial theory. In January of that year he had presented to the Geological Society a paper on the drifts of Norfolk in which he admitted that the only other place where he had seen a deposit comparable to the Norfolk tills (it was in this paper that he introduced both the terms 'drift' and 'till') was in the terminal moraines of the Swiss glaciers.[43] Having made this pertinent observation, however, he then immediately veered away from any explanation involving the former existence of English glaciers. Instead, he resorted to one of his favourite theories, and explained the deposit as being the result of melting icebergs dropping their dedris during the recent submergence of the region. Not until his excursion into Glen Prosen with Buckland in October 1840, did Lyell realise how close he had come to the truth in his Norfolk paper. When he spoke to the Geological Society in November he announced that his acceptance of the glacial theory had enabled him to solve many problems which had been puzzling him for some years. He also freely confessed that he had hitherto been in error in invoking a submergence and icebergs in explanation of the till and erratics of Strathmore. Even when confronted with drift sections revealing well bedded (glaciofluvial) sands and gravels, he now refused to call the sea to his aid. The following passage from the published proceedings of the Geological Society's meetings on 18th November and 2nd December explains Lyell's new position.

> Mr. Lyell objects to a general submergence of that part of Scotland, since the till and erratic blocks were conveyed to their present positions; as the stratified gravel is too partial and at too low a level to support such a theory; and he would rather account for the existence of the stratified deposits, by assuming that barriers of ice produced extensive lakes, the waters of which threw down ridges of stratified materials on the tops of the moraines.

E.D.—20

In his paper Lyell concluded that at the height of the glaciation most of Scotland must have been submerged under an ice-sheet so thick that only the highest mountain peaks were left uncovered. Scotland then, he explained, must have appeared very much as Antarctica does today. Buckland's paper on glaciers, although replete with detail, is an account of a reconnaissance journey through many parts of Scotland and England. Lyell's paper, on the other hand, is a circumstantial discussion of the glacial landforms present in one small area of the British Isles, and the paper is the earliest in that long line of regional glacial studies which have proved so popular with more recent generations of British geomorphologists.

Agassiz evidently delayed his departure from England so that he could be present to hear Buckland and Lyell complete the reading of their papers at the Geological Society on 2nd. December. When he finally left for Neuchâtel a day or so later, he departed very well satisfied with his achievements in Britain. His 2000-mile tour of England, Scotland and Ireland had not only confirmed his belief in an Ice Age, but had greatly enhanced his understanding of the bygone glaciers. 'It was in Scotland', he later admitted, 'that I acquired precision in my ideas regarding ancient glaciers'.[44] Equally important was the fact that he had now brought the glacial theory to the notice of the most august body of geologists to be found anywhere in the world – the Geological Society of London. Directly through his own paper, and indirectly through the papers of his two influential converts, he convinced many British geologists that henceforth they must take the ice of bygone glaciers into account in explaining the surface morphology of lands lying far beyond the modern glacial limits. In December 1840, immediately after his return to Neuchâtel, Agassiz wrote a jubilant letter to Humboldt containing the following passage:

> J'ai retrouvé les mêmes surfaces polies qu'en Suisse, les mêmes moraines latérales et terminales, la même disposition rayonnante du centre des chaînes de montagnes vers la plaine, les lacs partout également protégés contre le remplissage par les glaciers qui en occupaient le fond. . . . J'ai accumulé tant de preuves que personne en Angleterre ne doute maintenant que les glaciers n'y aient existé. . . .[45]

Here, as we will see, Agassiz rather underrated the obstinacy of Murchison and certain other leading British geologists, but as he sailed away from England Agassiz must have known that he left behind him dozens of geologists whose eyes had been opened and who were now eager to take the field in search of the reliquiae of the British glaciers. One other point concerning Agassiz's tour merits mention. During their journeyings, he and Buckland had of course examined numerous drift sections in search of stratigraphical evidence of former glaciation, but they focused their attention chiefly upon such morphological features as moraines, glacial pavements, and roches moutonnées. Agassiz and Buckland in 1840 were thus the first to take a geomorphic excursion through the British Isles.

The Reception of the Glacial Theory

At the Geological Society, in the autumn of 1840, the initial reaction to the glacial theory was hostile. When the subject was debated by the society on 18th November, there was not one Fellow prepared to take his stand squarely alongside Agassiz, Buckland, and Lyell, although the trio did receive some tacit support from Smith of Jordanhill who reminded the meeting of his discovery of an arctic fauna in the estuarine clays of the Firth of Clyde.[46] Most of those present clearly found the idea of former British glaciers something of a strain upon their imagination, and they certainly had difficulty in comprehending the scale of events during Agassiz's postulated Ice Age. George Greenough, for example, facetiously asked Agassiz whether he believed that Lake Geneva had once been filled with glacier ice 3000 feet thick. He was astonished when Agassiz took the question seriously, and replied that 3000 feet must be regarded merely as a minimum thickness for the ice in the area. To this Greenough could only retort that the glacial theory was the 'climax of absurdity in geological opinions'. Murchison, who had spoken against the theory at Glasgow and, doubtless, by the shores of the Moray Firth, was also inclined to employ ridicule rather than reasoned argument. If some scratches are explained as the work of former glaciers, he asked mockingly, then why not attribute all scratches to the same mechanism? The day will soon come, he observed, when

> Highgate Hill will be regarded as the seat of a glacier, and Hyde Park and Belgrave Square will be the scene of its influence.

At the Geological Society, only William Whewell offered much reasoned opposition to the theory. He made four chief points. Firstly, he asked what possible quirk of Nature could have converted the

287

British Isles and northern Europe into an icy wilderness. Secondly, with a site such as Moel-tryfan in mind, he asked the glacialists to explain how it came about that some of their supposed glacial tills contained fragments of marine shells. Thirdly, he claimed that drift deposits are widespread throughout the world, and that their diversity of type and environment is alone sufficient to preclude the glacial concept as a general theory of drift formation. Finally, Whewell confessed his inability to understand how glaciers could once have existed on a lowland such as the North European Plain. His knowledge of glaciers was clearly limited to the valley glaciers of mountain regions, and Agassiz's concept of an ice-sheet glaciation of lowland areas was evidently entirely lost upon Whewell.

Similar criticism of the glacial theory was heard outside the quarters of the Geological Society. William Conybeare, for example, protested that the theory was 'a glorious example of hasty unphilosophical & entirely insufficient induction',[47] and in December 1840 he wrote to Buckland as follows:

> . . . you see a few scratches on the face of a rock & a heap of gravil at its base – & then by an argument *per saltum* get at yr. Q.E.D. However it will make a fine *new slide in our raree* show Geological magic lantern . . . we shall have Jamaica *covered with trackless snow.* . . .[48]

On 11th February 1841 the Geological Society of Edinburgh discussed the glacial question and recorded in its minutes that

> . . . the Glacier Theory of Agassiz is not applicable to Scotland, at least in general.[49]

Three years later the society re-examined the question and again arrived at precisely the same conclusion. Even such an able field-geologist as J. B. Jukes for long remained blind to glacial phenomena. He joined the Geological Survey in 1846, and had spent four years mapping in North Wales before he first began to notice glacial landforms during the course of a tour of south-western Ireland made with De La Beche about 1851.[50]

Despite such early opposition and want of perception, there were many British geologists who were entirely convinced by the logic of Agassiz's argument. His visit had ended the years of indifference to the glacial theory, and glacial studies suddenly became the rage. In Scotland between 30th December 1840 and 13th January 1841 Charles Maclaren devoted four leading articles in *The Scotsman* to the discussion of glacial landforms, and on 27th January 1841 the paper carried a long and detailed article by Maclaren on the glacial morphology of that prominent Edinburgh landmark, Arthur's Seat.[51] Jameson was not to be outdone; between January 1840 and December 1843 he published more than thirty papers and notes on glacial topics in the *Edinburgh New Philosophical Journal*. Even a small society such as the Galashiels Geological Society felt the impact of Agassiz's visit; as early as 29th January 1841 the Society was addressed by one William Kemp on the subject of a series of moraines that he had identified around the town.[52] Edward Forbes was hardly exaggerating when on 13th February 1841 he wrote from Edinburgh to Agassiz in Neuchâtel:

> You have made all the geologists glacier-mad here, and they are turning Great Britain into an ice-house.[53]

Southward of the Scottish border it was Wales which attracted most attention, perhaps because Agassiz himself had left it as a glacial *terra incognita*. Buckland was at work amidst the glacial riches of Snowdonia in 1841, and in December of that year he reported his findings to the Geological Society of London.[54] He was particularly impressed by the glacial relics in the Pass of Llanberis where 150 years earlier Edward Llwyd had mused over the history of the innumerable boulders that litter the valley floor. Buckland knew the answer; everywhere he looked in the valley he saw the work of a former glacier. He nevertheless still clung to his belief in a late glacial submergence of the British Isles, and he claimed the shelly gravels of the Vale of Clwyd and Moeltryfan as a Welsh legacy dating from this event. Similarly, he regarded the Scottish erratics in North Wales as debris deposited by ice-bergs which had floated southwards over the waters of the submergence. This was not the only use that he made of these postulated ice-bergs. On the coast at Dinas Dinlleu, five miles south-south-west of Caernarvon, he had seen disturbances (cryoturbation) in the drift, and following a suggestion made by Lyell the previous year in his paper on the Norfolk drifts, Buckland suggested that the disturbances were the result of icebergs thrusting into the drift as they grounded.

Darwin soon followed Buckland's tracks into the Principality, and in 1842 he wrote a letter from North Wales containing the following passage:

Yesterday (and the previous days) I had some most interesting work in examining the marks left by *extinct* glaciers. I assure you, an extinct volcano could hardly leave more evident traces of its activity and vast powers. . . . The valley about here and the site of the inn at which I am now writing must once have been covered by at least 800 or 1,000 feet in thickness of solid ice! Eleven years ago I spent a whole day in the valley where yesterday everything but the ice of the glaciers was palpably clear to me, and I then saw nothing but plain water and bare rock. These glaciers have been grand agencies.[55]

Only a few months later Darwin was in print with his own contribution to the glacial history of the region.[56]

In Ireland interest in the glacial theory was less marked, but even there C. W. Hamilton read a paper on glacial phenomena to the Geological Society of Dublin as early as 9th March 1842. His subject was the glacial markings to be seen on the rocks around Bantry Bay and on Keeper Mountain, Co. Tipperary, but unfortunately his paper was never published.[57] Soon afterwards Ireland passed through the traumatic experience of the Famine, and it was 1850 before the first significant paper on the glaciation of Ireland appeared in print. This was John Ball's paper 'On the former existence of small Glaciers in the County of Kerry' which he read to the Geological Society of Dublin in November 1849.[58] In the paper he showed that small corrie glaciers had recently existed in south-western Ireland, and he thus became the first to recognise the relics of an event which much later generations of glacial geologists came to know as the Lesser Cork-Kerry Glaciation.

By no means all of this new British interest in glaciers was focused upon the events of the Ice Age. Soon after 1840 British geologists, led by J. D. Forbes and John Tyndall, joined their continental colleagues in making careful studies of existing glaciers, and these studies were soon shedding much valuable light upon the swollen glacial giants of the Pleistocene. By 1850 it was generally admitted that glaciers were geological agents of considerable importance, and henceforth they received careful attention in most of the texts which continued to pour from the presses in response to the sustained public interest in the geological science. Indeed, the glacial theory was now so firmly entrenched that some geologists began to look for evidence of glaciations other than that dating from the Pleistocene. Thus in 1855 A.C. Ramsay adduced the presence of striated and polished boulders in a Permian breccia as evidence of a Permian glaciation of Britain,[59] and in his

Climate and Time of 1875 James Croll discussed the evidence of no less than nine glaciations of pre-Pleistocene date.

Initially, some British geologists found the idea of one glacial period, let alone a multiplicity of glaciations, in sharp conflict with their preconceived notion that the Earth had been cooling steadily ever since its origin as a flaming, nebulous ball. The evidence of the existence of former glaciers was nevertheless so strong that nothing was allowed to hinder the general acceptance of the glacial theory. During the 1840s and 1850s only two British geologists of repute consistently refused to concede that the world's glaciers had once reached far beyond their present limits. Amazingly, one of these geologists was Sir Charles Lyell; more predictably the other was that hidebound conservative Sir Roderick Murchison.

Lyell's case is at once both interesting and puzzling. After his excursion into Glen Prosen in October 1840 he gave the glacial theory an enthusiastic welcome as the key to a multitude of problems that had long perplexed him. In the following month he expounded the theory before the Geological Society, claimed that Scotland had recently been submerged beneath an ice-sheet, and specifically renounced his former belief in the marine origin of even the bedded drifts of Strathmore. Agassiz must have left Britain in December 1840 convinced that he had secured the conversion of the man who without doubt was the world's most influential geologist. Almost immediately, however, Lyell began to entertain doubts about his new faith, and he was soon guilty of apostasy. This is only too apparent from the second edition of his *Elements of Geology* published in 1841 and bearing a preface dated 10th July 1841 – a date barely seven months after the presentation of his glacial paper to the Geological Society. In the book, while admitting that small glaciers might once have existed in Scotland, he dismisses the glacial theory of till formation and reverts to his old theory that the Scottish drifts are submarine deposits formed by melting icebergs. He even went so far as to claim that glaciers could never have carried the Alpine erratics to the Jura because the gradient between the two points is only 2°, and this he now regarded as too gentle a slope for glacial flowage.

In the first edition of volume III of the *Principles* published in 1833,[60] Lyell had claimed that the Jura erratics had been transported from the Alps frozen into ice-rafts which were swept northwards by debacles. Now, in 1841, he preferred this fanciful theory to Agassiz's notion of

291

transportation by swollen Alpine glaciers, and it remained Lyell's explanation of the Jura erratics down to 1858. In that year a tour of Switzerland – it was by no means his first such tour – finally convinced him of the validity of Agassiz's theory, and Lyell admitted his re-conversion in a letter written in January 1858.

> I came to the conclusion that Agassiz, Guyot, and others are right in attributing a great extension to the ancient Alpine glaciers, and that floating ice which I believe has done much in Great Britain, Scandinavia, and the United States will not aid, as I formerly suspected, in explaining the transfer of Alpine erratics to the Jura.[61]

This, however, was only a limited re-conversion. Not until the publication of his *Antiquity of Man* in 1863 did he again admit that Scotland had experienced an ice-sheet glaciation, and twenty-three years thus elapsed before his return to the position which he had originally taken up in 1840. Even in 1863 he was still far from accepting Agassiz's theory *in toto*. He still believed that many drift phenomena in Scotland were the result of a deep glacial submergence, and down to his death in 1875 he remained convinced that many of the tills of lowland regions were marine rather than glacial in origin.

It is difficult to understand Lyell's renunciation of the land-ice theory and his preference for a marine submergence. Two points nevertheless deserve to be made. Firstly, after the initial impact of the glacial relics around Kirriemuir had been forgotten, Lyell perhaps concluded that a marine submergence was a much more likely occurrence than a glacial inundation. This would certainly have been an understandable decision for a strict uniformitarian. No part of the British Isles is far removed from marine influences, and large areas would be placed under water by nothing more drastic than a rise of a few hundred feet in sea-level, or by a subsidence of the land of similar magnitude. To suppose that Britain had once been shrouded in an ice-sheet, on the other hand, involved the supposition that major changes had occurred in the natural order. Agassiz's theory, after due consideration, may to Lyell have looked dangerously like a new variety of catastrophism. Secondly, as we have seen, long before Agassiz brought the glacial theory to Britain, Lyell was explaining the Earth's topography in terms of wave and current erosion during a recent submergence. He was never very precise as to the date and duration of this submergence, but it was obviously very simple to suppose that a large proportion of the world's drift deposits had been laid down during the same inundation that had been instrumental in shaping so many of the world's landscapes. Lyell was thus to some extent able to explain two sets of phenomena – topography and the drift – in terms of a single natural event.

Murchison's opposition to the glacial theory is easier of explanation. Capable field-geologist though he may have been, he had a mind that was closed to innovations, and he opposed the glacial theory just as steadfastly as he opposed the fluvial doctrine. Where fluvialism was concerned he did have some weighty arguments to support his disbelief, but his opposition to the glacial theory rested upon nothing more substantial than his own cussed refusal to see the glacial relics around him. He rarely let slip an opportunity of denouncing the theory, and after succeeding Buckland in the chair of the Geological Society in 1841, he made vigorous attacks upon the theory in his anniversary addresses to the society in 1842 and 1843. In his 1842 address he laid emphasis upon the shelly gravels, the raised beaches, and Darwin's explanation of the Parallel Roads of Glen Roy, and sought to show that the supposed glacial drifts were really all marine in origin.[62] As a neodiluvialist he similarly held that most erratics had been carried to their present sites as a result of the combined efforts of icebergs and waves of translation, and that the debris swept along by the waves had, together with grounded icebergs, been responsible for cutting all striations except those lying in the immediate vicinity of modern glaciers.[63] His innate conservatism emerges from the following passage in his 1842 address.

> Once grant to Agassiz that his deepest valleys of Switzerland, such as the enormous chasm of the lake of Geneva, were formerly filled with solid snow and ice, and I see no stopping-place. From that hypothesis you may proceed to fill the Baltic and Northern Seas, cover Southern England, and half of Germany and Russia with similar icy sheets, on the surfaces of which all the northern boulders might have been shot off.

In his address the following year he came dangerously close to implying that both Buckland and Darwin were either incompetent as geologists, or else guilty of a gross misrepresentation of the Welsh evidence, because he now informed the Fellows of the society that during a recent tour of Snowdonia he had failed to find so much as a single moraine.[64]

Even an Alpine excursion with de Charpentier in 1848 failed to convince Murchison of the validity of the glacial theory, and not until

293

1851 did he first freely admit that glaciers might once have existed within the British Isles. In that year, in papers read to the Geological Society and the British Association, he conceded that during the recent submergence glaciers might perhaps have occurred in those parts of the British Isles which were sufficiently lofty to rise above the diluvial waters, although he still protested that icebergs and waves of translation were entirely adequate to account for all the supposed glacial phenomena.[65] Thus, as befitted an officer who was proud to have served under Sir John Moore at Corunna, Murchison's was a very guarded withdrawal, and not until 1862 did he confess that he had been mistaken in doubting the former presence of glaciers amidst the mountains of his native Scotland. In that year in a letter to Agassiz, Murchison wrote:

> I have had the sincerest pleasure in avowing that I was wrong in opposing as I did your grand and original idea of my native mountains. Yes! I am now convinced that glaciers did descend from the mountains to the plains as they do now in Greenland.[66]

This was very magnanimous, but in his study at Harvard Agassiz must have wondered why it had taken twenty-two years for the scales to fall from Murchison's eyes.

The Glacial Submergence

Agassiz and Murchison represent the two opposing poles of thought on the subject of the glacial epoch. The Swiss geologist envisaged the Ice Age as a period when ice-sheets and valley glaciers had reigned supreme as the chief geological agents at work. If a glacial submergence had occurred, it was for Agassiz an event of only trivial significance. Murchison, on the other hand, would have none of this. He denied the former existence of ice-sheets, relegated even valley glaciers to a role of only minor importance, and, as the leading neo-diluvialist, he claimed that during the so-called glacial period, waves of translation had in fact been the chief formative influence at work upon the world's landscapes. During the middle years of the nineteenth century most British geologists occupied a position lying midway between these two extremes. Like Agassiz, they admitted the onetime extent and importance of valley glaciers, but at the same time they rejected his concept of a former European ice-sheet. In place of Agassiz's deluge of ice, they substituted the neo-diluvial concept of a recent submergence by water.

This unfortunate compromise version of the glacial theory found favour for five chief reasons. Firstly, the idea of a recent submergence was already deeply embedded in British geological thought, whether in the guise of catastrophic neo-diluvialism, or in the uniformitarian concept of an inundation which had allowed waves and currents to mould the Earth's topography. Ever since the early years of the century such a submergence had regularly been invoked to explain the super-ficial deposits of northern Europe, and the glacial theory did not effect the immediate demise of this older theory of drift formation. Secondly, the existence of shelly tills in so many parts of the British Isles seemed to confirm that the drifts were indeed marine rather than glacial in origin. Thirdly, investigation showed that while many drift sections reveal the debris to be angular and unsorted, there are other sections displaying well rounded, stratified sands and gravels containing current bedding and other like phenomena. Later generations learned that such deposits are of glaciofluvial origin, but in the middle years of the nineteenth century these deposits seemed ample proof of the marine origin of the drifts as a whole. Fourthly, the notion of a vast ice-sheet covering most of northern Europe was too much for the comprehension of most British geologists, and this deficiency in their imaginative powers continued to plague glacial geology for more than twenty years. It was James Croll who later wrote:

> Few things have tended more to mislead geologists in the interpreta-tion of glacial phenomena than inadequate conceptions regarding the magnitude of continental ice.[67]

This point emerged clearly from the discussion of the glacial theory at the Geological Society in the autumn of 1840. British geologists could understand how valley glaciers such as those of the Alps might once have extended down their troughs to reach the adjacent lowlands, but the concept of lowland ice-sheets of vast thickness and extent was much less easily digested in an age when Greenland and Antarctica were virtually unknown. It seemed that no glacier capable of forming the drifts of the North European Plain could ever have existed, and it was therefore concluded that the deposits could only be marine in origin. Finally, it was noticed that instead of being dispersed radially from mountain centres, some erratic material has actually been trans-ported into mountain regions from the neighbouring lowlands. As James Smith observed before the Geological Society in 1845:

> I never yet saw or heard of an erratic block in the valley of the Clyde, where its course could be traced, that did not come in an opposite direction to the flow of the river.[68]

Presented with such a seemingly anomalous distribution of erratics, and with ice-sheets ruled out of court, it seemed entirely reasonable to suggest that the erratics must have been imported into the mountains by icebergs during a time of at least partial submergence.

As we have seen, Agassiz had himself admitted the possibility of a glacial submergence in his address to the British Association at Glasgow in September 1840, and two months later Buckland made use of the idea in the earliest of the glacial papers which he presented to the Geological Society. For the next twenty-five years this blend of the glacial and submergence theories dominated the British glacial scene. In 1851, in his *Geological Observer*, De La Beche went so far as to include a map showing the British Isles submerged in the glacial sea to the level of the present 1000-foot contour, and Lyell included a series of similar maps in his *Antiquity of Man* of 1863. Most tills outside mountain regions were interpreted as debris deposited from melting icebergs, and even within mountain areas a man of the standing of Thomas Oldham, then the Local Director of the Irish Geological Survey, could in 1846 regard the stadial moraines in the Wicklow Glens as 'valley bars' formed when the glens were arms of the sea.[69] Striations were believed to be abrasions formed as icebergs grounded, and the parallelism of the scratches over wide areas was attributed to marine currents which were supposed to have swept the icebergs along well-marked and regular paths. Similarly, the Irish eskers were generally regarded as shingle bars formed by marine currents in the glacial sea. In 1863, obviously with Darwin's work on coral reefs in mind, G. H. Kinahan, a Senior Geologist of the Irish Geological Survey, divided the Irish eskers into three types: Fringe Eskers, Barrier Eskers, and Shoal Eskers.[70] Some geologists nevertheless saw a perplexing problem in the eskers; why had such delicate features not been flattened by the swarms of icebergs which must have been swept southwards over Ireland during the period of the glacial submergence?

The wedding of the submergence theory to the glacial theory is well illustrated in Lyell's writings. In 1841, a few months after renouncing his short-lived belief in a Scottish ice-sheet, Lyell set out for North America, and there he satisfied himself that the so-called glacial phenomena were much better explained in terms of a submergence. He concluded, for example, that the striations to be seen around Boston could only have been cut by grounding icebergs because, he claimed, the city is too far removed from any mountains ever to have been reached by glaciers. Similarly, when in Nova Scotia, he was delighted to learn that some furrows which he had discovered on a wave-cut platform had indeed been formed by ice-flows during the previous winter.[71] While on passage to and from the New World he kept a close watch for Atlantic icebergs because he was anxious to see whether they really did carry sufficient debris to striate the sea-bed. His vigil was rewarded in June 1846 while travelling from Boston to Liverpool at the conclusion of his second American tour. Then, from the decks of the Cunard liner *Britannia*, he saw a large iceberg carrying what he took to be a cargo of boulders.[72] Convinced as he was that the present is the key to the past, he doubtless now felt satisfied that if such heavily armed icebergs exist today, then it was entirely reasonable to suppose that similar monsters had at some date in the past scraped their way across the drowned continental surfaces.

One problem connected with the marine origin of the drifts did worry Lyell. Granted the existence of some shelly tills, why is it that marine fossils are not to be found in all the deposits supposed to have been formed during the glacial submergence? In 1841 he suggested that meltwater from the icebergs of the submergence period had perhaps so chilled and freshened the water of the inundation as to make normal marine life impossible.[73] Later he had to abandon this view because zoological studies in Arctic waters showed that life abounded even amidst the ice-floes, and this discovery did finally raise some doubts in his mind about the marine origin of the drifts. This matter came to a head during his tour of Switzerland in 1857, and it was the complete absence of marine shells from the Swiss tills which then resulted in Lyell's partial reconversion to the glacial theory.

In Lyell's eyes, however, the British drifts were an entirely different matter from those of Switzerland. Even after his grudging 1863 admission that Scotland must have experienced a recent ice-sheet glaciation, he still remained convinced that the English drifts were nothing more than morainic debris dropped by melting icebergs, that most erratics had been similarly transported by ice-rafts, and that at least some striations were the result of icebergs grounding. In 1863 he visited Moel-tryfan where new sections had recently been excavated in the shelly till, and he found twenty different species of shells, many of them characteristic of present-day Arctic regions. After the visit he wrote in a letter:

> These shells show that Snowdon and all the highest hills which are in the neighbourhood of Moel Tryfaen were mere islands in the sea at a comparatively late period. . . .[74]

Lyell's thinking on the subject of the drifts progressed no further. Down to his death in 1875 he remained convinced that the waters of the glacial submergence had been at least as important as the glaciers in giving the Earth its present drift morphology. Even in the final edition of the *Principles*, published in the year of Lyell's death, we find him explaining that unsorted drifts are the result of icebergs melting in still water so that the debris falls directly to the sea-bed, whereas bedded drifts are formed when icebergs melt in waters affected by marine currents so that the debris is sorted as it sinks.

By the 1850s, research in the drifts had proceeded sufficiently far to allow the formulation of the earliest chronologies of the British Pleistocene, and the glacial submergence featured prominently in all such schemes. Ramsay was responsible for one of the first of the chronologies, and he divided the glacial period in North Wales into three stages.[75] Firstly, he recognised a period when valley glaciers had flowed out of the Welsh mountains producing striations and roches moutonnées on the valley floors. This episode was followed by a period of submergence during which the shelly tills were deposited on Moel-tryfan, and he claimed to have traced similar drifts up the flanks of Carnedd Dafydd and Carnedd Llywelyn to a height of 2300 feet. He therefore regarded this height as a minimum value for the depth of the glacial submergence of the region. Throughout the period of the submergence, marine currents swept a stream of icebergs south-westwards across Anglesey, and the frequent grounding of these ice-masses caused the island's rocks to become heavily striated in a north-east to south-west direction. Further to the east a few rogue icebergs left the main stream and floated into the Welsh valleys to deposit Scottish and other northern erratics. Finally, the waters abated, and the small valley glaciers which had lingered in the highlands throughout the submergence, now pushed down their valleys to form numerous moraines lying in the newly emerged areas below the 2300-foot contour. Had he substituted a great Irish Sea glacier for his glacial submergence, Ramsay's chronology would have been a close, if crude, approximation to the truth.

In the 1860s Jamieson devised a similar but rather more complex glacial chronology for Scotland.[76] There he recognised the following five stages:

    1. Scotland is raised to a height far above its present level, and because of its altitude, the country is soon covered by extensive glacial systems which striate and polish the rocks and leave a widespread mantle of till.

    2. The country sinks beneath the ocean and a strand-line is established at about the present 3,000-foot contour. New drifts, characterised by the presence of arctic shells, are formed as a result of deposition from melting icebergs.

    3. The region is elevated to a level slightly higher than that of the present day, and valley glaciers push down from the mountains towards the lowlands.

    4. A renewed period of submergence to about the level of the modern 40-foot contour. The glaciers now disappear, while the carse clays are formed in the firths of Forth and Tay.

    5. A slow emergence raises Scotland to its present level.

Such chronologies show a growing awareness of the complexity of Pleistocene history, and by the 1870s a tripartite division of the British drifts had obtained wide currency. Firstly, and at the bottom of the series, there was the Lower Boulder Clay supposed to have been left by an ice-sheet or by massive valley glaciers. Secondly, there were the Middle Sands and Gravels dating from the glacial submergence. Finally, there was the Upper Boulder Clay left by relatively small late-glacial valley glaciers. While generalisation is impossible, this tripartite division of the drifts did to some extent correspond to the distinction which was later drawn between the Older Drift, the out-wash gravels, and the Newer Drift. The southern English counties were of course excluded from all such chronological schemes; it was early recognised that southern England had escaped both the glaciation and the glacial submergence, although many must have wondered how it came about that a marine transgression in Wales to the level of the 2300-foot contour had left southern Britain entirely unaffected.

During the 1860s the reality of the glacial submergence seemed so securely established that attention was turned to the examination of the possible causes of the submergence. Here two explanations were widely discussed. The first of these made its earliest appearance in Britain in

The Reader on 2nd September 1865 when Croll suggested that the submergence had probably resulted from the displacement of the Earth's centre of gravity following the development of thick polar ice-sheets. He employed mathematics to prove that the existence of an ice-sheet 7000 feet thick in the northern hemisphere would so upset the Earth's present equilibrium that middle latitudes in the hemisphere would be drowned extensively as a result of a 1000-foot rise in sea-level. This theory attained some popularity, although in the same periodical a week later – on 9th September – Searles V. Wood junior replied to Croll raising a fundamental objection to his theory:

> . . . it seems to me that a result, precisely the opposite of a general submergence would be the consequence; since, as the sea is the source of all water, whether in the vaporous, liquid, or solid form, the abstraction of so large a proportion from the fluid state, and its accumulation in a solid form over the higher latitudes, must necessarily have reduced the general sea-level, and left great areas of its shallower parts in the state of land.

Thus the important glacio-eustatic theory was born in Wood's mind.

The second theory seeking to account for the glacial submergence was also first promulgated in 1865. Then, in a paper to the Geological Society, Jamieson suggested that during the glacial period the continents must have sunk under the weight of their heavy ice burdens, and that later the continents must have rebounded as the glaciers waned.[77] He thus originated the isostatic theory, and the theory was further developed by Charles Ricketts in two presidential addresses to the Geological Society of Liverpool in 1871 and 1872,[78] and by Jamieson himself in 1882.[79]

During the 1870s the glacial submergence theory fell into disrepute as British geological opinion gradually swung in favour of the land-ice theory which Agassiz had propounded more than thirty years earlier. This development will be examined later in the chapter, but it should be noted that the concept of a glacial submergence died a very slow death. In 1881, in the second edition of his great memoir on the geology of North Wales, Ramsay still retained his belief in a submergence which had placed the glacial shoreline about the level of the present 2000-foot contour in Snowdonia.[80] In 1893 Joseph Prestwich still believed in a late glacial sea-level 1000 feet higher than modern sea-level,[81] and in 1910, Edward Hull, the former Director of the Geological

Survey of Ireland, still regarded the deposits on Moel-tryfan as proof of a marine transgression.[82] Even as late as 1920 Professor J. W. Gregory of the University of Glasgow was still prepared to argue that a glacial submergence had been responsible for the formation of the Irish eskers.[83]

Glaciers as Agents of Erosion

From the earliest days of the glacial theory it had been recognised that glaciers produce such features as striations and *roches moutonnées* and great importance had always been attached to such erosional phenomena in delimiting the extent of the former glaciers. Early writers on the theory had even explained exactly how ice performs its erosional work. In his paper to the Geological Society in June 1840, for example, Agassiz attributed striations and rock-polish to the presence of quartz grains at the sole of a glacier. Slightly later Buckland explained to one of his classes the means whereby a glacier acquires its load of debris, and he continued:

> Every glacier is thus thickly set with fragments fixed firmly in the ice, like the teeth of a file.[84]

Similarly, in 1845, J. D. Forbes adduced rock flour as evidence of glacial abrasion and he emphasised that it was the glacial load which performs the abrasion, and not the ice itself:

> ... ice is only the *setting* of the harder fragments, which first round, then furrow, afterwards polish, and finally scratch the surface over which it moves.[85]

Another erosional effect associated with glaciers was described by John Tyndall after an Alpine excursion in 1858. He had noticed that some rocks which had recently emerged from beneath glaciers were spalling, and he wrote:

> The action of the glacier appeared to resemble that of the break [*sic*] of a locomotive upon rails, both being cases of exfoliation brought about by pressure and friction.[86]

Thus during the years before 1859 the erosional work of ice certainly did not pass unnoticed, but it was only the micro-features produced by glacial erosion which then claimed attention. Not until after 1859 did it begin to be appreciated that glaciers are among Nature's most powerful tools, and that many major landforms are entirely the result of glacial erosion. A few works published in Britain before 1859 do nevertheless show their authors to have possessed some inkling of the ability of ice to remodel a landscape.

Among these early writers on glacial erosion pride of place must be given to James Hutton. A tantalisingly vague passage taken from his *Theory* of 1795, and seeming to relate to glaciers as agents of erosion, was quoted earlier (p. 169), and one wonders whether as a young man Hutton ever took time off from his medical studies in Paris to visit the glacial landscapes of the Alps. The second author deserving of mention is Jens Esmark. His paper published in *The Edinburgh New Philosophical Journal* in 1826 contains the following passage which seems to leave little doubt that Esmark had a sound understanding of the significance of ice in the shaping of Norwegian topography.

> . . . that the Norwegian mountains have been covered with ice down to the level of the sea, and therefore that the sea itself must have been frozen, we may from this find the reason why the Norwegian mountains in general are so steep, I may say perpendicular, on the sides which hang over the valleys, not only in the valleys which are high above the level of the sea, but in those from the bottom of which the waters run into the Norwegian Fiords. Ice, or glaciers, by their immense expanding powers, must, beyond doubt, have produced this change in their original form, from this circumstance, that they are continually sliding downwards from the higher mountains to the lower districts, and, by this progressive motion, carried with them the masses of stone which they had torn from the mountains. It is easy to explain why no trace of these masses thus separated is to be found immediately below the precipices thus formed.[87]

Thirdly, there is Robert Mallet, an Irish engineer and keen amateur geologist. In January 1838 he read a paper on glacial motion to the Geological Society of Dublin, and he observed that

> . . . the bed of a glacier is in continual process of degradation, or *deepening* by the resistless passage of these vast masses of ice and rocks over it . . . .[88]

Finally, there is Colonel George Greenwood. Attention has always focused upon him as a fluvialist, but in his *Rain and Rivers* of 1857 we find the following passage relating to a glaciated landscape.

> That the beds of the valleys of glaciers suffer denudation like those of other valleys, I think we may infer for two reasons. First, the grooves or lines *scratched* on the sides of the valleys by the passage of the glaciers are

discovered far above their present level. Next, if this were not so, there would be an abrupt break or step from the glacier valley to the valley in prolongation of it. [89]

In further support of his belief in glacial erosion he drew attention to the turbid melt-water streams draining from glacial snouts, and to the immense quantities of debris contained in glacial moraines.

There was thus nothing novel in the concept of erosion by glaciers, but Ramsay nevertheless found himself at the centre of a major controversy when in 1859 he was bold enough to suggest that the Pleistocene glaciers had played a major role in shaping the Earth's present landscapes. Ramsay became interested in the effects of glacial erosion as a result of spending his honeymoon in Switzerland in the summer of 1852, and Archibald Geikie informs us that when Ramsay first saw the Alps he 'opened his eyes so wide that he feared they never would close again'.[90] It was nevertheless 1859 before Ramsay first discussed the significance of glacial erosion in print. In that year he wrote an essay for the Alpine Club on the subject of 'The Old Glaciers of Switzerland and North Wales',[91] and in the essay he boldly asserted 'that all glaciers must deepen their beds by erosion'. In discussing this subject he made a pertinent observation which was to become a commonplace in so many later geomorphic texts. He contrasted the rugged, serrated skyline of the Alps with the even, catenary form of the Alpine valleys. During the Ice Age, he suggested, the Alpine peaks rose above the valley glaciers and were 'scarred by rending frosts', while at lower elevations the topography beneath the ice was abraded and smoothed by the ponderous glaciers. Still more startling was his proposition concerning rock basins. He pointed out that such basins are today found chiefly in areas of recent glaciation, and he went on to draw the obvious inference. Such a correlation, he claimed, must exist because it was the bygone glaciers which excavated these remarkable concavities. Strangely, however, he made no attempt to explain how glaciers came to possess such singular powers of excavation.

Perhaps few members of the Alpine Club were in a position to appreciate the significance of Ramsay's essay, but in 1860 the essay was published separately as a booklet, and Ramsay immediately began to win converts. In that year, for example, Edward Hull expressed his indebtedness to Ramsay and adopted his theory by suggesting that many of the tarns in the English Lake District had been formed 'by the scooping action of glacier ice'.[92] It was nevertheless 1862 before Ramsay's theory really made its impact upon the geological world. It was then, on 5th March, that Ramsay presented his famous paper on the glacial origin of rock basins to the Geological Society of London.[93] In the paper he demonstrated that the lake basins occurring in a region such as the Alps are not structural features associated with synclines, nor are they areas of local crustal subsidence or gaping fissures lying along lines of faulting. Having successfully disposed of such earlier explanations, he went on, more by inference than by logical deduction, to present his conclusion that the basins could only be the product of glacial erosion. The actual sites of the lakes, he suggested, must have been determined by factors such as the thickness of the former glaciers and the presence of local weaknesses in the rocks, but he still made no real attempt to explain exactly how the glaciers had performed their erosive work. In adopting this policy he was perhaps wise; even a century later we are still far from a complete understanding of the amazing excavating power of glacier ice.

Ramsay was much less timorous when it came to offering examples of glacially excavated basins. He claimed that in the Alps lakes Geneva, Thuner, Lucerne, Zurich, Constance, Maggiore, Como, Lugano and many others all occupy ice-scooped basins, while in the British Isles he adduced the lakes of North Wales, the Lake District, and the west of Ireland as examples of similar glacial features. It was to his native Scotland, however, that he turned for his finest British examples of rock basins. There ribbon lakes such as lochs Ness, Morar, and Shiel seemed apposite illustrations of his theory, and on a somewhat smaller scale he also maintained that the innumerable rock basins in the areas of knock-and-lochan topography in the West Highlands and Outer Isles were similarly the work of glacier ice. Indeed, so convinced was he of the validity of his theory that he boldly claimed that ice-gouging was the process responsible for the formation of both the sea-lochs of Scotland and the fiords of Norway.

The paper provoked a lively discussion among the Geological Society's Fellows. Many dismissed the notion of pliable glaciers having excavated deep basins in some of the world's toughest rocks as too absurd to stand in need of serious refutation, but among the Fellows Ramsay did immediately find one stalwart champion in T. H. Huxley. Soon others came over to take their stand alongside Ramsay and Huxley, and before the year was out Ramsay was able to write about his paper with some satisfaction.

When it was read Dr. Falconer of Indian-fossil-elephant celebrity made an onslaught on it of forty minutes. I observe that most of the men older than myself repudiate it, while most of the younger bloods accept it. Lyell rejects, but then I have Darwin, Hooker, Sir William Logan, Jukes, and Geikie.[94]

The theory had thus acquired influential support, but within the Geological Society the opposition was so strong that the Council was in two minds about publishing the paper. On 21st September 1862 Darwin wrote to J. D. Hooker saying:

What presumption as it seems to me, in the Council of Geological Society that it hesitated to publish the paper.[95]

In the event the paper did appear in the society's *Quarterly Journal*, but Ramsay was convinced that the Council assented to its publication only because he happened to be the society's president at the time. The paper was evidently better received in the New World because it was reprinted in full in *The American Journal of Science*.[96]

Opposition to Ramsay's glacial erosion theory persisted for many years, with Lyell as the theory's most influential critic.[97] Lyell admitted that there was a close correlation between the existence of major lakes and regions of former glaciation, but he refused to allow that ice was capable of excavating any rock basins larger than those on the floor of corries. The following were his chief objections to Ramsay's theory.

1. Valleys which have been abandoned by glaciers within historic times contain no rock basins.

2. Crevasses repeatedly form in a glacier at the same point, thus proving that the glacier is incapable of performing sufficient erosion to remove even those minor underlying irregularities which cause crevasses to develop.

3. Ice, like water, needs a slope down which to move, and a glacier therefore could never have flowed up a gradient in order to escape from a rock basin. Thus the ice within such a basin could only stagnate, and such stagnant ice is clearly incapable of erosion.

4. Although many areas of former glacial activity do contain rock basins, the correlation between the two sets of phenomena is incomplete. Heavily glaciated regions such as the Caucasus and the southern flanks of the Himalayas, for example, are largely devoid of rock basins.

5. Some lakes, such as Lake Nyasa, are identical in appearance to the lakes of glaciated regions, but are clearly themselves not glacial in origin.

6. Lakes have existed at all periods of the earth's history and under a wide variety of climatic conditions. They are therefore not features found only after periods of glacial activity.

7. There is no conceivable mechanism which would give glaciers the power to excavate basins hundreds of feet deep and miles in length.

Having dismissed Ramsay's theory, Lyell tried to devise an alternative explanation to account for at least the Alpine rock basins. Here he made use of his uniformitarian concept of slow crustal warping, and he suggested that immediately before the glacial period, the Alps had been gently uplifted by several thousands of feet. During the uplift the rivers degraded vigorously to form deep valleys, but then, with the onset of glacial conditions, the degradation ceased and the Alps began to subside. This subsidence, Lyell claimed, affected a zone 80 to 100 miles in width. At the middle of the zone, along the axis of the movement, the rate of downwarping was five feet per year, but away from the axis, to the north and south, the rate of subsidence diminished steadily. As a result of this differential movement, the upper reaches of the deep pre-glacial valleys sank more rapidly than the lower reaches, and eventually the gradients on the valley floors were reversed to form a whole series of rock basins along the margins of the flexure. According to his theory the sole function of the glaciers was to occupy the newly formed rock basins and thus prevent them from being filled in with alluvium and other debris, and in this way he sought to explain the correlation between areas of lakes and areas of recent glaciation. Rock basins must have been formed as a result of differential warping in other parts of the world, he suggested, but there, in the absence of glaciers, the basins have been obliterated during the accumulation of thick alluvial mantles. Lyell believed that his theory could explain the origin of the Scandinavian lakes and fiords, but, as Ramsay pointed out in 1865,[98] it could never hope to explain the profusion of lakes in the Scottish Highlands without going to the absurd extreme of relating almost every lake to a separate flexure. Lyell nevertheless steadfastly refused to accept Ramsay's glacial erosion theory, and in 1875, in the twelfth and final edition of the *Principles*, he still dismissed the entire subject of glacial erosion in nothing more than a few sentences.

It goes almost without saying that Murchison was another who was resolutely opposed to Ramsay's theory. He was convinced that the rock basins were nothing less than gigantic crustal fissures, and in 1864 he launched a forceful attack upon the glacial erosion theory both in a series of letters written to the editor of *The Reader*, and in his anniversary address to the Royal Geographical Society.[99] This outburst gave Ramsay the excuse for a reply through the pages of *The Philosophical Magazine*, and in this rejoinder he made two good points.[100] Firstly, he observed that his opponents were making an elementary mistake and thus complicating the whole problem presented by the rock basins. They were allowing the greatly exaggerated vertical scale of sections drawn across the basins to give them a completely false impression of the magnitude of the features. In reality, Ramsay emphasised, the basins are not the deep, vertical sided and abruptly terminated trenches which figure in the cross-sections. Did they really possess such a form, he admitted, it would indeed be difficult to see how they could have been excavated by glaciers. Secondly, he emphasised that the exit of the glaciers from the rock basins presented no real problem because, despite their false representation in cross sections, the gradients at the downstream ends of basins are commonly comparatively gentle, and glaciers can flow up such a slope provided there is a sufficient pressure of ice to the rear. Murchison was nevertheless unrepentant. He returned to the attack upon the glacial erosion theory in 1870 when he asked:

> . . . where in any icy tract is there the evidence that any glacier has by its advance excavated a single foot of solid rock? In their advance, glaciers striate and polish, but never excavate rocks.[101]

John Ruskin also regarded the theory as absurd. He wrote in *The Reader* on 12th November 1864 that

> . . . the idea of the excavation of valleys by ice has become one of quite ludicrous untenableness.

Similarly, J. W. Tayler protested in 1870 that glaciers are impotent as erosional agents because in many places today they can be seen to rest upon beds of till. If the modern glaciers are so patently incapable of disturbing such unconsolidated deposits, he asked, what justification have we for believing that the Pleistocene glaciers were capable of gouging out hundreds of feet of some of the world's toughest rocks?[102]

Ramsay himself was philosophical about all this opposition. As he remarked to James Geikie:

> When a man does anything really in advance he may be well pleased if in 10 to 14 years he gets a fair proportion of the best men on his side.[103]

He nevertheless had the satisfaction of knowing that many of 'the best men' were indeed coming around to his way of thinking, and in May 1864 he wrote to Archibald Geikie expressing his general satisfaction with the way the glacial erosion theory was progressing.[104] Archibald Geikie, then a young officer of the Geological Survey of Scotland, was one of Ramsay's converts; in his *Scenery of Scotland* of 1865 he came out strongly in favour of Ramsay's theory as the only adequate explanation of the origin of the Scottish lake basins. Jukes, on the staff of the Geological Survey of Ireland, was equally convinced of the validity of Ramsay's views, and in 1863 he applied the glacial erosion theory to the corrie basins of Co. Kerry.[105]

Perhaps Ramsay's most fervent disciple was John Tyndall. In 1858 he and Ramsay had travelled together during the Alpine tour which had led Ramsay to formulate his glacial erosion theory of rock basins, but Tyndall was soon carrying the theory to what even Ramsay regarded as an unjustified extreme. In 1862, for example, Tyndall claimed that the ancient Swiss glaciers had done much more than excavate a few rock basins; they had, he maintained, gouged and fretted the pre-glacial topography to such an extent that the present-day landscape is almost entirely glacial in origin.[106] Darwin, convert though he was to Ramsay's theory, felt Tyndall's views to be preposterous. In November 1862 he wrote to Hooker:

> For Heaven's sake instil a word of caution into Tyndall's ears. I saw an extract that valleys of Switzerland were wholly due to glaciers. He cannot have reflected on valleys in tropical countries. The grandest valleys I ever saw were in Tahiti. Again, if I understand, he supposes that glaciers wear down whole mountain ranges; thus lower their height, decrease the temperature, and decrease the glaciers themselves. Does he suppose the whole of Scotland thus worn down?[107]

Ramsay was much disturbed. He felt it necessary to publish an immediate disclaimer of Tyndall's extremist views because he thought 'it a pity to let it be supposed that my theory led to such extravagance'.[108]

Only two years later, however, Ramsay was coming around to Tyndall's point of view,[109] and by 1870 it was increasingly being recognised that glaciers have indeed wrought profound changes upon the world's landscapes. In January 1874, when J. Clifton Ward read a paper to the Geological Society on the glacial origin of rock basins in the Lake District, Ramsay confessed in the ensuing discussion that he

... was so accustomed to meet with papers such as this, confirming his original views, that he was almost becoming weary of the subject.[110]

Shortly after, in 1876, when J. W. Judd dared to question the glacial excavation theory of rock basins in the pages of *The Geological Magazine*, seven geologists – five of them Geological Survey men – immediately rushed for their pens in order to defend the theory. The sole geologist to come to Judd's aid was T. G. Bonney who was soon to be appointed to the chair of geology at University College, London, and who for the next forty years consistently refused to concede that glaciers had played a major role in shaping many of the world's landscapes. E. J. Garwood later joined Bonney in his stand, and these two geologists became the leading British exponents of what came to be known as the glacial protection theory. They and their handful of followers held that the erosive potential of rain and rivers far exceeds that of glacier ice, and they claimed that the presence of glaciers in a region protects the landscape from the much more effective fluvial attack. Surprisingly, such views received a favourable airing in an influential British geomorphic text published as recently as 1937,[111] but in reality the protectionist school never achieved much popularity in Britain. Long before Ramsay died in 1891, it had become generally accepted that glacier ice is one of Nature's most powerful geomorphic tools.

The Land-Ice Theory

While Ramsay's glacial erosion theory was steadily gaining ground during the 1860s, another parallel and related development was taking place in the field of glacial studies – the glacial submergence theory was gradually yielding to the land-ice theory. After 1840, as we saw earlier, British geologists thought of the glacial period as a time when valley glaciers had existed in high mountain regions, and when lower areas had been submerged beneath a deep glacial sea. During the 1860s, however, opinion became increasingly critical of the view that all the lowland drifts, erratics, striations, rock-polish, and roches moutonnées were the product of icebergs floating over the waters of the transgression. Despite Lyell's observation in mid-Atlantic in 1846, it was now beginning to be appreciated that icebergs are generally composed of relatively clean ice; they carry neither sufficient tools to striate the sea-bed, nor sufficient debris to form the tills which mantle so much of the northern hemisphere. In any case, was it realistic to suppose that whole swarms of icebergs had possessed such a remarkable uniformity that individual bergs drew just sufficient water to allow them to touch the sea-bed, yet insufficient water to check their forward progress and thus prevent them from scouring the drowned continents? Again, were droves of icebergs swept forward by powerful currents really a plausible explanation of the parallelism of striations over wide areas? How could such icebergs have cut the striations which in many areas can be traced up the slopes of a hill, over the crest, and down the opposite side? How could ice-rafting explain the presence of erratics at elevations far higher than the source from which the rocks had been quarried? The erratics of Antrim chalk resting high on Moel-tryfan were here a case in point, and similarly was it realistic to suppose that icebergs had carried blocks of Shap granite, up and over the Stainmore fells, and thence down to the Yorkshire coast? In 1848 Darwin had grappled with this particular problem but his efforts to explain the phenomenon in terms of ice-pushing along the shores of a slowly sinking land-mass was not entirely convincing.[112] Perhaps the most serious of the many problems which the submergence theory had to face, however, was the unsorted character of so much of the drift. If the lowland tills really had been formed on the sea-bed from debris dropped by melting icebergs, then why did the deposits so rarely display any stratification?

It was because of difficulties such as these that British geological opinion gradually swung away from the glacial submergence theory and towards the land-ice theory. Lowland features which hitherto had been explained in terms of submergence and icebergs, now began to be explained instead in terms of massive ice-sheets. Of the leading British geologists only Lyell and Murchison failed to move with the times, but even their more enlightened contemporaries deserve little credit for their tardy acceptance of the land-ice theory. It was a theory which they should have adopted at least twenty years earlier. Agassiz had advocated the theory in Britain in 1840, and three years later, in a paper written for Jameson and published in *The Edinburgh New Philosophical Journal*, Agassiz had pictured the British Isles, the North Sea, Norway, Sweden, Russia, Germany and France as all having been enveloped by a single, gigantic ice-sheet.[113] It is therefore remarkable that it was the 1860s before the land-ice theory began to make any headway in Britain. As we have seen, it was in 1857 that Lyell at last conceded that it was a former glacier which had carried the erratics from the Alps to the Jura, and we have Ramsay's confession that he personally entertained no doubts about the ice-rafted origin of the Jura erratics until his third visit to the Alps in 1860.[114] It is highly significant that Ramsay was revising his views on the subject of the Jura erratics at precisely the same time as he was formulating his glacial-erosion theory of rock basins, because the land-ice and glacial erosion theories were very closely connected. If, as Ramsay claimed, rock basins in lowland areas in many parts of the world had all been excavated by ice, then that ice could only have been land-ice. As he wrote in his paper of 1862:

> An iceberg that could float over the margin of a deep hollow would not touch the deeper recesses of the bottom. I am therefore constrained to return, at least in part, to the theory many years ago strongly advocated by Agassiz, that, in the period of extremest cold of the Glacial epoch, great part of North America, the north of the Continent of Europe, great part of Britain, Ireland, and the Western Isles, were covered by sheets of true glacier-ice in motion, which moulded the whole surface of the country, and in favourable places scooped out depressions that subsequently became lakes.

Ramsay was not the first to appreciate that icebergs were unlikely to be effective agents of erosion. Only a week before he read his paper on rock basins to the Geological Society, Jamieson had presented to the same society a paper in which he argued very effectively that features such as striations, glacial polish, and roches moutonnées, no matter where they occur, could only be the work of land-ice such as exists today in Greenland and Antarctica.[115] Another influential convert to the land-ice theory was Archibald Geikie. He abandoned the rival sea-ice theory in 1861, and two years later he published a soundly reasoned paper demonstrating that land-ice afforded the only rational explanation of the glacial phenomena of Scotland.[116] In the first edition of his *Scenery of Scotland*, published in 1865, Geikie observed that the old theory relating striations to the abrasive action of passing icebergs was dying rapidly, and little was heard of the sea-ice theory after the publication of James Croll's powerful advocacy of land-ice in his *Climate and Time* of 1875.

Even now, however, the long delayed acceptance of the land-ice theory did not result in the immediate demise of the belief in a glacial submergence. Most of the unsorted drifts might have been deposited by valley glaciers and ice-sheets, but there remained the shelly tills and the stratified drifts, both of which continued to be widely regarded as marine deposits. Even the most strenuous advocates of the land-ice theory still felt it necessary to find room for a glacial submergence within their chronologies of the Pleistocene period. Jamieson in 1862, for example, and Archibald Geikie in 1865, both still retained a belief in a late-glacial submergence of the British Isles to at least the level of the present 2000-foot contour. Similarly, as we saw earlier, Ramsay in the 1880s still believed that a deep glacial submergence was necessary to explain the presence of the shelly till on Moel-tryfan. The true, ice-dredged origin of the shelly deposits was first appreciated by James Croll in 1870 when he suggested that the shelly tills of Caithness, Orkney, and Shetland had all been dredged from the bed of the North Sea by glacier ice flowing westwards from Scandinavia.[117] Later this idea was carefully developed by Benjamin Neeve Peach and John Horne in two memorable papers presented to the Geological Society in 1879 and 1880.[118] Soon even that incubus, the Moel-tryfan deposit, was being viewed in a clearer light; it was really nothing more than a till incorporating debris dredged from the sea-bed by a great southward moving glacier which had formerly occupied the Irish Sea basin. James Geikie virtually abandoned the sea-ice theory during the interval between the publication of the first edition of his *Great Ice Age* in 1874 and the appearance of the second edition three years later, but not all his contemporaries were possessed of Geikie's perception, and resistance to the land-ice theory lingered long in some quarters. By 1878, however, icebergs and the glacial submergence were no longer factors of importance in British geomorphic thought.

The final establishment of the land-ice theory, and the growing appreciation of the erosional and depositional work of glaciers, were events of great moment in the history of geomorphology. Scarcely less important than the recognition of the geomorphic role of glaciers, however, was the impact which the glacial theory had upon the fluvial doctrine. In the middle years of the last century geologists were slowly becoming aware of the complexity of the events which have shaped the continents, and after 1859 it was increasingly recognised that many of the topographical features which hitherto had been brought into court to testify against the fluvial doctrine, in reality had no relevance to the case being tried. The hanging valleys, for example, which had seemed to nullify Playfair's defence of fluvialism through his law of accordant junctions, were now seen to be extraneous to the issue; they were merely the result of differential glacial erosion. The steps present in so many mountain valleys could no longer be adduced as evidence against fluvialism; they too were the result of differential glacial action. Similarly, the aggradational work of innumerable modern rivers in mountain regions is no proof that rivers cannot excavate valleys; it merely proves that we are observing the rivers at an exceptional moment in their history and immediately after the disappearance of the glaciers which overdeepened the valleys in question and left the modern rivers as misfits. Above all else, Ramsay's glacial erosion theory of rock basins removed that most serious of all the obstacles to the acceptance of fluvialism – the limnological objection. We cannot do better than conclude this chapter with a quotation from a paper written by Archibald Geikie late in the 1860s. With reference to Ramsay's theory Geikie wrote:

> It removes in the most simple way a difficulty which has long perplexed all who have tried to trace the history of the present outlines of land-surfaces, and who have felt that the existence of rock-basins presented an anomaly which they could not satisfactorily explain.[119]

Thus Ramsay in 1859 opened the way to an acceptance of the thoroughgoing fluvialism which Hutton had advocated more than seventy years earlier.

REFERENCES

1. See Jules Marcou, *Life, Letters, and Works of Louis Agassiz* (New York 1896); F. J. North, 'Centenary of the Glacial Theory', *Proc. Geol. Ass., Lond.*, LIV (1943), pp. 1–28; Helmut Hölder, *Geologie und Paläontologie in Texten und ihrer Geschichte*, pp. 319–347 (Freiburg and München 1960); A. V. Carozzi, 'Agassiz's Amazing Geological Speculation: the Ice Age', *Studies in Romanticism*, V (1966), pp. 57–83.
2. James G. Playfair, *The Works of John Playfair, Esq.*, I, p. xxix (Edinburgh 1822).
3. *Edinb. New Philos. Journ.*, XL (1846), p. 99f.
4. *Jameson Papers*, Gen. 121.
5. Georges Cuvier, *Essay on the Theory of the Earth. With Geological Illustrations, by Professor Jameson*, fifth edition, pp. 352–354; 482 (Edinburgh 1827).
6. *Edinb. New Philos. Journ.*, II (1827), pp. 107–121.
7. *Ibid.*, XXI (1836), pp. 210–222.
8. *Ibid.*, XXII (1837), pp. 27–36.
9. *Ibid.*, XXIV (1838), pp. 176–179.
10. *Ibid.*, XXIV (1838), pp. 364–383.
11. *Ibid.*, XXVII (1839), pp. 383–390.
12. Elizabeth O. Gordon, *The Life and Correspondence of William Buckland*, p. 141 (London 1894).

13. William J. Sollas, 'The Influence of Oxford on the History of Geology', *Sci. Progr.*, VII (1898), pp. 22 and 23; *The Age of the Earth*, p. 246 (London 1905).
14. *Proc. geol. Soc.*, III (1840–41), No. 72, pp. 332 and 333.
15. Dowager Duchess of Argyll, *Georges Douglas, Eighth Duke of Argyll*, I, pp. 350 and 351 (London 1906).
16. *Proc. geol. Soc.*, III (1839), No. 63, pp. 118 and 119.
17. *The Scotsman* (7th October 1840); Robert Cox and James Nicol, *Select Writings . . . of the late Charles Maclaren*, II, p. 65 (Edinburgh 1869).
18. *Proc. geol. Soc.*, III (1840), No. 71, pp. 321 and 322.
19. Louis Agassiz, *Geological Sketches*, second series, p. 3 (Boston 1890).
20. Letter in the archives of the Geological Society of London.
21. G. L. Davies, 'The Tour of the British Isles made by Louis Agassiz in 1840', *Ann. Sci.*, in the press.
22. *Rep. Brit. Ass., Glasgow 1840*, pt. 2, pp. 113 & 114.
23. *The Athenæum* (17th October 1840), No. 677, p. 824.
24. Archibald Geikie, *Life of Sir Roderick I. Murchison*, I, p. 307 (London 1875).
25. *The Scotsman, loc. cit.*; Cox and Nicol, *op. cit.* (1869), II, p. 65.
26. *Edinb. New Philos. Journ.*, XXXIII (1842), p. 228.
27. Elizabeth C. Agassiz, *Louis Agassiz*, I, p. 307 (London 1885).
28. Gordon, *op. cit.* (1894), p. 141.
29. *Edinb. New Philos. Journ.*, XXXIII (1842), p. 222.
30. *A Tour in Scotland. MDCCLXXII*, Pt. II, pp. 394–396 (London 1776).
31. Francis Darwin and Albert C. Seward, *More Letters of Charles Darwin*, II, p. 188 (London 1903).
32. See also Cox and Nicol, *op. cit.* (1869), II, p. 65.
33. Agassiz, *op. cit.* (1890), pp. 70 & 71.
34. Agassiz, *op. cit.* (1885), I, p. 309.
35. *Ibid.*, I, p. 310.
36. Marcou, *op. cit.* (1896), I, p. 171.
37. *The Pollok-Morris MSS: Robert Jameson's Letter Book and Diary, 1840–1845.* The MSS. are in the possession of Mrs. Seton Dickson of Symington, Ayrshire.
38. *The Scotsman* (2nd January 1841); Cox and Nicol, *op. cit.* (1896), II, p. 73.
39. *Proc. geol. Soc.*, III (1840–41), No. 72, pp. 327–332; *The Athenæum* (21st November 1840), No. 682, pp. 927 & 928; *Phil. Mag.*, N.S. XVIII (1841), pp. 569–574.
40. *Proc. geol. Soc.*, III (1840–41), No. 72, pp. 332–337; *The Athenæum* (28th November 1840), No. 683, pp. 948 & 949, and (19th December 1840), No. 686, pp. 1012 & 1013; *Edinb. New Philos. Journ.*, XXX (1840–41), pp. 194–198, 202–205; *Phil. Mag.*, N.S. XVIII (1841), pp. 574–579, 587–590.
41. Sollas, *op. cit.* (1898), p. 23.
42. *Proc. geol. Soc.*, III (1840–41), No. 72, pp. 337–345; *The Athenæum* (12th December 1840), No. 685, pp. 991–993; *Edinb. New Philos. Journ.*, XXX (1840–41), pp. 199–202; *Phil. Mag.*, N.S. XVIII (1841), pp. 579–587.
43. *Proc. geol. Soc.*, III (1840), No. 67, pp. 171–179; *Phil. Mag.*, N.S. XVI (1840), pp. 345–380.
44. *Edinb. New Philos. Journ.*, XXXIII (1842), p. 240.
45. Marcou, *op. cit.* (1896), I, pp. 170 & 171.

46. *The Midland Naturalist*, VI (1883), pp. 225–229; Horace B. Woodward, *The History of the Geological Society of London*, pp. 138–144 (London 1907).

47. F. J. North, 'Dean Conybeare, Geologist', *Trans. Cardiff Nat. Soc.*, LXVI (1933), pp. 15–68.

48. *The Sir Henry De La Beche Papers* in the National Museum of Wales, Cardiff, 34.492 G4.

49. *Trans. Edinb. geol. Soc.*, XIII (1931–38), p. 262.

50. *Geol. Mag. Lond.*, IV (1867), p. 41.

51. See also Cox and Nicol, *op. cit.* (1869), II, pp. 67–102.

52. *Phil. Mag.*, N.S. XVIII (1841), pp. 337–343.

53. Agassiz, *op. cit.* (1885), I, p. 338.

54. *Proc. geol. Soc.*, III (1841–42), No. 84, pp. 579–584.

55. Agassiz, *op. cit.* (1885), I, p. 342 & 343.

56. *Phil. Mag.*, N.S. XXI (1842), pp. 180–188; *Edinb. New Philos. Journ.*, XXXIII (1842), pp. 352–363.

57. MS. *Minutes of the General Meetings of the Geological Society of Dublin* in the library of the Department of Geology, Trinity College, Dublin. See also *Journ. geol. Soc. Dublin*, II (5) (1843), pp. 10–13; IV (1848–50), p. 186.

58. *Journ. geol. Soc. Dublin*, IV (1848–50), pp. 151–154.

59. *Quart. J. geol. Soc. Lond.*, XI (1855), pp. 185–205.

60. *Principles of Geology*, III, pp. 148–150 (London 1833).

61. Katharine M. Lyell, *Life, Letters and Journals of Sir Charles Lyell*, II, p. 279 (London 1881).

62. *Proc. geol. Sec.*, III (1842), No. 86, pp. 637–687.

63. See also Roderick I. Murchison *et alia*, *The Geology of Russia in Europe*, I, pp. 507–556 (London 1845).

64. *Proc. geol. Soc.*, IV (1) (1843), No. 93, p. 94.

65. *Quart. J. geol. Soc. Lond.*, VII (1851), pp. 349–398; *Rep. Brit. Ass., Ipswich 1851*, pt. 2, pp. 66 & 67.

66. Agassiz, *op. cit.* (1885), I, p. 341.

67. James Croll, *Climate and Time*, p. 385 (New York 1875).

68. *Quart. J. geol. Soc. Lond.*, II (1846), pp. 33–37.

69. *Journ. geol. Soc. Dublin*, III (1849), pp. 197–199.

70. *Ibid.*, X (1862–64), pp. 109–112.

71. *Travels in North America*, I, pp. 9, 173, 174 (London 1845).

72. *A Second Visit to the United States of North America*, II, pp. 366–370 (London 1849).

73. *Elements of Geology*, pp. 248–256 (London 1841).

74. Lyell, *op. cit.* (1881), II, p. 380.

75. *Quart. J. geol. Soc. Lond.*, VIII (1852), pp. 371–376; *Rep. Brit. Ass., Liverpool 1854*, pt. 2, pp. 94 & 95; John Ball, *Peaks, Passes, and Glaciers* (London 1859).

76. *Quart. J. geol. Soc. Lond.*, XVI (1860), pp. 347–371; XXI (1865), pp. 161–203.

77. *Ibid.*, XXI (1865), pp. 161–203.

78. *Proc. Lpool. geol. Soc.*, II, Session 13 (1871–72), pp. 4–35; II, Session 14 (1872–73), pp. 3–30.

79. *Geol. Mag. Lond.*, N.S. Dec. II, IX (1882), pp. 400–407; 457–466.

80. 'The Geology of North Wales', *Mem. geol. Surv. U.K.*, p. 278 (London 1881).

81. *Philos. Trans.*, CLXXXIV (1893), pp. 903–984.

82. *Reminiscences of a Strenuous Life*, pp. 111 & 112 (London 1910).

83. *Philos. Trans.*, Series B, CCX, pp. 115–151.

84. Gordon, *op. cit.* (1894), p. 143.

85. *Travels through the Alps of Savoy*, p. 47 (Edinburgh 1845).

86. *The Glaciers of the Alps*, pp. 143 & 144 (London 1860).

87. *Edinb. New Philos. Journ.*, II (1827), pp. 118 & 119.

88. *Journ. geol. Soc. Dublin*, I (1838), p. 332.

89. *Rain and Rivers*, p. 100 (London 1857).

90. Archibald Geikie, *Memoir of Sir Andrew Crombie Ramsay*, p. 198 (London 1895).

91. Ball, *op. cit.* (1859).

92. *Edinb. New Philos. Journ.*, N.S. XI (1860), pp. 31–44.

93. *Quart. J. geol. Soc. Lond.*, XVIII (1862), pp. 185–204.

94. Geikie, *op. cit.* (1895), p. 272.

95. Darwin and Seward, *op. cit.* (1903), II, p. 155.

96. *Amer. J. Sci.*, Series 2, XXXV (1863), pp. 324–345.

97. *The Geological Evidence of the Antiquity of Man*, Chap. XV (London 1863); *Elements of Geology*, sixth edition, pp. 169–174 (London 1865); *Student's Elements of Geology*, pp. 151–164 (London 1871).

98. *Phil. Mag.*, Series 4, XXIX (1865), pp. 285–298.

99. *J. Roy. Geogr. Soc.*, XXXIV (1864).

100. *Phil. Mag.*, Series 4, XXVIII (1864), pp. 293–311.

101. *J. Roy. Geogr. Soc.*, XL (1870), p. clxxiii.

102. *Ibid.*, XL (1870), pp. 228–230.

103. Marion I. Newbigin and John S. Fleet, *James Geikie*, p. 60 (Edinburgh 1917).

104. Geikie, *op. cit.* (1895), p. 281.

105. *Mem. geol. Surv. Irel.*, 'Explanation of Sheets 160, 161, 171, and Part of 172', p. 8f (Dublin 1863).

106. *Phil. Mag.*, Series 4, XXIV (1862), pp. 169–173.

107. Darwin and Seward, *op. cit.* (1903), I, p. 471.

108. *Phil. Mag.*, Series 4, XXIV (1862), pp. 377–380.

109. *Ibid.*, Series 4, XXVIII (1864), p. 303.

110. *Quart. J. geol. Soc. Lond.*, XXX (1874), p. 104.

111. Sidney W. Wooldridge and Ralph S. Morgan, *The Physical Basis of Geography. An Outline of Geomorphology*, Chap. XXII (London 1937).

112. *Quart. J. geol. Soc. Lond.*, IV (1848), pp. 315–323.

113. *Edinb. New Philos. Journ.*, XXXV (1843), pp. 1–29.

114. *Quart. J. geol. Soc. Lond.*, XVIII (1862), pp. 185–204.

115. *Ibid.*, XVIII (1862), pp. 164–184.

116. *Trans. geol. Soc. Glasg.*, I(2), (1863), pp. 1–190.

117. *Geol. Mag. Lond.*, VII (1870), pp. 209–214; Croll *op. cit.* (1875), Chap. XXVII.

118. *Quart. J. geol. Soc. Lond.*, XXXV (1879), pp. 778–812; XXXVI (1880), pp. 648–663.

119. *Trans. geol. Soc. Glasg.*, III (1868–69), p. 181.

Chapter Nine

Fluvialism Revived
1862-1878

It seems to me, therefore, that the time is come when geologists should study a little more closely this problem of the mode of production of the surface of the land, and determine exactly the method of the formation of those variations in its outline which we call mountains, hills, table-lands, cliffs, precipices, ravines, glens, valleys, and plains.

> J. B. Jukes
> in his Presidential Address to
> Section C of the British Associa-
> tion at Cambridge, Thursday 2nd
> October 1862.

THE year 1862 was a turning-point in the history of British geomorphology. For sixty years the subject had been the neglected Cinderella of the Earth-sciences, but now early in the second half of the nineteenth century, geomorphology suddenly came into its own as geologists began to devote serious attention to landforms for the first time since the days of Playfair and de Luc. In seeking an explanation for this sudden change in the fate of geomorphology four factors need to be borne in mind. Firstly, old men were passing from the scene; De La Beche died in 1855, Buckland in 1856, Murchison in 1871, Sedgwick in 1873, and Lyell in 1875. As we saw earlier, these geologists of the first half of the nineteenth century had for a variety of reasons preferred to focus attention upon the purely geological aspects of Earth-history, rather than upon the more recent, geomorphic aspects. Many of these reasons, however, were no longer influential with the shrewd professional geologists of the Geological Survey who formed the spearhead of British geology during the 1860s and 1870s.

Secondly, by 1862 most of the more obvious fundamental problems in geology had been answered in at least a rudimentary fashion, but one major question remained unresolved. How had the continents acquired their present configuration? Was the prevailing marine erosion theory really adequate as an explanation of the form of most of the Earth's topography? Such questions assumed great significance as the concept of a glacial submergence faded from the scene, taking with it the marine waves and currents which hitherto had been invoked in explanation of so many landforms.

Thirdly, there was the seminal impact of discoveries made in the western United States of America. During the first half of the nineteenth century the true geomorphic significance of rain and rivers was much better appreciated in the United States than in Great Britain. In consequence, when the American geologists began to explore the western territories during the middle years of the century, they had no hesitation in invoking the fluvial processes in explanation of the scenic grandeur which they found there. The published reports of these expeditions, embodying fluvialistic interpretations of topography, circulated widely, and in Britain they served to stimulate interest in geomorphology in general and in the fluvial doctrine in particular. The earliest of these reports to attract attention in Britain was the memoir on the Colorado river published by Joseph Christmas Ives in 1861 and containing a long account of the region's geology and morphology from the pen of John Strong Newberry.[1] Later, in the following decade, there appeared two other extremely influential memoirs; John Wesley Powell's famous report on the Colorado river of 1875, and Grove Karl Gilbert's equally famous report of 1877 on the Henry Mountains of Utah.[2] At least one British geologist – Archibald Geikie – was so impressed by these studies emanating from the American west that in 1879 he visited the United States in order to see the remarkable western landscapes for himself. It nevertheless needs to be emphasised that the rekindling of the flame of fluvialism in Britain cannot be attributed to the arrival of these American geological reports; that flame was already beginning to burn brightly several months before the earliest of the invigorating breaths was received from across the Atlantic.

Finally, the revival of interest in geomorphology was intimately connected with the formulation of Ramsay's glacial erosion theory. The theory was sufficiently startling to capture widespread attention and it brought into being a situation the like of which had not been seen since the early years of the century; it brought a geomorphic issue to the centre of the British geological forum. Even more important than this is the fact that the adoption of Ramsay's theory opened the way for a return to the fluvialistic interpretation of topography. Perhaps in the 1840s and 1850s there were many geologists who felt tempted to explore geomorphic problems, but who demurred from such research because they felt uneasy about the prevailing marine erosion theory of landscape, and because the only plausible alternative theory – the fluvial doctrine – had to be ruled entirely out of court in view of the numerous and seemingly damning objections which it had to face. The most vital of these objections was of course the old limnological objection. Indeed, this objection had grown in force with the passage of years as catastrophic interpretations of topography waned in popularity. By the middle of the century few were prepared to argue, as had Hutton and Playfair, that the lake basins might all be the result of crustal collapse post-dating the fluvial excavation of the valleys in which the basins lie. In 1862, however, Ramsay's glacial excavation theory of rock basins had at one blow eliminated the limnological objection to fluvialism, and it was soon seen that the glacial erosion theory was equally capable of explaining the origin of those other landforms whose existence had hitherto seemed a serious obstacle to the acceptance of fluvialism. Thus through his paper of March 1862 Ramsay both stimulated a renewed interest in gemorphology and made possible the acceptance of the fluvial doctrine.

The revival of geomorphology in Britain was largely the work of a small group of Geological Survey officers. Three of these men were outstanding in their contribution, and very appropriately they were each associated with a different part of the United Kingdom. There was Joseph Beete Jukes in Ireland, Andrew Crombie Ramsay himself in England and Wales, and Archibald Geikie in Scotland. The contribution of these three men to the fluvialistic revival must be our chief concern in this final chapter.

Joseph Jukes and the Rivers of Southern Ireland

Joseph Beete Jukes, the son of a prosperous manufacturer, was born in 1811 at Summerhill on what was then the fringe of Birmingham.[3]

319

As a child he used to visit the home of an aunt who, in the fashion of her day, had assembled a small collection of fossils, and these specimens early gave Jukes his interest in geology. On his schoolboy rambles in the country around Dudley he searched for fossils to add to his own cabinet, but when the time came for him to choose a career, geology seemed to offer few prospects. Apart from such academics as Buckland, Jameson, and Sedgwick, the only professional geologists of the day were surveyors like Robert Bakewell, John Farey, and the illustrious William Smith, and while such men were very able in their chosen field, they hardly belonged to the social class to which the Jukes family aspired. Joseph therefore went up to Cambridge as an Exhibitioner at St John's College with the intention of preparing himself for ordination. His interest in geology nevertheless remained strong, and despite the opposition of tutors who regarded the subject as an unsuitable study for a prospective ordinand, he joined Sedgwick's geology class, and there he soon fell completely under his master's spell. Indeed, the feeling between the two men was mutual, and Sedgwick always recollected Jukes as one of the most able students ever to have passed through his hands. As he sat in the geology classes week after week, enthralled by Sedgwick's lectures, all thought of a clerical career melted from Jukes's mind. When he came down from Cambridge in 1836, instead of following the high calling which his family had expected of him, he became a geological itinerant. He wandered through the English countryside, a collecting-bag over his shoulder and a hammer in his hand, and he paid his living expenses from the fees he earned by delivering geological lectures to the many local societies which then flourished throughout the kingdom.

In 1839 Jukes was appointed Geological Surveyor of Newfoundland, and he spent the next year in the arduous task of exploring the geology of what was then a remote and little-known colony. He returned to England in December 1840, and soon after he was invited to become the naturalist to a scientific expedition which was preparing to leave for Australian waters. The expedition cleared Falmouth in H.M.S. *Fly* in April 1842, and was away from England for more than four years. By the time the *Fly* returned to Britain in June 1846 Jukes was a geologist possessed of considerable field experience, and as might have been expected, he was offered an appointment on the staff of the newly expanded Geological Survey of Great Britain. He was delighted at the opportunity of joining the small but enthusiastic band of 'hammerers' who were working under the direction of Sir Henry De La Beche, and he reported for duty with the Survey in August 1846, his initial salary being nine shillings per day for a six day week. For the next four years he was engaged in the Survey's mapping programmes in North Wales and the English Midlands, and these years, spent in the congenial company of colleagues such as Ramsay and De La Beche himself, were the happiest of Jukes's life.

In 1850 De La Beche offered Jukes the Local Directorship of the Geological Survey of Ireland, a post which had fallen vacant as a result of Thomas Oldham's appointment as Superintendent of the Geological Survey of India. At first Jukes was loath to accept the Irish post. He was reluctant to give up the care-free life of a field-geologist and he was sad at the thought of losing the close companionship of his English Survey colleagues. On the other hand, after four years of field-mapping on the One Inch maps then being used by the Geological Survey in Britain, he was intrigued at the opportunities presented in Ireland where all field-work was being carried out on the Ordnance Survey's magnificent new maps drawn on a scale of six inches to the mile. In the event, however, it was financial considerations which decided the matter. Despite his salary of nine shillings per day, Jukes was impecunious. He had recently married, and his finances were in so depleted a condition that his father-in-law had had to pay off Jukes's debts (they included the expenses he had incurred when fitting himself out for the *Fly* expedition seven years earlier) and lend him several hundred pounds so that he might start his married life in a state of solvency.[4] The prospect of increasing his salary by accepting the Irish post was therefore attractive, the more so since De La Beche assured him that he and his wife would be able to live very much more cheaply in Ireland than in Britain. After some hesitation Jukes made his decision, and on 20th September 1850 he wrote to De La Beche from Llangollen accepting the Irish appointment. In one respect Jukes was disappointed. Oldham had been both Local Director of the Geological Survey and Professor of Geology in the University of Dublin, but when Jukes applied for the vacant chair he was passed over in favour of an Irish geologist, the Carlow-born Samuel Haughton.

Jukes proved to be a good Local Director in the sense that he expedited the geological mapping of Ireland, and One Inch geology maps covering more than half of the country were published during his tenure of office. A study of the correspondence preserved in the

Geological Survey Office in Dublin nevertheless reveals that Jukes's relations with his subordinates were not entirely happy. To some extent this may have been because his Irish staff resented the appointment of an Englishman as their Local Director, but Jukes himself was evidently sadly deficient in those qualities necessary to smooth the troubled waters. His skill as a field-geologist was never in question, but he proved to be a tactless and niggling bureaucrat who was constantly reprimanding his staff for their petty infringements of Survey regulations. On occasion he was charming to his colleagues, but the picture which emerges from the Dublin correspondence is that of a conscientious man whose heart lay in the field and who was completely overwhelmed by the many facets of an administrative post for which he was ill-suited.

Jukes was never happy in Ireland. He arrived in Dublin during the aftermath of the Great Famine, and he was appalled by the conditions which then prevailed in the country. After spending only seven months in Ireland he wrote to a friend:

> I hardly know whether it is the air of Ireland, or the nature of the work here, or what; but certainly much of the zest of life has departed, and nothing but duty and business remain. . . .[5]

In his letters he complains repeatedly about the rigours of the Irish climate, about the hazards of travelling long distances in open cars while inspecting the field-work of his staff, and about the dirtiness of the accommodation he had to use when examining territory in the country's remote western regions. During the Fenian troubles there was an element of risk involved in being a government official in Ireland, and some officers of the Geological Survey were authorised to carry firearms when working in the field. Jukes himself evidently never took the danger very seriously because in March 1867 we find him writing:

> There is a strong rumour that we are all to get up murdered next Monday morning, nevertheless Augusta, C., and Miss B. are making preparations to dance at St. Patrick's ball on Monday night. . . .[6]

In a letter written at Castlebar, Co. Mayo, on 29th April 1867 Jukes observed that 'I shall be heartily glad when I can escape with a pension'.[7] Sadly, that day never came. On 27th July 1864 at Kenmare, Co. Kerry, Jukes had slipped on the staircase of the inn in which he was staying while on a tour of inspection, and he fell headlong on to the flagstone

322

floor beneath. The accident caused a slight concussion of the brain, and he was forced to take six-months' leave of absence in order to try to regain his health. A change of scene was clearly imperative, and he therefore set out on a continental tour, but at the end of September, while at Coblenz, he suffered a seizure and became partly paralysed.[8] Later his health improved sufficiently to allow him to resume his duties in Dublin, but he never fully recovered from the accident at Kenmare and it was directly responsible for the fatal illness which struck him down in July 1869.

Jukes had always been interested in geomorphology, but this interest was suddenly quickened in 1855 when Murchison succeeded De La Beche as the Director-General of the Geological Survey and decreed that henceforth each published One Inch geology map was to be accompanied by an explanatory booklet which would include a discussion of the region's topography. The editing of these memoirs was one of Jukes's tasks in Dublin, and many of the Irish memoirs in fact contain sections on the form of the ground written by Jukes himself. He was happy that Irish geomorphic problems were constantly being brought to his attention during the course of this work, and it was in this way that the question of the southern Irish drainage first came to his notice.

In 1861 Jukes had to prepare a memoir on the region around Cappoquin, Co. Waterford, and in this area one geomorphic feature above all others arrested his attention – the anomalous course of the River Blackwater.[9] This river is one of the subsequent streams in the Ridge and Valley Province of the south of Ireland. It rises on the borders of counties Kerry and Cork and flows eastwards along a narrow, limestone-floored syncline overlooked to the north and south by ridges of hills and mountains developed upon anticlines of Old Red Sandstone. The river follows its strike course for more than 50 miles, and by the time its waters reach Cappoquin they are within 11 miles of the sea at Dungarvan Harbour. Between Cappoquin and Dungarvan there is an eastward continuation of the river's synclinal limestone valley, but, amazingly, the river spurns this easy and obvious route to the sea. Instead, the Blackwater turns abruptly southwards at Cappoquin and leaves the syncline to cut through four ridges of anticlinal Old Red Sandstone hills in order to reach the sea at Youghal Bay. Not only does this course of the Blackwater between Cappoquin and Youghal present the river with much more formidable geological obstacles than

323

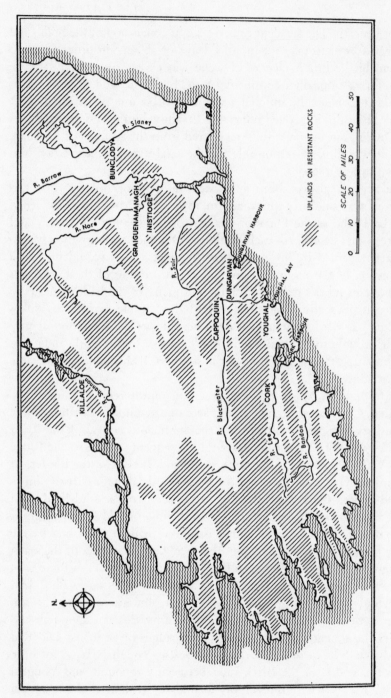

UPLANDS ON RESISTANT ROCKS

SCALE OF MILES

0   10   20   30   40   50

R. Slaney

BUNCLODY

R. Barrow

GRAIGUENAMANAGH

INISTIOGE

R. Nore

R. Suir

DUNGARVAN HARBOUR

CAPPOQUIN

DUNGARVAN

YOUGHAL

YOUGHAL BAY

KILL ALOE

Shannon R.

R. Blackwater

CORK

R. Lee

CORK HARBOUR

R. Bandon

Fig. 3 *Juke's problem area: the Topography and Drainage of Southern Ireland*

it would encounter if it persisted in its strike course below Cappoquin, but the river's preferred course from Cappoquin to Youghal is actually several miles longer than the very much simpler course from Cappoquin to Dungarvan (Figure 3).

While musing over the problem presented by the course of the Blackwater, Jukes observed that it is by no means unique in its anomalous behaviour. Two other eastward flowing subsequent rivers in the south of Ireland – the Lee and the Bandon – both suddenly abandon their strike courses and turn southwards to enter the sea via north to south reaches cut across a series of anticlinal ridges, although in the case of the Lee this north to south reach has now been drowned to form Cork Harbour. Another related but slightly different anomaly in the Irish drainage also arrested Jukes's attention. He noticed that after flowing southwards for many miles over the Carboniferous Limestone plain of the Irish midlands, certain major Irish rivers then cut their way across uplands developed upon inliers rising through the Carboniferous strata, or else they enter narrow defiles which carry the rivers across the pre-Carboniferous rocks of Ireland's mountainous rim. The Shannon, for example, cuts across the Old Red Sandstone and Silurian inlier of the Arra and Slieve Bernagh Mountains at Killaloe, while in southeastern Ireland the Slaney, Barrow, and Nore pass through the granite, schist, and Ordovician strata of the Leinster Mountain chain via three narrow and very picturesque valleys at Bunclody, Graiguenamanagh, and Inistioge.

In his explanatory memoir of 1861 Jukes confessed his inability to account for the strange behaviour of the Blackwater at Cappoquin, but early in the following year the light suddenly dawned upon him as he considered the broader issue of the southern Irish drainage as a whole. On 28th April 1862 he wrote to Murchison, his chief:

> I am preparing a paper on the formation of the river valleys of the S. of Ireland and should like to read it to the London Geological Society; before making any definite arrangements for that purpose however I write according to instructions for your permission to do so.[10]

The necessary permission was soon granted, but it was the members of the Geological Society of Dublin and not the Fellows of the Geological Society of London who first had the pleasure of hearing Jukes expound his novel theory. It was on 14th May 1862 that the members of the Dublin society gathered in the newly completed Museum Building of

325

Trinity College to hear Jukes deliver his now famous paper 'On the Mode of Formation of some of the River-Valleys in the South of Ireland'. At its meeting a week later the Council of the Geological Society of Dublin ordered the printing of Jukes's paper in the society's Journal but in the event nothing more than an abstract and a review appeared.[11] The president later explained that the society, which was in permanent financial straits,[12] had been unable to publish the paper in full because of the cost of printing its accompanying maps and diagrams. Fortunately the paper was not lost to the world. In accordance with his original plan, Jukes crossed to England and read his paper to the Geological Society of London on 18th June, and it was published in the London society's *Quarterly Journal* for 1862.[13]

In his paper Jukes claimed that virtually the whole of Ireland had been submerged during the Carboniferous period, and he argued that the transgression had left the island with a thick and extensive coat of Carboniferous strata.

> I am, in fact, unable to escape the conviction that at the close of the Carboniferous Period one great plain of Coal-measures extended horizontally over all Ireland, with the exception perhaps of the loftier peaks of Connemara, Donegal, Down, and Wicklow, even if any parts of those mountains remained uncovered by the highest Coal-measure beds.

He went on to explain that Nature had sculpted modern Ireland out of this Carboniferous-mantled block; but this view in itself was by no means novel. Ever since his arrival in Ireland Jukes had repeatedly voiced the opinion that the numerous Irish plateaux developed upon outliers of Upper Carboniferous strata were merely the last lingering remnants of a former extensive Upper Carboniferous cover.[14] For more than a decade he had regarded Ireland as a gigantic denudation, but the winter of 1861–62 saw a major advance in his thinking upon this subject. Hitherto he had always regarded the sea as the agent responsible both for the removal of the Irish Upper Carboniferous strata and for the moulding of the island's present topography. Quotations illustrative of the great importance which he attached to marine erosion were offered in an earlier chapter, but now, in 1862, he firmly rejected the marine erosion theory of landscapes and claimed instead that Ireland's topography could only have been shaped by the long-continued action of the fluvial processes. More especially he argued cogently that only river erosion could have produced valleys such as that of the Shannon around Killaloe or that of the Blackwater below Cappoquin.

Although Jukes now emerged as a thorough-going fluvialist, it is interesting to note that he still seriously underestimated the potency of the fluvial processes. He no longer doubted their ability to effect major landscape changes, but he remained convinced that such changes could only be produced as a result of the continuous operation of the fluvial processes throughout aeons of time. Thus he believed that except for the brief interlude of the glacial submergence, the greater part of Ireland had stood above sea-level ever since the Carboniferous, and he held that the reduction of the Irish Upper Carboniferous strata to their present restricted outcrops must of necessity have occupied the fluvial processes for the whole of post-Carboniferous time, a period, incidentally, whose duration he greatly exaggerated. It is now known that some 240 million years have elapsed since the close of the Carboniferous, but Jukes was one of those who thought that Darwin had made a mistake in 1861 when in the third edition of the *Origin of Species* he withdrew his previous claim that post-Mesozoic time alone amounted to not less than 300 million years.[15]

Jukes's belief that the greater part of Ireland had stood above sea-level throughout the Mesozoic and the Tertiary was of course related to the fact that in his day post-Carboniferous marine sediments were unknown in southern Ireland. So impressed was he with the slowness of action of the fluvial processes, however, that he never even considered the possibility that post-Carboniferous strata might have been deposited in the area only to be completely stripped off by later denudation. In this respect Jukes unfortunately set a pattern for later geologists. Numerous palaeogeographical atlases show Ireland as standing above sea-level throughout recent geological time, and it was not until 1959 that the discovery of an outlier of Cretaceous chalk near Killarney opened the eyes of geologists to the reality of at least one post-Carboniferous marine transgression in southern Ireland.

Despite his emphasis upon the geomorphic importance of the fluvial processes, Jukes in his theory did reserve one important role for the sea. The Irish Carboniferous rocks were folded during Permo-Carboniferous times by the earth-movements of the Armorican orogeny, but rivers such as the Shannon, Slaney, Barrow, and Nore display little or no relationship to the Armorican fold-structures. Jukes therefore reasoned that the folds must have lost their topographical expression before Ireland's Permo-Carboniferous emergence allowed the ancestors of the present Irish rivers to take their birth. He suggested that the emergence must have occurred in so leisurely a fashion that wave action had sufficient time to bevel a great plain across the whole of Ireland, thus effacing all the recently formed anticlinal ridges. It was upon this plain of marine denudation that he believed the Irish drainage to have originated. When the uplift was complete, he claimed, the plain stood higher than the loftiest of Ireland's present mountain peaks, and its surface dipped gently southwards so that a series of southward flowing (consequent) rivers developed. Slowly the fluvial processes denuded the plain and a variegated topography developed as the relatively weak Carboniferous rocks were peeled away to expose the much more resistant Devonian and Lower Palaeozoic rocks. The major rivers nevertheless persisted in their original directions of flow. They were let down off the Carboniferous cover so gradually that they were able to superimpose themselves on to the older rocks beneath, and in this way Jukes explained the anomalous courses of the Shannon, Slaney, Barrow, and Nore.

Jukes believed that in the far south of Ireland the presence of a series of closely spaced east to west trending Armorican anticlines and synclines had given the drainage of counties Waterford and Cork a somewhat more complex history than that possessed by the remainder of the Irish drainage. He claimed that in the two counties, as elsewhere in Ireland, the original drainage had flowed southward over a plain of marine erosion, but later, he argued, tributaries to these southward flowing rivers etched out the east to west strike of the folded rocks to yield what in modern terminology is known as a trellised drainage pattern. In 1862 he had little to say about the precise mode of development of the longitudinal or strike streams, but in the following year, with the southern Irish drainage still fresh in his mind, he observed that once the north to south transverse valleys were well established, then

> . . . other small streams would flow into them from their sides; and these, acting on the softer or more easily destructible bands of rock that run parallel to the length of the mountain chain, would commence the formation of the longitudinal valleys of the mountain chain. The lateral [transverse] valleys then would be the primary or first eroded river valleys, and the longitudinal would be the secondary valleys.[16]

In this manner Jukes explained the strange behaviour of the Blackwater, Lee, and Bandon. He regarded the north to south reaches of the
three rivers as surviving elements of the ancient transverse streams,
while he viewed the west to east reaches of the rivers as longitudinal
tributaries which, thanks to the region's geological structure, have now
become the dominant elements in the drainage pattern. He explained
the persistence of the north to south reaches of the rivers in the following passage from the 1862 paper:

> When once a lateral valley has succeeded in cutting a sufficiently deep
> channel, the waters of the longitudinal valleys that are afterwards
> poured into it cannot cross it, or overmount the walls opposite to their
> own mouths, because they are inevitably deflected down the lateral
> valley, and help to excavate it deeper and deeper below their junction
> with it. Hence the longitudinal valley and the part of the lateral one
> below their junction may be equally wide and deep, and appear to be
> the result of one action, whilst the lateral valley above the junction may
> be a narrow and broken ravine with a much smaller and apparently in-
> significant brook, although in reality the prime mover, the 'fons et
> origo' of the whole excavation.

Jukes added a postscript to his paper suggesting that the lessons which
he had learned in the south of Ireland were presumably of universal
application. He urged, for example, that the Wealden Denudation is
probably the work of rain and rivers, and that the streams which break
through the North and South Downs via impressive water gaps are
really transverse streams similar in character to the north to south
reaches of the southern Irish rivers. Here he was striking an important
blow in the cause of fluvialism because it will be remembered that the
existence of these water gaps had long been regarded as a serious
objection to the fluvial doctrine. Further afield he drew attention to the
course of the Rhône above Lake Geneva. He regarded that portion of
the river which drains north-westwards between Martigny-Ville and
the lake as transverse in character, and therefore analogous to the Blackwater below Cappoquin, while he regarded the south-westward
draining portion of the Rhône above Martigny-Ville as a longitudinal
river, the genetic equivalent of the Blackwater above Cappoquin. He
went on to formulate a general rule which was to become familiar to
all later generations of geomorphologists; streams transverse to the
geological structure are normally older than their longitudinal or
strike tributaries. Jukes was certainly under no misapprehensions as to

the significance of the views he was propounding. If the theory is correct, he observed at the conclusion of his paper

> ... atmospheric denudation or degradation will then have to be taken into account as one of the most important geological agencies in the production of the 'form of the ground' on all the dry lands of the globe.

Jukes's theory in explanation of the anomalous features of the southern Irish drainage has been widely accepted for more than a century. Later workers have modified the theory in only one important respect; with their greater understanding of both the rate of denudation and the general stratigraphical history of the British Isles, they have recognised that the Irish drainage is much more likely to have originated upon a sedimentary surface of Cretaceous Chalk than upon a Permo-Carboniferous plain of marine denudation cut in the Coal Measures. Perhaps Jukes's theory is today approaching the end of its useful scientific life, but in the historical context the paper will retain its significance for all time.

In the eyes of the historian the paper is important for four reasons. Firstly, it was a pioneer study in the evolution of a regional drainage pattern. Secondly, it introduced the useful concept of drainage super-imposition. Thirdly, it advanced the idea that rivers not only excavate their valleys, but that during the degradation they also adjust their courses to fit the underlying geological structures. (After a pilgrimage to the bend of the Blackwater at Cappoquin in 1911, William Morris Davis, the great American geomorphologist, paid tribute to Jukes for his recognition of this vital principle.[17] When Davis adopted the term 'subsequent' for any tributary developed in response to structural controls, he gave acknowledgment to Jukes for his earlier use of the word with reference to the strike valleys of the south of Ireland, although Jukes himself admittedly never intended to employ the word in a technical sense.[18]) Finally, Jukes's paper represents a convincing application of the fluvial doctrine to the geomorphic problems of a specific region and it marked the beginning of the fluvialistic revival.

When he read his paper to the Geological Society of Dublin in May 1862, Jukes was a very recent convert to fluvialism. Indeed, so recent was his conversion that in the second edition of his *Students' Manual of Geology* published in 1862 we find him still proclaiming the marine erosion theory of topography and still emphasising that rivers exist because there are valleys, and not valleys because there are rivers. Jukes later admitted that his conversion had taken place early in 1862,[19] and it must have occurred sometime before the end of April because on 28th of that month, in the letter already cited, he wrote to Murchison requesting permission to read his paper on the southern Irish rivers. Now it was on 5th March 1862 that Ramsay read his paper on the glacial excavation of rock basins to the Geological Society of London, and the present writer believes that the close time relationship between Ramsay's elimination of the limnological objection to fluvialism and Jukes's adoption of fluvialism can hardly have been fortuitous. Jukes had certainly not been oblivious to the problem posed by the existence of the rock basins because he observed in 1862:

> The formation of lakes lying in 'rock-basins' . . . had always been a complete puzzle to me until I read Professor Ramsay's paper in the last Number of the Geological Journal.[20]

Thus, on his own admission, Jukes first appreciated the true nature of rock basins as a result of the publication of Ramsay's paper. It is nevertheless difficult to believe that he remained entirely unaware of Ramsay's theory prior to its appearance in print. He may not have been in London to hear the paper read, but he and Ramsay were close friends, and they must surely have discussed the glacial erosion theory in their correspondence during the first three months of 1862. It is therefore not unreasonable to suggest that Jukes's conversion to fluvialism was a result of his suddenly appreciating the fluvialistic implications of Ramsay's glacial erosion theory. The paper which Jukes read to the Dublin Geological Society in May 1862 may perhaps be regarded as a natural outgrowth of the paper which Ramsay presented to the Geological Society of London two months earlier.

In October 1862 the British Association for the Advancement of Science met at Cambridge, and Jukes returned to his *alma mater* as the president of Section C (Geology). A presidential address was of course expected of him, and in the flush of his new-found enthusiasm for the fluvial doctrine he chose to devote his prelection to a discussion of the processes responsible for shaping the Earth's topography.[21] The address falls readily into two sections. The first section was obviously directed at Murchison and William Hopkins, and it was a well reasoned demonstration of the fact that topography is almost never the direct result of seismic convulsions. Jukes protested at those 'educated men' who believed in the existence of valleys of elevation and who adduced the cliffs on either side of the Straits of Dover as proof that the straits had been torn open by some violent convulsion. In the second part of the address Jukes affirmed his belief that almost all of the Earth's topography is not the result of the sudden unleashing of titanic forces inside the globe, but is rather a cumulative response to the slow but prolonged activities of the external forces of denudation. He admitted that in some places valleys and scarps may be developed along faults, but this, he observed, is only because denudation is commonly guided by pre-existing lines of structural weakness. He went on to draw a distinction which is all too often forgotten by geologists even today – the distinction between fault scarps which are formed directly as a result of earth-movements, and fault-line scarps which are the result of differential denudation along some ancient dislocation. Interestingly, Jukes commented that it was the detailed field-mapping in Ireland on the Six Inch maps which had satisfied him that topography normally owes little to the underlying fold and fault structures.[22]

Those geologists at the Cambridge meeting who had already heard Jukes's paper on the southern Irish rivers must have perceived the end towards which their president was working in his address. When it came, however, Jukes's expected reaffirmation of his fluvialistic faith was in much less emphatic terms than might have been anticipated. Indeed, by October 1862 Jukes seems to have withdrawn somewhat from the position which he had taken up five months earlier. He still believed in the geomorphic significance of rain and river action, but it seems that his mental pendulum was now swinging back in favour of the marine erosion theory of topography. He observed, for example, that the ocean is 'the grandest' of the geomorphic agents, and he reverted to his pre-1862 attitude by claiming that marine agencies have shaped innumerable inland cliffs and escarpments, and that the waves were even responsible for blocking out the broad outline of the world's mountain ranges. He still maintained that the fluvial processes had played the major part in stripping many hundreds of feet of rock off the region today occupied by the Irish midlands, but otherwise he believed that rain and rivers had merely added the detail to a topography which was essentially the product of wave and current action. Thus by October 1862 Jukes's geomorphic thinking was virtually back where George Poulett Scrope's had been in the 1820s, and Jukes's relapse from fluvialism between May and October 1862 is reminiscent of Lyell's relapse from the glacial theory during the early months of 1841.

The reasons for Jukes's change of opinion are far from clear. Perhaps something said during the discussion following the reading of his paper at the Geological Society of London caused him to have second thoughts upon the subject. Unlike Lyell, however, Jukes rapidly recovered from his relapse. As his Survey colleagues rallied around the standard which he had raised, Jukes's flagging faith in the fluvial doctrine was speedily restored. One event which helped to dispel such lingering doubts as remained in his mind was his receipt in February 1863 of a complimentary copy of Lieutenant Ives's great memoir on the Colorado river. The magnificent illustrations contained therein, together with Newberry's excellent discussion of the Colorado landscapes in terms of the fluvial processes, were sufficient to convince Jukes that his Irish paper of the previous year had not been merely idle speculation. Nevertheless, as late as 1866 he was still concerned about the evident slowness of modern changes wrought by rain and rivers, and we find him wondering whether rates of denudation might not have been very much more rapid in the geological past at a time before grasses had arrived upon the scene to bind and protect the Earth's superficial mantle.[23] (Interestingly, this notion has been revived by some geomorphologists of the last two decades.) By 1866, however, Jukes was again a confirmed fluvialist, and in that year as he looked back over the period of his conversion he observed:

> This conclusion, to which I found myself unconsciously and almost reluctantly brought, acted on me like a sudden revelation. It connected together and explained to me all that had been mysterious in the 'form of ground' in Wales and England, and other parts of the world, during my observations of the last thirty years. . . .[24]

Between 1862 and 1863 Jukes thus became the first British geologist since the days of Hutton and Playfair to subscribe unreservedly to the fluvial doctrine. In tracing the history of British geomorphology we have encountered many epoch-making works, but Jukes's paper of May 1862 marks the end of one era and the beginning of another more clearly than does any other single contribution to the subject.

Andrew Ramsay and the Wealden Denudation

Andrew Crombie Ramsay's geomorphic studies have earned him repeated mention in the two previous chapters. He was born in Glasgow in 1814 and he is therefore entitled to a place among that sturdy band of Scots who from the days of Hutton and Playfair onwards have done so much to further the progress of the Earth-sciences.[25] The death of his father in 1827 left the family in somewhat straitened circumstances, and Ramsay was in consequence forced to take employment with a firm of Glasgow linen merchants. He found no happiness in the world of commerce, however, and about 1836 he began to take solace in the study of geology using Lyell's *Principles* as his guide. Family holidays spent in Arran gave him some field experience and he soon began to construct both a map and a model of the island to illustrate its geological structure. Now came the turning-point in his career. In 1840 the British Association met in Glasgow and as the local authority on the geology of Arran, Ramsay was asked to read a paper to Section C upon the subject and to exhibit his model of the island. In addition he was scheduled to conduct a party of the Association's members on a field excursion to Arran, but to his everlasting regret he overslept on the morning in question and he missed the boat across to the island. Nowadays perhaps few excursion leaders' reputations could survive such a fiasco, but his model of Arran, together with his explanatory discourse, had already made a profound impression upon the leading lights of the geological world who were present at the meeting that year, and this impression was by no means effaced as a result of Ramsay's temporary failure as an excursion leader. Murchison in particular was delighted with the young man's obvious enthusiasm for geology, and in the following year he used his influence to secure Ramsay a post on the staff of the Geological Survey as an Assistant Geologist. Ramsay joined De La Beche at Tenby for training in April 1841 and within only four years he had shown such outstanding aptitude for the work that he had been promoted to the office of Local Director for Great Britain. Two years later, in 1847, Ramsay added academic responsibilities to his Survey duties when he was appointed Professor of Geology at University College, London, and when Murchison died in 1871, Ramsay was his obvious successor as the Survey's Director-General. He remained Director-General until 1881 when, after receiving the customary knighthood, he retired to live at Beaumaris in North Wales, close to the place where as a young Survey officer he had first met the Welsh girl who was now Lady Ramsay. There he died in 1891, and he lies buried in the ancient Anglesey churchyard of Llansadwrn. His grave looks out across the Menai Straits and towards a magnificent panoramic view of the mountains of Snowdonia where Ramsay toiled for so many years. Very appropriately his headstone is a glacial erratic.

Ramsay early displayed a keen interest in landforms, and readers will recollect the memoir which he published in 1846 explaining the morphology of South Wales as being largely the work of marine erosion. His geomorphic interests were further stimulated as a result of a series of Alpine excursions, the earliest of which took place in 1852. As a geomorphologist, and as a close friend of Jukes, Ramsay was fascinated by Jukes's paper on the Irish rivers. In May 1864 Ramsay wrote to Archibald Geikie:

> Have you brooded patiently for six months without ceasing over that passage at the end of Jukes's memoir on the Irish rivers, in which he discusses the valley of the Rhône above the Lake of Geneva? It is admirable and true, and by'r lakins! he never saw the location![26]

By the time he wrote this letter the fluvial doctrine had already replaced the marine erosion theory in Ramsay's mind, and Jukes's paper, as Ramsay himself admitted, had clearly played a major role in effecting this transformation. It is one of the strange facts of history that it was evidently Ramsay's paper on the glacial excavation of rock basins which made it possible for Jukes to adopt a fluvialistic interpretation of the Irish landscapes, and that it was Jukes's paper which in its turn played a major role in causing Ramsay to abandon the marine erosion theory in favour of the fluvial doctrine.

The shift in Ramsay's thinking upon the subject of denudation first became apparent during a course of six working men's lectures that he delivered at the Royal School of Mines in London during January and February 1863. The lectures were published in March of the same year as Ramsay's very successful *Physical Geology and Geography of Great Britain* which ran through six editions, the last being published in 1894. The book of the lectures is primarily geological in character, but Ramsay's deep concern for geomorphic problems is evident throughout the work. At the time when he delivered the six lectures his conversion to fluvialism remained incomplete and he retained a vestige of his earlier belief in the geomorphic importance of marine erosion. For example he still believed that the Oolitic escarpment of southern England was a fossil sea-cliff, and he observed:

> The sea waves on the cliffs by the shore are the only power I know that can denude a country, so as to shave it across and make a plain either horizontally or slightly inclined.[27]

335

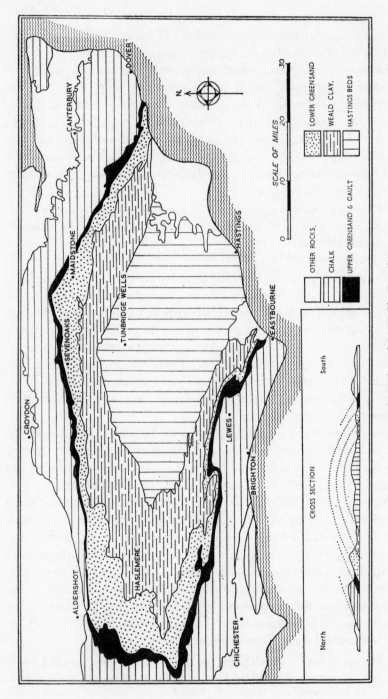

Fig. 4 *The Geology of the Wealden Denudation*

Strangely, Thomas Burnet and many another early naturalist had during the seventeenth century possessed a very much better understanding of the base levelling potential of rain and rivers than did Ramsay in 1863.

By 1863 Ramsay did nevertheless attach considerable geomorphic significance to the fluvial processes as the following passage reveals.

> Water running over the surface wears away the ground over which it passes, and carries away detrital matter, such as pebbles, sand, and mud, and if this goes on long enough over large areas, there is no reason why any amount of matter should not in time be removed.[28]

His new-found fluvialistic faith emerges most clearly in the third of his six lectures when he offered a discussion of the Wealden Denudation which is really a development of the ideas which Jukes had tentatively proffered the previous year in the postscript to his Irish paper. Ramsay observed that the Chalk, Upper Greensand, Gault Clay, Lower Greensand, and Weald Clay (Figure 4) must all once have extended in continuous sheets over the crest of the Wealden pericline, and he confessed that, following Lyell, he had hitherto attributed the removal of all these rocks to marine processes, working at a time of high sea-levels. Now, however, he admitted to entertaining grave doubts about this traditional interpretation of the Denudation. He still thought that the sea might have played some small part in the initial breaching of the pericline, but he was now satisified that the greater part of the work, together with the task of shaping 'all the present details, great and small, of the form of the ground', had been performed by the fluvial processes alone.

Ramsay went on to offer six convincing arguments against the marine erosion theory of the Wealden denudation.

1. If the sea really had worked over the Weald, then surely it would have produced a smooth plain of marine denudation and not a topography diversified with scarps and vales.

2. If the Weald has only recently emerged from the sea, then why are the region's vales not floored with marine sediments?

3. If the sea cut the Wealden escarpments, then the sea-level must once have stood at about the position of the present 400-foot contour. Such a submergence, however, would bring into existence a very strange pattern of islands and narrow straits quite unlike anything to be found in the modern world.

337

4. If the Weald was once flooded, then the narrow straits between the cuesta-islands would provide insufficient fetch to allow the generation of the powerful waves demanded by the marine erosion theory.

5. If the great Wealden escarpments were formed by wave action, then why did the waves not cut similar escarpments on the dip slopes of the cuestas?

6. If the escarpments are former sea-cliffs, then how does it come about that the level of their base varies from region to region instead of maintaining a constant altitude slightly below that of the former sea-level?

Then, having rejected the marine erosion theory of the Denudation, Ramsay proceeded to outline his own simple denudation chronology of the region. He suggested that south-eastern England had been folded into a dome following the deposition of the Cretaceous Chalk and that marine erosion had then removed the crest of the dome to reveal the pre-Cretaceous rocks in a series of concentric outcrops. Next, the region emerged from the sea and a series of transverse rivers developed draining radially down the flanks of the dome at right angles to the geological strike. These rivers, the ancestors of such modern rivers as the Stour, Medway, Ouse, and Arun, cut the impressive water gaps which today break the continuity of the chalk uplands forming the rim of the Denudation. Finally, a number of longitudinal rivers developed within the Denudation and these etched out the strike of the weaker formations to yield a series of broad vales separated from each other by cuestas developed upon the more resistant strata. Thus did Ramsay display his sound understanding of Jukes's concept of the adaptation of drainage to structure.

Early in 1863 Ramsay perhaps still harboured more doubts about the fluvial doctrine than he cared to admit to publicly. Like Jukes he was certainly still puzzled at the evident slowness with which the fluvial processes perform their work, and inwardly he may even have entertained some slight misgivings about the fluvial origin of the Wealden Denudation. By the end of the year, however, these final shreds of doubt were fast disappearing and in November we find him writing to Archibald Geikie as follows:

> By the way, I think I have given up the marine denudation of the Weald. Atmosphere, rain, and rivers must ha' done it. I'm coming to that, I fear and hope, and hoping, fearing, trembling, regretfully triumphant, and tearfully joyous with the alloy of despair at my heart, and the balm of

a truthful Gilead spread upon the struggling soul, bursting the bonds of antique prejudice, I yet expect to moor the tempest-tossed bark of Theory in the calm moral waters of Assurance.[29]

Here we have an insight into the anguish that Ramsay suffered as he transferred his allegiance from the marine erosion theory of landscape to the fluvial doctrine. It is a mark of his intellectual stature that he had the insight and courage to reject a theory which he had cherished for more than twenty years, but by the end of 1863 Ramsay's climacteric was over and the fluvial doctrine had secured its second important British convert.

Archibald Geikie and the Scenery of Scotland

The third mid-nineteenth-century British geologist to become a convert to fluvialism was Archibald Geikie, and through his fluent and prolific writings he soon became the theory's most effective champion. The remarkable similarities in the careers of Geikie and Hutton were noted earlier. Geikie was born in Edinburgh in 1835,[30] and his interest in geology was first aroused as a schoolboy when he discovered a fossil in the Burdiehouse Limestone to the south of Edinburgh. He immediately began a serious study of the subject, and although he was soon familiar with the works of De La Beche and Lyell, he always maintained that he had found his greatest early inspiration in Hugh Miller's *Old Red Sandstone*. He was soon so enamoured of geology that when in 1851 he was given the opportunity of spending his summer holidays either in London visiting the Great Exhibition, or in that geological treasure chest which is the Isle of Arran, he hesitated scarcely a moment before accepting the latter alternative. Geikie was an instinctive writer, and his geologising in Arran that summer resulted in the publication of two newspaper articles. These happened to catch the eye of Hugh Miller himself and he was so impressed by the young man's efforts that he arranged a meeting with the sixteen year old author. Despite the great disparity in their ages, geology was a common bond which held the two men in a firm friendship until Miller's tragic suicide in 1856.

After leaving school Geikie went to work in the office of a Writer to the Signet, but, like Hutton a century earlier, he soon discovered that he had no interest in legal affairs. He thereupon abandoned his legal career and entered the University as a student of the humanities. Surprisingly, he never attended classes in geology, and he later explained that he had seen no point in enrolling for this particular course because it was being taught to an antiquated syllabus by the senile Robert Jameson. He nevertheless persisted with his private geological studies, and in 1853, when Ramsay visited Edinburgh to see whether the work of the Geological Survey could be extended into Scotland, it was Geikie who conducted the visitor on an excursion to Arthur's Seat. As a result of this encounter with Ramsay, and backed by a warm recommendation from Hugh Miller, Geikie was in 1855 appointed to the staff of the Geological Survey.

The excellence of Geikie's work soon attracted the attention of Murchison, the Director-General, and when in the autumn of 1860 Murchison decided to make a geological tour of the Scottish Highlands he invited Geikie to be his travelling companion. Together they sailed from Greenock to Islay and Jura, and they travelled thence through much of the Western and Grampian Highlands, the return journey being made via Blair Atholl where they paused awhile in order to make a pilgrimage to Hutton's intrusive granite body in near-by Glen Tilt. This Scottish tour is still remembered today because the erroneous interpretation of Highland geology which Murchison and Geikie then arrived at laid the foundations for the famous Highland Controversy, but the tour also has significance in our present context. Geikie had always possessed an interest in topography, but he tells us that it was in the Highlands in the autumn of 1860 that he first began to appreciate the extent of the circumdenudation necessary to give Scotland its present configuration. This lesson was powerfully reinforced during the following year when Geikie toured Auvergne and Haute-Loire with Scrope's book on the extinct volcanoes of central France as his guide. Soon after this continental excursion Geikie wrote of the area around Clermont Ferrand:

> It is not without an effort, and after having analysed the scene, feature by feature, that the geologist can take it all in. But when he has done so, his views of the effects of subaerial disintegration become permanently altered, and he quits the district with a rooted conviction that there is almost no amount of waste and erosion of the solid frame-work of the land which may not be brought about in time by the combined influence of springs, frost, rain, and rivers.[31]

This passage, it should be noted, was published early in 1862 at just about the same time that Ramsay, Geikie's friend and colleague, read his paper on the glacial excavation of rock basins to the Geological Society of London. We know from the passage cited earlier that Geikie had long been perplexed by the existence of the rock basins, and he was certainly a very early convert to Ramsay's theory. Here again, therefore, we may have further evidence of a causal relationship between Ramsay's removal of the limnological objection to the fluvial doctrine, and the adoption of the doctrine by another of the younger British geologists.

As soon as Geikie returned from France in 1861 he began to collect material for a book on the scenery of Scotland and this task occupied most of his spare time for the next four years. He gave the subject a preliminary airing in October 1864 when he read a paper to the Geological Society of Glasgow on the origin of the Scottish valleys.[32] In the paper he was primarily concerned to demonstrate that the valleys in question could not be seismic fissures, and it is difficult to avoid the conclusion that he was really seeking to controvert the catastrophic theories which he had doubtless heard Murchison expounding during their Scottish tour four years earlier. Interestingly, some of the arguments which Geikie now employed to challenge the few remaining catastrophists and to support his own fluvialistic interpretations of topography, were the very arguments which Playfair had employed in the same situation more than sixty years earlier.

Geikie's book entitled *The Scenery of Scotland viewed in Connexion with its Physical Geology* was completed in the following year; the Preface is dated 2nd June 1865 at Ayr where Geikie was then stationed on Survey work. The book is written in his usual lucid and highly readable style (his qualities as an author perhaps owed much to the fact that his formal education was in the classics rather than in science) and it is an able exposition of the thesis that Scotland owes most of its present form to the long-continued action of rain and rivers. He admitted that while developing this theme he had found valuable inspiration in the writings of Hutton, Playfair, and Ramsay, but it was to Jukes that Geikie expressed a very special debt of gratitude.

> Although I have long held the belief of Hutton, that our valleys are mainly the work of atmospheric waste, the history of the process of their excavation was but dimly understood by me until the appearance of the admirable paper by my colleague, Mr. J. B. Jukes, on the River-Valleys of the South of Ireland.[33]

As we will see, the influence of Jukes's 1862 paper is evident throughout Geikie's book.

341

By 1865 Geikie's geological studies had made him thoroughly familiar with both the Scottish Highlands and the Southern Uplands, and one feature of the country's topography had profoundly impressed him. He had noticed that so often the climber attaining the summit of a Scottish mountain is greeted not by a panorama of rugged peaks rising to widely differing altitudes, but by a remarkably even skyline formed as a result of the gently convex crests of the surrounding mountains all rising to approximately the same level. Geikie claimed that this general accordance of summit level proves that Scotland is a dissected plateau, and he regarded the convex summits of the Scottish mountains as preserving the last surviving fragments of an ancient plain of marine denudation similar to the wave-cut surface on which Jukes supposed the Irish rivers to have originated. Geikie claimed that in Scotland the relics of this ancient platform today stand at altitudes varying from 1000 to 4000 feet.

Here we must note the existence of a difference between the theories of Jukes and Geikie. Jukes held that the Irish plain of marine denudation was the initial land surface upon which all the major Irish rivers had taken their origin. Geikie, on the other hand, regarded Scotland as what Hutton would have termed a 'compound mass', and he believed that the summit surface of the Scottish mountains was merely an exhumed feature exposed during the reduction of Scotland from some still higher level. He maintained that the surface had been formed while Scotland was undergoing a slow submergence during Upper Silurian and Devonian times, and that it had been cut partly as a result of wave action and partly as a result of the abrasive work of grounding icebergs. Then, when the submergence was complete, Devonian and Carboniferous rocks were laid down over the marine erosion platform, and he suggested that the Scottish Old Red Sandstone consists largely of debris worn from the country as the waves cut their way across the slowly sinking land mass. Finally, at some unspecified time, the whole region was again uplifted and the former sea-floor was exposed to the sub-aerial attack which has given Scotland its present morphology.

Geikie discussed the early history of the newly emerged sea-floor in the following passage.

> That surface is not a mere dead level, so that when rain falls upon it drainage necessarily sets in from the highest parts down to the shores. The rain gathers into runnels, following the inequalities of the sea-worn slopes, and widening into brooks and rivers; or the moisture falls in the

form of snow, and glaciers grind a path for themselves from the high grounds to the shore. Thus begins the scooping out of a system of valleys diverging from the higher parts of the rising land. These depressions are slowly dug deeper and wider, until at last the ancient elevated sea-bed is worn into a system of hills and mountains, valleys and glens.[34]

Where these valleys and glens are concerned, Jukes's influence upon Geikie is again clearly in evidence. Geikie believed that soon after the emergence two types of valley began to form. Firstly, there were the Scottish equivalents of the Shannon, Slaney, Barrow, and Nore; a series of transverse streams flowing either north-westwards or south-eastwards at right angles to the country's Caledonian strike and developed in response to the topographical dips of the youthful Scottish plateau. For his best examples of modern rivers descended from these ancient transverse streams he turned to the Southern Uplands where he cited the Clyde, Doon, Annan, Nith, and Dee as streams which are markedly discordant in relation to the underlying geological structures. Secondly, he drew attention to the existence of Scottish counterparts to the Blackwater, Lee, and Bandon; longitudinal streams which have etched out the geological strike and which in Scotland flow either north-eastwards or south-westwards. As examples of features of this type he offered glens Mor and Carron, the troughs occupied by lochs Awe, Linnhe, and Tay, and the valleys of the upper Tweed, the Yarrow Water, the Ettrick Water, and the Teviot. Thus Geikie, like Ramsay, had fully accepted Jukes's concept of the adaptation of drainage to structure.

Implicit in the idea of drainage adaptation to structure is of course the notion that valleys are the result of fluvial degradation, and in his book Geikie again emphasised that valleys are not crustal fissures. His Survey mapping had satisfied him that while many Scottish valleys are aligned along the regional strike, very few of them are developed along faults.

> But that our valleys and ravines are not mere cracks, would seem to be put beyond dispute by the fact that for one valley which happens to run along the line of a dislocation, there are, I dare say, fifty or a hundred which do not.[35]

He held that as a general rule surface morphology bears little direct relationship to the underlying tectonic structures, and he was at some pains to dispose of the notion that anticlines give rise to mountains while synclines form valleys. Indeed, he stressed that in many parts of

343

Scotland denudation has produced an inversion of relief so that the synclines now underlie the mountains while the adjacent valleys are developed over anticlines.

One of the outstanding features of Geikie's book is an excellent discussion of the subaerial processes, and within this section of the work the high-light is to be found where he conducts his readers on an imaginary journey down a river valley in order to examine the landforms produced there by the fluvial processes. Here he employed his narrative powers to their fullest effect in presenting a most graphic picture of a valley excavated by a swift-flowing torrent and opened out as a result of landslides and rain-wash. Few of his readers can have remained unmoved by this sound presentation of the subject, and following on from this microcosmic example, he proceeded to demonstrate that it must have been the action of the very same fluvial processes which, assisted by glacier ice, has given the whole of Scotland its present configuration. The country's mountains he urged, are nothing more than a circumdenuded residue existing today only because Nature's destructive forces have as yet not completed their appointed task. His vivid portrayal of Scotland as a decaying ruin is one that would have horrified and disgusted any of the teleological naturalists of the previous century.

By 1865 Geikie had completely rejected the marine erosion theory of topography, but he nevertheless remained convinced of the potency of wave attack. Strangely, even in his *Scenery of Scotland* we find 36 pages given over to a discussion of the work of the sea, but only 24 pages devoted to a consideration of the fluvial processes, and in 1865 he certainly still believed that only the sea was capable of planating a land mass. As we will see, however, his growing appreciation of the true power of the fluvial processes was soon to lead him to a revision of his views on the relative importance of submarine as opposed to subaerial processes.

In the Preface to his book, after acknowledging the inspiration which he had found in the geomorphic principles first enunciated by Hutton and Playfair, Geikie continued:

> I can claim nothing more than to have tried in some detail to develop these principles in an inquiry into the origin of the existing scenery of Scotland. The views to which I have been led, however, run directly counter to what are still the prevailing impressions on this subject, and I am therefore prepared to find them disputed, or perhaps thrown aside as mere dreaming.[36]

Here he perhaps misjudged the temper of his age because by 1865 the fluvialistic revival had already made considerable headway, but when he wrote this passage he must surely have had Murchison very much in mind. Murchison remained an unrepentant catastrophist, and it was therefore a somewhat provocative act upon Geikie's part when he chose to dedicate his book to no less a person than Sir Roderick himself. Those less charitably inclined towards Geikie might regard this dedication as a blatant attempt to curry favour in high places, and if it was so intended, it could be regarded as having paid off handsomely. Geikie expected a rejoinder from the Director-General, but none ever came and the book did nothing to damage the friendship which had developed between the two men during their Scottish tour of 1860. In 1867 Murchison made Geikie the Local Director of the Geological Survey of Scotland and four years later he was instrumental in securing Geikie's election to the newly established chair of geology in the University of Edinburgh. By the time Murchison died later in 1871 Geikie's star was set firmly in the ascendant. In 1881 he succeeded the ailing Ramsay as Director-General of the Geological Survey, in 1891 he was knighted, and in 1901 he retired from the Director-Generalship only to begin a new career as secretary and later president of the Royal Society. He died at his home at Haslemere in Surrey in November 1924 full of years and replete with honours. He was the last of that long line of great British geologists whose names had been household words among men of science throughout the world.

Fluvialism Triumphant

Jukes, Ramsay, and Geikie had all struck powerful blows on behalf of the fluvial doctrine and there ensued a major controversy as the opponents of the doctrine tried to stem the rising tide of its influence. The significance of this resistance must nevertheless not be exaggerated. By the 1860s the objections which had been raised against the doctrine in earlier decades were fast disappearing, if they had not already been completely eliminated, and after 1862 the fluvialists had to face little that can be called reasoned criticism. The controversy which took place during the 1860s was really little more than a rear-guard action fought by those older geologists who were determined to remain loyal to geomorphic theories which had been popular earlier in the century.

The adherents of three other rival theories of landscape evolution were involved in this final struggle with fluvialism. Firstly, there were a few lingering neo-diluvialists; secondly, there was a somewhat larger group of cataclysmic geologists; and finally, there were the numerous geologists who were still anxious to interpret topography in terms of marine erosion. Of these three groups the neo-diluvialists may be dismissed briefly. Apart from Murchison, Joseph Prestwich of Oxford was the only neo-diluvialist of any standing after 1862, and although he for several decades continued to maintain that many landscapes had been moulded by torrents draining off the continents after some recent inundation, few deemed his theory worthy of serious consideration.

Murchison was still the most prominent figure among the cataclysmic geologists, still holding aloft his now faded and tattered banner of catastrophism, and still convinced that eventually geological opinion would swing around to his point of view. In October 1864 he was labouring under a strange delusion as to the real state of British geological thought when he wrote:

> In common with what I hold to be the opinion of by far the greater number of practical geologists, I believe that most of the valleys in mountain-chains owe their first traces not only to fractures, but often to great and rapid convolutions, or foldings of the strata, which have left depressions where sharp synclinal lines or narrow troughs have been formed between up-raised masses of rock.[37]

Almost five years later he was still trying to convince the Fellows of the Royal Geographical Society that the cliffs on either side of the Straits of Dover and the Irish Sea afforded ample proof that France, Britain, and Ireland had been torn asunder by some violent convulsion.[38] In his address to the British Association at Cambridge in 1862 Jukes had specifically protested at the absurdity of such an interpretation of marine cliffs, but Murchison was deaf to the reasoned pleas of his Local Director in Ireland. Jukes himself must have been chagrined to know that even among his own Survey staff in Dublin Murchison had a stalwart supporter in the burly form of George Henry Kinahan. In 1869, the year of Jukes's death, we find Kinahan maintaining, as had de Luc almost a century earlier, that a landscape is immune from the attack of rain and rivers once it has acquired a mantle of soil and grass,[39] and in 1878 he went so far as to claim that

> ... in Ireland, in general, the rivers are due to the valleys, not the valleys to the rivers; the valleys occupying dykes of fault-rock or lines of breaks or other shrinkage fissures in the strata or accumulations in which they are situated.[40]

Although Murchison himself never published a reply to Geikie's *Scenery of Scotland*, Geikie did not escape attack from the camp of the catastrophists. In both a paper read to the Geological Society of London in 1868, and a presidential address to the same body five years later, the Duke of Argyll sought to show that rain, rivers, and ice had played only an insignificant role in shaping the Scottish landscape and that in Argyll, at least, the mountains and valleys are essentially the result of recent differential earth-movements.[41]

The refusal of the catastrophists to be weaned away from their favourite doctrine caused the fluvialists little discomfort. Murchison and his disciples were few in number, and the fluvialists recognised that the catastrophic creed was merely a lingering survival from a bygone age. They could afford to leave the antiquated doctrine to die a natural death. Much more serious was the opposition to fluvialism offered by those who subscribed to the marine erosion theory of topography. Theirs was a modern theory dating no further back than the 1830s, and through the influential writings of Lyell the theory had achieved currency throughout the world. After 1862, however, Lyell left the defence of the marine erosion theory to pens other than his own, and in Britain the theory's chief advocate was now Daniel Mackintosh.

Mackintosh was yet another geologist of Scottish origin. He was born at Blairgowrie in 1815 and after moving to England about 1845 he became a very successful lecturer on geology and other scientific subjects. He was elected to Fellowship of the Geological Society of London in 1861, and later, after taking up residence in Birkenhead, he served as the president of the Liverpool Geological Society. His most important research lay in the field of the English glacial drifts, but in 1869 he published a book on the scenery of England and Wales which, like Geikie's *Scenery of Scotland*, was dedicated to Murchison.[42] Mackintosh was nevertheless no catastrophist; he believed that almost every topographical feature of England and Wales, from the lynchets of southern England to the corries of Snowdonia and the surface of the Kinder Scout Plateau in the Peak District, had all been shaped by marine agencies during the glacial submergence. Mackintosh had a remarkably detailed knowledge of the configuration of southern Britain (he was still climbing Welsh mountains after he had passed his seventieth birthday) and his book is replete with a wealth of detailed evidence all purporting to support the marine erosion theory. Indeed he evinces a much greater concern for the minutiae of landscapes than

347

does Geikie in his *Scenery of Scotland*, and Mackintosh was clearly anxious that his book should be regarded as an effective reply to Geikie's work. Geikie offered no rejoinder, but some significance may perhaps be read into an event that occurred more than twenty years later. Mackintosh died in July 1891, but when the president of the Geological Society of London presented his anniversary address to the society in the following February, Mackintosh's name was conspicuously absent from the list of deceased Fellows to whom the president accorded obituary notices. The president at the time was none other than Sir Archibald Geikie himself.

When the fluvialistic revival began, Lyell was in his sixty-fifth year and it was perhaps too much to expect of an old man that he should renounce his long-cherished marine erosion theory of topography in favour of the fluvial doctrine. In the sixth edition of his *Elements of Geology*, published in 1865, he curtly dismissed Ramsay's arguments concerning the Weald, and adhered to his old theory that the transverse valleys were fissures, that the escarpments were former sea-cliffs, that the dry valleys in the chalk had been cut by torrents draining off the region as it finally rose out of the sea, and that the greater part of the topography within the Denudation was the result of marine attack.[43] As a general rule, Lyell observed,

> . . . the principal valleys in almost every great hydrographical basin in the world, are of a shape and magnitude which imply that they have been due to other causes besides the mere excavating power of rivers.

By 1867 he had modified his thinking somewhat because in the summer of that year he wrote to William Whitaker as follows:

> I have long ago modified my opinions on denudation, and I now agree with you in considering that the escarpments round the Weald are not inland cliffs, as I formerly supposed, although at some points the sea may have entered through transverse valleys and modified parts of them.[44]

This was as close as Lyell ever came to an acceptance of the fluvial doctrine, and in 1871, when he published a modified fluvialistic interpretation of the Wealden Denudation, he continued to insist that the sea must have played a major part in the excavation of the region's vales.[45]

Despite all this opposition, the fluvial doctrine rapidly gained ground among the younger British geologists after 1862, and here we need do little more than highlight some of the most important contributions indicating the doctrine's return to favour. Pride of place must be given to a paper which was presented to the Geological Society of London in May 1865 by two young officers of the Geological Survey – Clement le Neve Foster and William Topley.[46] Their subject was the Weald and they convincingly demonstrated the validity of the suggestions made concerning the region's history by Jukes in 1862 and by Ramsay in 1863; the Wealden Denudation was indeed the product of the prolonged action of rain and rivers. Next, in May 1867 William Whitaker, another officer of the Geological Survey, read an important paper to the Geological Society demonstrating that the escarpments of south-eastern England could only be subaerial in origin.[47]

Yet another Geological Survey man to enter the fray on behalf of the fluvial doctrine was James Croll, the Secretary in the Survey's Edinburgh office. In 1868 he published a paper which did much to dispel any remaining doubts concerning the efficacy of Nature's subaerial processes.[48] Using estimates of the amount of debris annually carried down to the Gulf of Mexico by the Mississippi, he calculated that the river must be lowering its drainage basin at the rate of one foot every 4566 years. If denudation continues at this rate, he noted, then the entire basin will be reduced to sea-level within a period of about four and a half million years. Such a figure doubtless delighted his fluvialist colleagues, but Croll arrived at a still more heartening conclusion when he compared the estimated rates of marine and subaerial denudation over the entire globe. He considered the work of marine erosion first. Taking 116 531 miles as the total length of the world's coastlines, and assuming the average height of the coast to be 25 feet and the average rate of coastal recession to be one foot every hundred years, he arrived at the conclusion that marine erosion will remove 15 382 500 000 cubic feet of rock from the continents every century. Now the continental surfaces exposed to subaerial denudation amount, he claimed, to a total of 57 600 000 square miles, and assuming, in the light of the calculation for the Mississippi basin, that the continents are lowered by one foot every six thousand years, then we arrive at the conclusion that 26 763 000 000 000 cubic feet of rock must be removed from the continents every century by the action of rain and rivers alone. In other words, Croll claimed, the subaerial processes are destroying the continents 1740 times more rapidly than are the marine processes, and for the sea to achieve the same rate of erosion as rain and rivers it would have to devour the continents at the average rate of 17 feet per year.

Such figures gave weighty support to the fluvialist cause and they served to demonstrate that Lyell and his followers had been in error in believing that only the sea possessed the power necessary to mould a landscape.

This paper of Croll's was instrumental in bringing one eminent scientist over into the fluvialist camp. That convert was Charles Darwin. Initially Darwin had been sceptical about the revival of fluvialism. In July 1865 he wrote to J. D. Hooker about the Parallel Roads of Glen Roy:

> You seem to have been struck with what most deeply impressed me at Glen Roy . . . viz. the marvellous manner in which every detail of surface of land had been preserved for an enormous period. This makes me a little sceptical whether Ramsay, Jukes, etc., are not a little over-doing sub-aërial denudation.[49]

After reading Croll's paper of 1868 these doubts melted away, and in September of that year Darwin wrote to Croll from Down House:

> I was formerly a great believer in the power of the sea in denudation, and this was perhaps natural, as most of my geological work was done near sea-coasts and on islands. But it is a consolation to me to reflect that as soon as I read Mr. Whittaker's [sic] paper on the escarpments of England, and Ramsay and Jukes' papers, I gave up in my own mind the case; but I never fully realised the truth until reading your papers just received. How often I have speculated in vain on the origin of the valleys in the chalk platform round this place, but now all is clear. I thank you cordially for having cleared so much mist from before my eyes.[50]

Jukes's accident in July 1864, and his subsequent years of ill health, prevented him from playing a leading role in the re-establishment of the fluvial doctrine. Both Geikie and Ramsay, however, remained very active in the fluvialist camp for more than a decade. Geikie's most important later contribution to the fluvialist cause was a paper entitled 'On Modern Denudation' which he read to the Geological Society of Glasgow in March 1868.[51] The paper bears obvious similarities to Croll's essay on the same subject published later in the same year and it is worth remembering that Geikie was Croll's chief in the Edinburgh office of the Geological Survey. Indeed, it was Geikie who in the previous year had used his influence to secure Croll's appointment to the Survey after the brilliant but self-taught mathematician-geologist had failed to pass the necessary Civil Service examination.[52] In his paper Geikie, like Croll, used the amount of sediment being transported by rivers as the basis for an estimate of current rates of denudation in various river basins. He concluded that in Europe the rates vary so that it takes 6846 years to erode one foot of rock from over the entire surface of the Danube basin, but only 729 years to remove a similar thickness of rock from the Po basin, and he suggested that the whole of Europe would be reduced to a featureless plain within about four million years.

We now know that Geikie – like Thomas Burnet almost two hundred years earlier – was grossly overestimating the potency of the subaerial processes, but he emerged from his studies entirely satisfied that William Thomson's newly revised appraisal of the Earth's age need hold no terrors for the geologist. In view of the rapidity of subaerial denudation, Geikie observed, Thomson's allowance of 100 million years must be regarded as ample to encompass all the events of Earth-history. Indeed, Geikie was now so convinced of the effectiveness of rain and rivers that he withdrew his earlier claim that only the sea had the power necessary to planate a land mass.

> Hence, before the sea, advancing at the rate of ten feet in a century, could pare off more than a mere marginal strip of land, between 70 and 80 miles in breadth, the whole land would be washed into the ocean by atmospheric denudation.

In the light of this realisation he revised his earlier interpretation of Scottish geomorphology by suggesting that the summit surface of the Highlands and Southern Uplands was in all probability subaerial rather than marine in origin. It was this paper from Geikie, together with Croll's similar paper published later in the same year, which sounded the death knell of the marine erosion theory as a serious contender for a place in modern geomorphology.

Ramsay's chief contribution to the advancement of the fluvial doctrine after 1865 is to be found in a series of three papers which he presented to the Geological Society of London in 1872, 1874, and 1876.[53] The first of this trio of papers was a study of the major English rivers, and Ramsay tried to trace their evolution just as Jukes had traced the evolution of the Irish rivers ten years earlier. The second paper was a similar study concerned with the history of the Rhine, while the third paper was a discussion of the development of the river Dee in North Wales. Sadly, these three papers cannot be numbered among Ramsay's most successful works. The reconstruction of ancient drainage patterns is admittedly a task fraught with difficulty, and Ramsay's essay into this field was in any case probably premature, but even after such allowances have been made, it still has to be admitted that the three papers are feeble, incoherent, and poorly reasoned presentations. It is therefore small wonder that when they were first read to the Geological Society all three papers received a somewhat hostile reception at the hands of the Fellows. The papers are nevertheless of significance because of the insight that they afford into the nature of Ramsay's conception of landscape history. In each case he believed that the rivers in question had originated upon a former land surface standing far above the level of the present topography, and he argued that the rivers had all been let down to their present positions as a result of the steady reduction of the British Isles and Europe under long continued fluvial attack. He thus fully appreciated the dynamic character of a landscape, and he certainly no longer entertained any doubts as to the ability of rain and river action to effect topographical changes upon a grand scale. By the early 1870s the majority of British geologists evidently shared Ramsay's view. His three papers were criticised at the Geological Society on points of detail, but, so far as can be judged from the published discussions, the Duke of Argyll was the sole critic of Ramsay's conception of landscape evolution.

In 1865 Geikie had expected to find his fluvialism 'disputed, or perhaps thrown aside as mere dreaming', but by 1879 the issue had been decided and he was able to look back upon the fluvial controversy as a closed chapter in the history of the Earth-sciences. It was in 1879 that he visited Colorado, Utah, Idaho, and Wyoming in order to see the remarkable landscapes which the Americans had been interpreting so effectively in the light of the fluvial doctrine, and while in the West he was delighted to meet two of the leading American geomorphologists – Major John Wesley Powell and Captain Clarence Edward Dutton. Geikie was profoundly impressed by what he saw in the New World, and as he looked back over the history of geomorphology in Britain he was saddened to think of all the effort and ingenuity which had been expended in supporting false theories merely because certain geologists had been blind to the geomorphic significance of rain and river action. After his return from the United States he wrote:

> Now, it is unquestionably true that had the birthplace of geology lain on the west side of the Rocky Mountains, this controversy would never have arisen. The efficacy of denudation instead of evoking doubt, dis-

cussion, or denial, would have been one of the first obvious principles of the science, established on the most irrefrangable basis of patent and most impressive facts.[54]

Geikie was here undoubtedly correct. The drift and vegetation-mantled landscapes of Europe are of complex origin, and they do not readily yield up those geomorphic secrets which are so plainly evident in the arid and semi-arid landscapes of the New World. It is nevertheless very fitting that a pair of topographical features in one of the classical geomorphic regions of the American West should have been named in honour of two of the British geologists who had done so much to establish the truth of the fluvial doctrine. The two features are butte-forming laccoliths in the Henry Mountains of Utah, and they were named after Joseph Beete Jukes and Archibald Geikie by Grove Karl Gilbert, who was perhaps the most outstanding of the early American geomorphologists.

When Geikie crossed to the New World in 1879 one of the best-selling scientific works in Britain was T. H. Huxley's newly published text entitled *Physiography: An Introduction to the Study of Nature*. The book was first published in 1877 and it was regularly reprinted at least once every year until 1885. In the book Huxley emerges as the complete fluvialist. He still believed the sea to be capable of planating a land mass, but otherwise he explained most landscapes in terms of erosion performed by glacier ice and the fluvial processes. He drew his readers' attention to the impressive fluvial landforms of the American West; he reproduced one of the splendid illustrations of the Grand Canyon taken from Powell's memoir of 1875; and he anticipated a day some five and a half million years hence by which time the fluvial processes will have removed all Britain's mountains and converted the region into a low-lying plain. The success of Huxley's book carried the fluvial doctrine into class-rooms and homes throughout Britain, and its appearance three hundred years after the publication of William Bourne's *Treasure for Travellers* marks the end of this survey of the first three centuries in the history of British geomorphology.

REFERENCES

1. *Report upon the Colorado River of the West* (Washington 1861).
2. *Exploration of the Colorado River of the West* (Washington 1875); *Report on the Geology of the Henry Mountains* (Washington 1877).

353

3. C. A. Browne, *Letters and Extracts from the Addresses and occasional Writings of J. Beete Jukes* (London 1871).

4. *The Sir Henry De La Beche Papers* in the National Museum of Wales, Cardiff.

5. Browne, *op. cit.* (1871), p. 462.

6. *Ibid.*, p. 556.

7. *Ibid.*, p. 558.

8. *Geological Survey of Ireland Letter Books.* Letter from Jukes to Murchison (31st October 1864).

9. *Explanations to Accompany Sheets 176 and 177 of the Maps of the Geological Survey of Ireland*, pp. 5–9 (Dublin 1861).

10. *Geological Survey of Ireland Letter Books.*

11. *Journ. geol. Soc. Dublin*, X (1862–64), pp. 51 & 52; 72–74.

12. G. L. Davies, 'The Geological Society of Dublin and the Royal Geological Society of Ireland', *Hermathena*, C (1965), pp. 66–76.

13. *Quart. J. geol. Soc. Lond.*, XVIII (1862), pp. 378–403.

14. *Journ. geol. Soc. Dublin*, VI (1853–55), pp. 178–181; VIII (1857–60), pp. 166 & 167.

15. Francis Darwin, *The Life and Letters of Charles Darwin*, II, p. 296 (London 1887).

16. *The School Manual of Geology*, p. 96 (Edinburgh 1863).

17. *Ann. Ass. Amer. Geogr.*, II (1912), pp. 75 & 76.

18. *Geogr. J.*, V (1895), p. 131.

19. *Rep. Brit. Ass.*, *Cambridge 1862*, pt. 2, p. 64.

20. *Ibid.*, p. 64.

21. *Ibid.*, pp. 54–65.

22. See also Jukes's letter in *The Reader* for 10th December 1864.

23. *Geol. Mag. Lond.*, III (1866), pp. 234 & 235.

24. *Ibid.*, p. 233.

25. Archibald Geikie, *Memoir of Sir Andrew Crombie Ramsay* (London 1895).

26. *Ibid.*, pp. 281 & 282.

27. *The Physical Geology and Geography of Great Britain*, second edition, p. 141 (London 1864).

28. *Ibid.*, pp. 22 & 23.

29. Geikie, *op. cit.* (1895), p. 280.

30. Archibald Geikie, *A Long Life's Work* (London 1924).

31. Francis Galton, *Vacation Tourists and Notes of Travel in 1861* (London 1862); Archibald Geikie, *Geological Sketches at Home and Abroad*, p. 104 (New York 1882).

32. *Trans. geol. Soc. Glasg.*, II (1864–67), pp. 4–12.

33. *The Scenery of Scotland*, p. 141 (London and Cambridge, 1865).

34. *Ibid.*, p. 88.

35. *Ibid.*, p. 9.

36. *Ibid.*, p. viii.

37. *The Reader* (22nd October 1864).

38. *J. Roy. Geogr. Soc.*, XXXIX (1869), pp. cxxxv–cxciv.

39. *Geol. Mag. Lond.*, VI (1869), pp. 109–115.

40. *Manual of the Geology of Ireland*, p. 314 (London 1878).

41. *Quart. J. geol. Soc. Lond.*, XXIV (1868), pp. 255–273; XXIX (1873), pp. li–lxxviii.

42. *The Scenery of England and Wales* (London 1869).

43. *Elements of Geology*, pp. 351–358 (London 1865).

44. *Geol. Mag. Lond.*, IV (1867), p. 449.

45. *The Student's Elements of Geology*, pp. 80 & 81 (London 1871).

46. *Quart. J. geol. Soc. Lond.*, XXI (1865), pp. 443–474.

47. *Geol. Mag. Lond.*, IV (1867), pp. 447–454; 483–493.

48. *Phil. Mag.*, Series 4, XXXV (1868), pp. 363–384. See also *Climate and Time*, Chap. XX (New York 1875).

49. Francis Darwin and Albert C. Seward, *More Letters of Charles Darwin*, II, p. 156 (London 1903).

50. *Ibid.*, II, p. 211.

51. *Trans. geol. Soc. Glasg.*, III (1868–69), pp. 153–190.

52. James C. Irons, *Autobiographical Sketch of James Croll* (London 1896).

53. *Quart. J. geol. Soc. Lond.*, XXVIII (1872), pp. 148–160; XXX (1874), pp. 81–95; XXXII (1876), pp. 219–229.

54. Geikie, *op. cit.* (1882), p. 183.

Postscript

THE history of British geomorphology, like the history of the Earth's surface itself, has been cyclical in character. Spells of intense interest in landforms have alternated with prolonged periods when the subject lay in limbo. Strangely, each of the cycles has been of about a century's duration. The first cycle began towards the close of the sixteenth century, and a waxing interest in Earth-history culminated in the numerous late seventeenth-century theories of the Earth. Then, about 1705, all this interest suddenly evaporated and a new cycle commenced with a long period during which geomorphology lay neglected and forgotten. This second cycle was in its turn completed by the great late eighteenth-century resurgence of interest in landforms which reached its zenith in the writings of Hutton and Playfair. Soon after the appearance of Playfair's *Illustrations* in 1802 this second cycle ended and a third cycle began as the other branches of geology swept vigorously forward leaving geomorphology trailing far behind. The third cycle reached its climax in the years following 1862 when the revival of the fluvial doctrine stimulated a renewed British interest in topography. This, however, was by no means the end of geomorphology's cyclical progression. Soon after 1878 history again repeated itself. While geomorphology flourished in the United States, geomorphology in Britain suddenly went into eclipse, thus initiating yet a fourth cycle in the subject's history. This fourth cycle remains current down to the present day, and it entered its second, apogaeic phase in the 1930s with the beginning of another great resurgence of interest in landforms. Currently there are almost four hundred British scientists researching within the realms of geomorphology and never in its history has the subject commanded greater attention. Strangely, however, this twentieth-century revival of landform studies has been the work not of British geologists, but of British geographers. Not since the days of Jukes, Ramsay, and Geikie has British geology taken any serious interest in the configuration of the Earth's surface.

356

Bibliography

The Bibliography is arranged under the following headings:

> I   Primary Sources
>   A. Works written before 1808
>   B. Works written after 1808
> II   Secondary Sources
> III   Periodicals

Hitherto no comprehensive bibliography of source material for the study of early British geomorphology has been available, and it therefore seemed desirable to list in Section A of the Primary Sources all those works written before 1808 which have been found to contain discussions of landforms. After 1808 the output of geological literature increased rapidly, and any attempt at a complete bibliography of nineteenth-century British geomorphology would assume impossibly large proportions. Section B of the Primary Sources therefore contains only a selection of the literature written after 1808, but all works of any significance have been included.

Part III of the Bibliography lists those scientific periodicals which have been searched and which contain a considerable body of primary geomorphic material.

I. *Primary Sources*

A. Works published before 1808

Arbuthnot, John, *An Examination of Dr. Woodward's Account of the Deluge, Etc.* (London 1697).

Beaumont, John, *Considerations on a Booke, Entituled The Theory of the Earth* (London 1693).

Black, Joseph, *The Joseph Black Papers*, in the Edinburgh University archives.

Black, Joseph, *Lectures on the Elements of Chemistry, delivered in the University of Edinburgh; by the late Joseph Black, M.D.*, edited by John Robison (Edinburgh 1803).

Borlase, William, *The Natural History of Cornwall* (Oxford 1758).

Bourne, William, *A Booke Called the Treasure for Traveilers, devided into five Bookes or Partes* (London 1578).

Buffon, George L. L., Comte de, *Natural History, general and particular . . . translated, . . . with . . . notes and observations by William Smellie*, second edition (London 1785).

Burnet, Thomas, *Telluris Theoria Sacra* (London 1681 and 1689).

Burnet, Thomas, *The Theory of the Earth* (London 1684 and 1690).

Burnet, Thomas, *An Answer to the late Exceptions made by Mr. Erasmus Warren against the Theory of the Earth* (London 1690).

Burnet, Thomas, *A Short Consideration of Mr. Erasmus Warren's Defence of his Exceptions against the Theory of the Earth* (London 1691).

Burnet, Thomas, *Reflections upon the Theory of the Earth, occasion'd by a Late Examination of it* (London 1699).

Camden, William, *Britain*, translated by Philemon Holland (London 1637).

Carpenter, Nathanael, *Geography Delineated Forth in Two Bookes* (Oxford 1625).

Catcott, Alexander, *A Treatise on the Deluge* (London 1761).

Chambers, Ephraim, *Cyclopædia* (London 1738).

Childrey, Joshua, *Britannia Baconica* (London 1661).

Clayton, Robert, *A Vindication of the Histories of the Old and New Testament*, Pt. II (Dublin 1754).

Cockburn, Patrick, *An Enquiry into the Truth and Certainty of the Mosaic Deluge* (London 1750).

Coetlogon, Dennis de, *An Universal History of Arts and Sciences* (London 1745).

Croft, Herbert, *Some Animadversions upon a Book intituled The Theory of the Earth* (London 1685).

Darwin, Erasmus, *The Botanic Garden* (1789 and 1791).

De Luc, Jean A., *Lettres Physiques et Morales, sur les Montagnes et sur l'Histoire de la Terre et de l'Homme* (The Hague 1778).

De Luc, Jean A., *Lettres Physiques et Morales sur l'Histoire de la Terre et de l'Homme* (Paris 1779).

De Luc, Jean A., a series of geological letters addressed to Dr. James Hutton in *The Monthly Review*, II, III, V (1790–1791).

De Luc, Jean A., twenty-one letters in *Rozier's Journal* (*Le Journal de Physique*), XXXVII–XL (1790–1792).

De Luc, Jean A., a series of geological letters addressed to Professor Blumenbach in *The British Critic*, II–V (1793–1795).

De Luc, Jean A., *Letters on the Physical History of the Earth, addressed to Professor Blumenbach*, edited by Henry de la Fite (London 1831).

Derham, William, *Physico-Theology* (London 1713).

Dickinson, Edmund, *Physica Vetus & Vera: sive Tractatus de Naturali veritate hexaëmeri Mosaici* (London 1702).

Drayton, Michael, *Noahs Floud* (1630).

Dugdale, Sir William, *The History of Imbanking and Drayning of Divers Fenns and Marshes, both in Forein Parts, and in this Kingdom* (London 1662).

Edinburgh University Archives, *The Natural History Society of Edinburgh*. Twelve volumes containing MS copies of the papers read before the Society between 1782 and ca. 1806.

Ferber, John J., *Travels through Italy, in the Years 1771 and 1772*, translated by Rudolf E. Raspe (London 1776).

Goldsmith, Oliver, *An History of the Earth, and Animated Nature* (London 1774).

Hakewill, George, *An Apologie or Declaration of the Power and Providence of God in the Government of the World*, third edition (Oxford 1635).

Hale, Sir Matthew, *The Primitive Origination of Mankind* (London 1677).

Hamilton, Sir William, *Campi Phlegræi* (Naples 1776).

Hamilton, William, *Letters Concerning the Northern Coast of the County of Antrim* (Dublin 1786).

Harris, John, *Remarks on some late Papers, Relating to the Universal Deluge : and to the Natural History of the Earth* (London 1697).

Harris, John, *Lexicon Technicum* (London 1704–1710).

Havers, George, *A General Collection of Discourses of the Virtuosi of France, upon Questions of all Sorts of Philosophy, and other Natural Knowledge* (London 1664).

Havers, George, and Davies, John, *Another Collection of Philosophical Conferences of the French Virtuosi, upon Questions of All Sorts* (London 1665).

Hooke, Robert, *Micrographia* (London 1665).

Hooke, Robert, *The Posthumous Works of Robert Hooke*, edited by Richard Waller (London 1705).

Howard, Philip, *The Scriptural History of the Earth and of Mankind* (London 1797).

Hutton, James, *Theory of the Earth, with Proofs and Illustrations*, volumes I and II (Edinburgh 1795); volume III, edited by Sir Archibald Geikie (London 1899).

Jameson, Robert, *An Outline of the Mineralogy of the Shetland Islands, and of the Island of Arran* (Edinburgh 1798).

Jameson, Robert, *Mineralogy of the Scottish Isles* (Edinburgh 1800).

Jameson, Robert, *System of Mineralogy* (Edinburgh 1804–1808).

Jameson, Robert, *A Mineralogical Description of the County of Dumfries* (Edinburgh 1805).

Jameson, Robert, *The Robert Jameson Papers*, in the Edinburgh University archives.

Jones, William, *The Theological and Miscellaneous Works of the Rev. William Jones, M.A.F.R.S.*, IV (London 1810).

Keill, John, *An Examination of Dr. Burnet's Theory of the Earth. Together with some Remarks on Mr. Whiston's New Theory of The Earth* (Oxford 1698).

359

Keill, John, *An Examination of the Reflections on the Theory of the Earth. Together with a Defence of the Remarks on Mr. Whiston's New Theory* (Oxford 1699).

King, William, 'Of the Bogs and Loughs of Ireland', *Philos. Trans.*, XV (1685), No. 170, pp. 948–960.

Kirwan, Richard, 'On the Primitive State of the Globe and its Subsequent Catastrophe', *Trans. R. Ir. Acad.*, VI (1797), pp. 233–308.

Kirwan, Richard, *Geological Essays* (London 1799).

Kirwan, Richard 'An Essay on the Declivities of Mountains', *Trans. R. Ir. Acad.*, VIII (1802), pp. 35–52.

Lovell, Archibald, *A Summary of Material Heads which may be Enlarged and Improved into a compleat Answer to Dr. Burnet's Theory of the Earth* (London 1696).

Mackaile, Matthew, *Terræ Prodromus Theoricus* (Aberdeen 1691).

Maillet, Benoît de, *Telliamed: or, Discourses Between an Indian Philosopher and a French Missionary* (London 1750).

Morton, John, *The Natural History of Northampton-shire* (London 1712).

Murray, John, *A Comparative View of the Huttonian and Neptunian Systems of Geology* (Edinburgh 1802).

Nares, Edward, 'A View of the Evidences of Christianity at the Close of the Pretended Age of Reason', *The Bampton Lectures* (Oxford 1805).

Nicholls, William, *A Conference with a Theist*, Pt. I and Pt. II (London 1696).

P., L., *Two Essays, sent in a Letter from Oxford, to a Nobleman in London, 1695*, in Lord John Somers, *A Collection of Scarce and Valuable Tracts*, XII, pp. 20–33 (London 1814).

Pettus, Sir John, *Fodinæ Regales* (London 1670).

Plattes, Gabriel, *A Discovery of Subterraneall Treasure* (London 1639).

Plattes, Gabriel, *A Discovery of Infinite Treasure, Hidden since the Worlds Beginning* (London 1639).

Playfair, John, *Illustrations of the Huttonian Theory of the Earth* (Edinburgh 1802).

Plot, Robert, *The Natural History of Oxford-Shire* (Oxford 1677).

Plot, Robert, *The Natural History of Stafford-Shire* (Oxford 1686).

Pluche, Noël A., *Spectacle de la Nature: or, Nature Display'd*, translated by Samuel Humphreys, third edition (London 1736–1739).

Pryce, William, *Mineralogia Cornubiensis* (London 1778).

Ray, John, *Observations made in a Journey Through Part of the Low-Countries, Germany, Italy, and France* (London 1673).

Ray, John, *The Wisdom of God Manifested in the Works of the Creation*, second edition (London 1692).

Ray, John, *Three-Physico-Theological Discourses*, second edition (London 1693).

Ray, John, *Philosophical Letters between the Late Learned Mr. Ray and Several of his Ingenious Correspondents*, edited by William Derham (London 1718).

Ray, John, *Memorials of John Ray*, edited by Edwin Lankester, Ray Society Series (London 1846).

Ray, John, *The Correspondence of John Ray*, edited by Edwin Lankester, Ray Society Series (London 1848).

Ray, John, *Further Correspondence of John Ray*, edited by Robert W. T. Gunther, Ray Society Series, CXIV (London 1928).

Richardson, William, 'Inquiry into the Consistency of Dr. Hutton's Theory of the Earth with the Arrangement of the Strata, and other Phænomena on the Basaltic Coast of Antrim', *Trans. R. Ir. Acad.*, IX (1803), pp. 429–487.

Richardson, William, 'A Letter on the Alterations that have taken place in the Structure of Rocks, on the Surface of the Basaltic Country, in the Counties of Derry and Antrim', *Philos. Trans.* (1808), pp. 187–222.

Robinson, Thomas, *The Anatomy of the Earth* (London 1694).

Robinson, Thomas, *New Observations on the Natural History of this World of Matter, and this World of Life* (London 1696).

Robinson, Thomas, *A Vindication of the Philosophical and Theological Exposition of the Mosaick System of the Creation* (London 1709).

Robinson, Thomas, *An Essay Towards a Natural History of Westmorland and Cumberland* (London 1709).

Rowlands, Henry, *Mona Antiqua Restaurata* (Dublin 1723).

St. Clair, Robert, *The Abyssinian Philosophy Confuted* (London 1697).

Shakelton, Francis, *A Blazyng Starre or Burnyng Beacon* (London 1580).

Sneyd, Ralph, *A Letter to Dr. Toulmin, M.D., Relative to his Book on the Antiquity of the World* (Lewes 1783).

Steno, Nicolaus, *The Prodromus to a Dissertation Concerning Solids Naturally Contained within Solids*, translated by Henry Oldenburg (London 1671).

Stillingfleet, Edward, *Origines Sacræ* (London 1663).

Strachey, John, 'An Account of the Strata in Coal-Mines, &c.', *Philos. Trans.*, XXXIII (1724–1725), No. 391, pp. 395–398.

Strachey, John, *Observations on the Different Strata of Earths, and Minerals* (London 1727).

Stukeley, William, *Itinerarium Curiosum* (London 1724).

Sulivan, Sir Richard J., *A View of Nature, in Letters to a Traveller among the Alps, with Reflections on Atheistical Philosophy, now Exemplified in France* (London 1794).

Swan, John, *Speculum Mundi or a Glasse representing the Face of the World* (Cambridge 1635).

Toulmin, George H., *The Antiquity and Duration of the World* (London 1780).

Toulmin, George H., *The Eternity of the World* (London 1785).

Toulmin, George H., *The Eternity of the Universe* (London 1789).

Townson, Robert, *Philosophy of Mineralogy* (London 1798).

Varenius, Bernhard, *Geographia Generalis*, edited by Isaac Newton (Cambridge 1672).

Verstegan [or Verstegen *alias* Rowlands], Richard, *A Restitution of Decayed Intelligence* (Antwerp 1605).

Walker, John, *Lectures on Geology*, edited by Harold W. Scott (Univ. Chicago Press, Chicago and London 1966).

Wallis, John, 'A Letter . . . Relating to that Isthmus, or Neck of Land, which is supposed to have joyned England and France', *Philos. Trans.*, XXII (1701), No. 275, pp. 967–979.

Walsh, Francis, *The Antediluvian World; or, a New Theory of the Earth* (Dublin 1743).

Warren, Erasmus, *Geologia: or, a Discourse Concerning the Earth before the Deluge* (London 1690).

Wesley, John, *A Survey of the Wisdom of God in the Creation* (London 1777).

Whiston, William, *A New Theory of the Earth* (London 1696).

Whiston, William, *A Vindication of the New Theory of the Earth from the Exceptions of Mr. Keill and Others* (London 1698).

Whiston, William, *A Second Defence of the New Theory of the Earth from the Exceptions of Mr. John Keill* (London 1700).

Whiston, William, *The Cause of the Deluge Demonstrated* (London 1714).

Whitehurst, John, *An Inquiry into the Original State and Formation of the Earth* (London 1778).

Williams, John, *The Natural History of the Mineral Kingdom* (Edinburgh 1789).

Wilson, Joseph, *A History of Mountains, Geographical and Mineralogical* (London 1807–1810).

Woodward, John, *An Essay Toward a Natural History of the Earth* (London 1695).

Woodward, John, *Naturalis Historia Telluris Illustrata & Aucta* (London 1714).

Woodward, John, *The Natural History of the Earth, Illustrated, Inlarged, and Defended*, translated by Benjamin Holloway (London 1726).

Worthington, William, *The Scripture-Theory of the Earth* (London 1773).

B. Works published after 1808

Agassiz, Louis, 'Glaciers, and the evidence of their having once existed in Scotland, Ireland, and England', *Proc. geol. Soc.*, III (1840–1841), No. 72, pp. 327–332.

Agassiz, Louis, 'The Glacial Theory and its recent Progress', *Edinb. New Philos. Journ.*, XXXIII (1842), pp. 217–283.

Ansted, David T., 'On some Phenomena of the Weathering of Rocks, illustrating the nature and extent of sub-aerial Denudation', *Trans. Camb. phil. Soc.*, XI (1871), pp. 387–395.

Argyll, Duke of, 'On the Physical Geography of Argyllshire in connexion with its Geological Structure', *Quart. J. geol. Soc. Lond.*, XXIV (1868), pp. 255–273.

Bakewell, Robert, *An Introduction to Geology*, third edition (London 1828).

Ball, John, 'On the Formation of Alpine Valleys and Alpine Lakes', *Phil. Mag.*, Series 4, XXV (1863), pp. 81–103.

Bonney, Thomas G., 'On the formation of "Cirques", and their bearing upon Theories attributing the Excavation of Alpine Valleys mainly to the action of Glaciers', *Quart. J. geol. Soc. Lond.*, XXVII (1871), pp. 312–324.

Bonney, Thomas G., 'Lakes of the North-Eastern Alps, and their bearing on the Glacier-Erosion Theory', *Quart. J. geol. Soc. Lond.*, XXIX (1873), pp. 382–395.

Bonney, Thomas G., 'Notes on the Upper Engadine and the Italian Valleys of Monte Rosa, and their Relation to the Glacier-Erosion Theory of Lake-Basins', *Quart. J. geol. Soc. Lond.*, XXX (1874), pp. 479–489.

Brande, William T., *Outlines of Geology* (London 1829).

Brown, Robert, 'Remarks on the Formation of Fjords and Cañons', *J. Roy. Geogr. Soc.*, XLI (1871), pp. 348–360.

Buckland, William, 'Description of the Quartz Rock of the Lickey Hill . . . with considerations on the evidences of a Recent Deluge afforded by the gravel beds of Warwickshire and Oxfordshire', *Trans. geol. Soc. Lond.*, V (1819), pp. 506–544.

Buckland, William, *Vindiciæ Geologicæ; or the Connexion of Geology with Religion* (Oxford 1820).

Buckland, William, 'On the Excavation of Valleys by diluvian Action . . .', *Trans. geol. Soc. Lond.*, N.S. I. (1822–1824), pp. 95–102.

Buckland, William, *Reliquiæ Diluvianæ* (London 1823).

Buckland, William, 'On the Formation of the Valley of Kingsclere and other Valleys by the Elevation of the Strata that enclose them', *Trans. geol. Soc. Lond.*, N.S. II (1829), pp. 119–130.

Buckland, William, 'A Memoir on the Evidences of Glaciers in Scotland and the North of England', *Proc. geol. Soc.*, III (1840–1841), No. 72, pp. 332–337 and 345–348.

Buckland, William, 'On the Glacia-Diluvial Phænomena in Snowdonia and the adjacent parts of North Wales', *Proc. geol. Soc.*, III (1841–1842), No. 84, pp. 579–584.

Campbell, John F., *Frost and Fire* (Edinburgh 1865).

Carr, J., 'Observations Suggested by the Geological Paper of Mr. John Farey', *Phil. Mag.*, XXXIII (1809), pp. 385–389.

Carr, J., 'On the Natural Causes which operate in the Formation of Valleys', *Phil. Mag.*, XXXIII (1809), pp. 452–459.

Carr, J., 'An Inquiry into the Terrestrial Phænomena produced by the Action of the Ocean', *Phil. Mag.*, XXXIV (1809), pp. 19–26.

Carr, J., 'On the Causes which have operated in the Production of Valleys' *Phil. Mag.*, XXXIV (1809), pp. 190–200.

Chambers, Robert, *Vestiges of the Natural History of Creation* (London 1844).

Chambers, Robert, *Ancient Sea-Margins, as Memorials of Changes in the Relative Level of Sea and Land* (Edinburgh 1848).

Chambers, Robert, 'On Glacial Phenomena in Scotland and Parts of England', *Edinb. New Philos. Journ.*, LIV (1853), pp. 229–281.

Close, Maxwell H., 'Notes on the General Glaciation of Ireland', *Journ. Roy. geol. Soc. Irel.*, I (1864–1867), pp. 207–242.

Close, Maxwell H., 'On Some Corries and their Rock-Basins in Kerry', *Journ. Roy. geol. Soc. Irel.*, II (1867–1870), pp. 236–248.

Conybeare, William D., and Phillips, William, *Outlines of the Geology of England and Wales* (London 1822).

Conybeare, William D., 'An Examination of those Phænomena of Geology, which seem to bear most directly on theoretical Speculations', *Phil. Mag.*, N.S. IX (1831), pp. 188–197 and 258–270.

Croll, James, 'On Geological Time, and the probable Date of the Glacial and the Upper Miocene Period', *Phil. Mag.*, Series 4, XXXV (1868), pp. 363–384.

Croll, James, *Climate and Time* (London 1875).

Cuvier, Georges L. C. F. D., *Essay on the Theory of the Earth. With Geological Illustrations, by Professor Jameson*, fifth edition (Edinburgh 1827).

Darwin, Sir Francis, and Seward, Albert C., *More Letters of Charles Darwin* (London 1903).

Daubeny, C., 'On the Diluvial Theory, and, on the Origin of the Valleys of Auvergne', *Edinb. New Philos. Journ.*, X (1831), pp. 201–229.

De La Beche, Sir Henry T., *The Sir Henry De La Beche Papers*, in the National Museum of Wales, Cardiff.

De La Beche, Sir Henry T., 'Notice on the Excavation of Valleys', *Phil. Mag.*, N.S. VI (1829), pp. 241–248.

De La Beche, Sir Henry T., *A Geological Manual* (London 1831).

De La Beche, Sir Henry T., *The Geological Observer* (London 1851).

de Luc, Jean A., *An Elementary Treatise on Geology*, translated by Henry de la Fite (London 1809).

de Luc, Jean A., *Geological Travels* (London 1810–1811).

de Luc, Jean A., *Geological Travels in Some Parts of France, Switzerland and Germany* (London 1813).

Dick, T. L., 'On the Parallel Roads of Lochaber', *Trans. roy. Soc. Edinb.*, IX (1823), pp. 1–64.

Farey, John, 'Observations on a late Paper by Dr. Wm. Richardson', *Phil. Mag.*, XXXIII (1809), pp. 257–263.

Farey, John, 'Geological Observations . . . on the Use and Abuse of Geological Theories', *Phil. Mag.*, XXXVII (1811), pp. 440–443.

Farey, John, 'An Account of the Great Derbyshire Denudation', *Philos. Trans.* (1811), Pt. II, pp. 242–256. Also *Phil. Mag.*, XXXIX (1812), pp. 26–35, with additional observations, pp. 93–106.

Farey, John, *General View of the Agriculture and Minerals of Derbyshire* (London 1811–1817).

Foster, Sir Clement le N., and Topley, William, 'On the Superficial Deposits of the Valley of the Medway, with Remarks on the Denudation of the Weald', *Quart. J. geol. Soc. Lond.*, XXI (1865), pp. 443–474.

Geikie, Sir Archibald, 'On the Phenomena of the Glacial Drift of Scotland', *Trans. geol. Soc. Glasg.*, I (2) (1863), pp. 1–190.

Geikie, Sir Archibald, *The Scenery of Scotland viewed in connexion with its Physical Geology* (London and Cambridge 1865).

Geikie, Sir Archibald, 'On Modern Denudation', *Trans. geol. Soc. Glasg.*, III (1868–1869), pp. 153–190.

Geikie, Sir Archibald, 'Earth Sculpture and the Huttonian School of Geology', *Trans. Edinb. geol. Soc.*, II (1869–1874), pp. 247–268.

Geikie, James, 'On Denudation in Scotland since Glacial Times', *Trans. geol. Soc. Glasg.*, III (1868–1869), pp. 54–74.

Geikie, James, *The Great Ice Age* (London 1874).

Greenough, George B., *A Critical Examination of the First Principles of Geology* (London 1819).

Greenwood, George, *The Tree-Lifter*, second edition (London 1853).

Greenwood, George, *Rain and Rivers; or, Hutton and Playfair against Lyell and all Comers* (London 1857).

Greenwood, George, *River Terraces* (London 1877).

Hall, J., 'On the Revolutions of the Earth's Surface', *Trans. roy. Soc. Edinb.*, VII (1815), pp. 138–211.

Hopkins, W., 'On the Elevation and Denudation of the District of the Lakes of Cumberland and Westmoreland', *Quart. J. geol. Soc. Lond.*, IV (1848), pp. 70–98.

Hopkins, W., 'Anniversary Address' [on the Glacial Theory], *Quart. J. geol. Soc. Lond.*, VIII (1852), pp. xxi–lxxx.

Hull, Edward, 'On the Physical Geography and Pleistocene Phænomena of the Cotteswold Hills', *Quart. J. geol. Soc. Lond.*, XI (1855), pp. 477–496.

Hull, Edward., 'Modern Views of Denudation', *Popular Science Review*, V (1866), pp. 453–461.

Hull, Edward, *The Physical Geology & Geography of Ireland* (London 1878).

Huxley, Thomas H., *Physiography* (London 1877).

Jamieson, Thomas F., 'On the Ice-Worn Rocks of Scotland', *Quart. J. geol. Soc. Lond.*, XVIII (1862), pp. 164–184.

Jamieson, Thomas F., 'On the Parallel Roads of Glen Roy, and their Place in the History of the Glacial Period', *Quart. J. geol. Soc. Lond.*, XIX (1863), pp. 235–259.

Jamieson, Thomas F., 'On the last stage of the Glacial Period in North Britain', *Quart. J. geol. Soc. Lond.*, XXX (1874), pp. 317–338.

Jukes, Joseph B., *The Student's Manual of Geology* (Edinburgh 1857).

Jukes, Joseph B., 'On the Mode of Formation of some of the River-Valleys in the South of Ireland', *Quart. J. geol. Soc. Lond.*, XVIII (1862), pp. 378–403.

Jukes, Joseph B., 'Address' [on the external features of the earth's surface], *Rep. Brit. Ass.*, *Cambridge 1862*, Pt. 2, pp. 54–65.

Jukes, Joseph B., *The Student's Manual of Geology*, third edition, edited by Archibald Geikie (Edinburgh 1872).

Kendall, J. D., 'The Formation of Rock Basins', *Manchr. geol. Min. Soc. Trans.*, XV (1878–1880), pp. 367–382.

Kidd, John, *A Geological Essay on the Imperfect Evidence in Support of a Theory of the Earth* (Oxford 1815).

Kinahan, George H., 'Ancient Sea Margins in the Counties Clare and Galway', *Geol. Mag. Lond.*, III (1866), pp. 337–343.

Kinahan, George H., *Valleys and their Relation to Fissures, Fractures, and Faults* (London 1875).

Kinahan, George H., *Manual of the Geology of Ireland* (London 1878).

Knight, William, *Facts and Observations towards forming a new Theory of the Earth* (Edinburgh 1819).

Lyell, Sir Charles, and Murchison, Sir Roderick I., 'On the Excavation of Valleys, as illustrated by the Volcanic Rocks of Central France', *Edinb. New Philos. Journ.*, VII (1829), pp. 15–48.

Lyell, Sir Charles, *Principles of Geology* (London 1830–1833).

Lyell, Sir Charles, *Elements of Geology* (London 1838).

Lyell, Sir Charles, 'On the Geological Evidence of the former existence of Glaciers in Forfarshire', *Proc. geol. Soc.*, III (1840–1841), No. 72, pp. 337–345.

Lyell, Sir Charles, *Travels in North America; with Geological Observations on the United States, Canada, and Nova Scotia* (London 1845).

Lyell, Sir Charles, *The Geological Evidences of the Antiquity of Man* (London 1863).

Lyell, Katharine M., *Life, Letters and Journals of Sir Charles Lyell, Bart.* (London 1881).

Macculloch, John, 'On the Granite Tors of Cornwall', *Trans. geol. Soc. Lond.*, II (1814), pp. 66–78.

Macculloch, John, 'On the Parallel Roads of Glen Roy' *Trans. geol. Soc. Lond.*, IV (1817), pp. 314–392.

Macculloch, John, *A Description of the Western Islands of Scotland, including the Isle of Man* (London 1819).

Macculloch, John, *A System of Geology* (London 1831).

Mackintosh, Daniel, 'The Sea against Rain and Frost; or the Origin of Escarpments', *Geol. Mag. Lond.*, III (1866), pp. 63–70.

Mackintosh, Daniel, 'The Sea against Rivers: or the Origin of Valleys', *Geol. Mag. Lond.*, III (1866), pp. 155–160.

Mackintosh, Daniel, 'Results of Observations on the Cliffs, Gorges, and Valleys of Wales', *Geol. Mag. Lond.*, III (1866), pp. 387–398.

Mackintosh, Daniel, *The Scenery of England and Wales* (London 1869).

Maclaren, Charles, *A Sketch of the Geology of Fife and the Lothians* (Edinburgh 1839).

Maclaren, Charles, *Select Writings, Political, Scientific, Topographical, and Miscellaneous, of the late Charles Maclaren*, edited by Robert Cox and James Nicol (Edinburgh 1869).

Martin, Peter I., *A Geological Memoir on a Part of Western Sussex* (London 1828).

Martins, C., 'Upon the Identity of the Marks of Glacial Action on the Rocks in the Environs of Edinburgh . . ., *Edinb. New Philos. Journ.*, L (1851), pp. 301–318.

Maw, George, 'Notes on the Comparative Structure of Surfaces produced by Subaërial and Marine Denudation', *Geol. Mag. Lond.*, III (1866), pp. 439–451.

Morgan, C. L., 'Physiography', *Geol. Mag. Lond.*, N.S. Dec. II, V (1878), pp. 241–254.

Murchison, Sir Roderick I., *The Silurian System* (London 1839).

Murchison, Sir Roderick I., 'On the Superficial Detritus of Sweden, and on the Probable Causes which have affected the Surface of the Rocks in the Central and Southern portions of that Kingdom', *Quart. J. geol. Soc. Lond.*, II (1846), pp. 349–381.

Murchison, Sir Roderick I., 'On the Distribution of the Flint Drift of the South-East of England, on the Flanks of the Weald, and over the Surface of the South and North Downs', *Quart. J. geol. Soc. Lond.*, VII (1851), pp. 349–398.

Murchison, Sir Roderick I., *Siluria*, fourth edition (London 1867).

Oldham, T., 'On the Geological Structure of a Portion of the Khasi Hills, Bengal', *Mem. geol. Surv. India*, I (1859), pp. 99–210.

Page, David, *The Philosophy of Geology* (Edinburgh and London 1863).

Prestwich, Sir Joseph, 'Theoretical Considerations on the Conditions under which the (Drift) Deposits containing the Remains of Extinct Mammalia and Flint Implements were accumulated, and on their Geological Age', *Philos. Trans.*, CLIV (1864), pp. 247–309.

Prestwich, Sir Joseph, 'On the Origin of the Parallel Roads of Lochaber and their Bearing on other Phenomena of the Glacial Period', *Philos. Trans.* CLXX (2), (1879), pp. 663–726.

Ramsay, Sir Andrew C., 'On the Denudation of South Wales and the adjacent Counties of England', *Mem. geol. Surv. U.K.*, I (1846), pp. 297–335.

Ramsay, Sir Andrew C., 'The Old Glaciers of Switzerland and North Wales', in John Ball, *Peaks, Passes, and Glaciers* (London 1859).

Ramsay, Sir Andrew C., 'On the Glacial Origin of certain Lakes in Switzerland, the Black Forest, Great Britain, Sweden, North America, and elsewhere', *Quart. J. geol. Soc. Lond.*, XVIII (1862), pp. 185–204.

Ramsay, Sir Andrew C., *The Physical Geology and Geography of Great Britain* (London 1863).

Ramsay, Sir Andrew C., 'On the Erosion of Valleys and Lakes', *Phil. Mag.*, Series 4, XXVIII (1864), pp. 293–311.

Ramsay, Sir Andrew C., 'Sir Charles Lyell and the Glacial Theory of Lake-Basins', *Phil. Mag.*, Series 4, XXIX (1865), pp. 285–298.

Ramsay, Sir Andrew C., 'On the River-Courses of England and Wales', *Quart. J. geol. Soc. Lond.*, XXVIII (1872), pp. 148–160.

Ramsay, Sir Andrew C., 'The Physical History of the Valley of the Rhine', *Quart. J. geol. Soc. Lond.*, XXX (1874), pp. 81–95.

Ramsay, Sir Andrew C., 'How Anglesey became an Island', *Quart. J. geol. Soc. Lond.*, XXXII (1876), pp. 116–122.

Ramsay, Sir Andrew C., 'On the Physical History of the Dee, Wales', *Quart. J. geol. Soc. Lond.*, XXXII (1876), pp. 219–229.

Reade, T. M., 'Presidential Address' [on rates of denudation], *Proc. Lpool. geol. Soc.*, III, Session 18 (1876–1877), pp. 211–235.

Richardson, William, 'On the Basaltic Productions, &c, of the county of Antrim', Appendix II in John Dubourdieu, *Statistical Survey of the County of Antrim* (Dublin 1812).

Ricketts, C. 'Presidential Address' [on denudation], *Proc. Lpool. geol. Soc.*, II, Session 13 (1871–1872), pp. 4–35.

Scrope, George Poulett, *Considerations on Volcanos* (London 1825).

Scrope, George Poulett, *Memoir on the Geology of Central France* (London 1827).

Scrope, George Poulett, 'On the Origin of Valleys', *Geol. Mag. Lond.*, III (1866), pp. 193–199.

Sedgwick, Adam, 'On the Origin of Alluvial and Diluvial Formations', *Ann. Philos.*, N.S. IX (1825), pp. 241–257 and N.S. X (1825), pp. 18–37.

Smith, James, 'On the Last Changes in the Relative Levels of the Land and

Sea in the British Isles, *Mems. Wern. Nat. Hist. Soc.*, VIII (1) (1837–1838), pp. 49–113.

Sorby, Henry C., 'On Yedmandale, as illustrating the excavation of some Valleys in the Eastern Part of Yorkshire', *Quart. J. geol. Soc. Lond.*, X (1854), pp. 328–333.

Topley, William, *The Geology of the Weald*, Memoirs of the Geological Survey (London 1875).

Townsend, Joseph, *The Character of Moses established for Veracity as an Historian* (Bath 1813).

Tyndall, J., 'On the Conformation of the Alps', *Phil. Mag.*, Series 4, XXVIII (1864), pp. 255–271.

Ure, Andrew, *A New System of Geology* (London 1829).

Ward, James C., 'The Origin of some of the Lake-Basins of Cumberland', *Quart. J. geol. Soc. Lond.*, XXX (1874), pp. 96–104.

Ward, James C., 'The Glaciation of the Southern Part of the Lake-District and the Glacial Origin of the Lake-Basins of Cumberland and Westmoreland', *Quart. J. geol. Soc. Lond.*, XXXI (1875), pp. 152–166.

Whitaker, William, 'On Subaërial Denudation, and on Cliffs and Escarpments of the Chalk and Lower Tertiary Beds', *Geol. Mag. Lond.*, IV (1867), pp. 447–454 and 483–493.

Whymper, Edward, *Scrambles amongst the Alps in the Years 1860–69* (London 1871).

Wynne, A. B., 'On Denudation with Reference to the Configuration of the Ground', *Geol. Mag. Lond.*, IV (1867), pp. 3–11.

Yates, J., 'Remarks on the Formation of Alluvial Deposites', *Edinb. New Philos. Journ.*, XI (1831), pp. 1–41.

Young, George, and Bird, John, *A Geological Survey of the Yorkshire Coast* (Whitby 1822).

II *Secondary Sources*

Adams, Frank D., *The Birth and Development of the Geological Sciences* (Baltimore 1938).

Allen, Don C., 'The Legend of Noah', *Illinois Studies in Language and Literature*, XXXIII (3 & 4) (1949).

Bailey, Sir Edward B., and Tait, D., 'Geology', pp. 63–99 in *Edinburgh's Place in Scientific Progress*, British Association (Edinburgh 1921).

Bailey, Sir Edward B., 'James Hutton: the Father of Modern Geology, 1726–1797', *Trans. Edinb. geol. Soc.*, XII (1927), pp. 183–186.

Bailey, Sir Edward B., 'James Hutton: Father of Modern Geology, 1726–1797', *Nature, Lond.*, CXIX (1927), p. 582.

Bailey, Sir Edward B., 'The Interpretation of Scottish Scenery', *Scot. geogr. Mag.*, L (1934), pp. 308–330.

Bailey, Sir Edward B., 'James Hutton, Founder of Modern Geology (1726–1797)', *Proc. roy. Soc. Edinb.*, LXIII B (4) (1950), pp. 357–368.

Bailey, Sir Edward B., *Geological Survey of Great Britain* (London 1952).

Bailey, Sir Edward B., *Charles Lyell* (London 1962).

Bailey, Sir Edward B., *James Hutton – the Founder of Modern Geology* (Amsterdam 1967).

Baulig, Henri, 'La Philosophie Géomorphologique de James Hutton et John Playfair', pp. 1–11 in Baulig's *Essais de Géomorphologie* (Paris 1950).

Bonney, Thomas G., *The Work of Rain and Rivers* (Cambridge 1912).

Bromehead, C. E. N., 'Geology in Embryo (up to 1600 A.D.)', *Proc. Geol. Ass., Lond.*, LVI (1945), pp. 89–134.

Cannon, W. F., 'The Uniformitarian – Catastrophist Debate', *Isis*, LI (1960), pp. 38–55.

Challinor, J., 'The Early Progress of British Geology – I. From Leland to Woodward, 1538–1728', *Ann. Sci.*, IX (1953), pp. 124–153; 'II. From Strachey to Michell, 1719–1788', *ibid.*, X (1954), pp. 1–19; 'III. From Hutton to Playfair, 1788–1802', *ibid.*, X (1954), pp. 107–148.

Charlesworth, John K., *The Quaternary Era* (London 1957).

Chorley, Richard J., Dunn, Antony J., and Beckinsale, Robert P., *The History of the Study of Landforms, or the Development of Geomorphology*, Vol. I (London 1964).

Clark, John W., and Hughes, Thomas McK., *The Life and Letters of the Reverend Adam Sedgwick* (Cambridge 1890).

Collier, Katharine B., *Cosmogonies of Our Fathers* (Columbia Univ. Press, New York 1934).

Crowther, James G., *Scientists of the Industrial Revolution* (London 1962).

Davies, G. L., 'Joseph Beete Jukes and the Rivers of Southern Ireland – A Century's Retrospect', *Ir. Geogr.*, IV (4) (1962), pp. 221–233.

Davies, G. L., 'Robert Hooke and his Conception of Earth-History', *Proc. Geol. Ass., Lond.*, LXXV (1964), pp. 493–498.

Davies, G. L., 'From Flood and Fire to Rivers and Ice – Three Hundred Years of Irish Geomorphology', *Ir. Geogr.*, V (1) (1964), pp. 1–16.

Davies, G. L., 'Early British Geomorphology 1578–1705', *Geogr. J.*, CXXXII (1966), pp. 252–262.

Davies, G. L., 'The Eighteenth-Century Denudation Dilemma and the Huttonian Theory of the Earth', *Ann. Sci.*, XXII (1966), pp. 129–138.

Davies, G. L., 'The Concept of Denudation in Seventeenth-Century England', *J. Hist. Ideas*, XXVII (1966), pp. 278–284.

Davies, G. L., 'George Hoggart Toulmin and the Huttonian Theory of the Earth', *Bull-geol. Soc. Am.*, LXXVIII (1967), pp. 121–124.

Davies, G. L. 'Another Forgotten Pioneer of the Glacial Theory: James Hutton', *J. Glaciol.*, VII (1968) pp. 115 & 116.

Davies, G. L., 'The Tour of the British Isles made by Louis Agassiz in 1840', *Ann. Sci.*, XXIV (1968), pp. 131–146.

Dickinson, Robert E. and Howarth, Osbert J. R., *The Making of Geography* (Oxford 1933).

'Espinasse, Margaret, *Robert Hooke* (London 1956).

Eyles, V. A., 'James Hutton (1726–1797) and Sir Charles Lyell (1797–1875)', *Nature, Lond.*, CLX (1947), pp. 694 & 695.

Fenneman, N. M., 'The Rise of Physiography', *Bull. geol. Soc. Am.*, CL (1939), pp. 349–359.

Fenton, Carroll L., and Fenton, Mildred A., *Giants of Geology* (New York 1952).

Fitton, W. H., 'Elements of Geology', *Edinb. Rev.*, LXIX (1839), pp. 406–466.

Garfinkle, N., 'Science and Religion in England, 1790–1800', *J. Hist. Ideas*, XVI (1955), pp. 376–388.

Geikie, Sir Archibald, *Life of Sir Roderick I. Murchison* (London 1875)

Geikie, Sir Archibald, *The Scottish School of Geology* (Edinburgh 1871).

Geikie, Sir Archibald, 'The Centenary of Hutton's "Theory of the Earth" ', *Rep. Brit. Ass., Edinburgh 1892*, pp. 3–26.

Geikie, Sir Archibald, *Memoir of Sir Andrew Crombie Ramsay* (London 1895).

Geikie, Sir Archibald, *The Founders of Geology*, second edition (London 1905).

Geikie, Sir Archibald, 'Lamarck and Playfair: A Geological Retrospect of the Year 1802', *Geol. Mag. Lond.*, N.S. Dec. V, III (1906), pp. 145–153 and 193–202.

Gillispie, Charles C., *Genesis and Geology* (Harvard Univ. Press, Cambridge, Mass. 1951).

Glacken, Clarence J., *Traces on the Rhodian Shore: Nature and Culture in Western Thought from Ancient times to the end of the Eighteenth Century* (Univ. California Press, Berkeley and Los Angeles, 1967).

Glass, Bentley, Temkin, Owsei, and Straus, William L., *Forerunners of Darwin: 1745–1859*(Johns Hopkins Press, Baltimore 1959).

Gordon, Elizabeth O., *The Life and Correspondence of William Buckland, D.D., F.R.S.* (London 1894).

Greene, John C., *The Death of Adam* (Iowa State Univ. Press, Ames 1959).

Gregory, H. E., 'A Century of Geology – Steps of Progress in the Interpretation of Land Forms', pp. 122–152 in *A Century of Science in America* (Yale Univ. Press, New Haven 1918).

Haber, Francis C., *The Age of the World* (Johns Hopkins Press, Baltimore 1959).

Hölder, Helmut, *Geologie und Paläontologie in Texten und ihrer Geschichte* (Freiburg and Munich 1960).

Hooykaas, R., *Natural Law and Divine Miracle: The Principle of Uniformity in Geology, Biology and Theology* (Leiden 1963).

Howorth, Sir Henry H., *The Glacial Nightmare and the Flood: a second appeal to common sense from the extravagance of some recent Geology* (London 1893).

Litchfield, E., 'Notes on Early References to Geology, Mostly Before 1800', *Proc. Geol. Ass., Lond.*, XI (1889–1890), pp. 187–193.

MacGregor, M., 'James Hutton, the Founder of Modern Geology', *Endeavour*, VI (1947), pp. 109–111.

MacGregor, M., 'Life and Times of James Hutton', *Proc. roy. Soc. Edinb.*, LXIII B (4) (1950), pp. 351–356.

McCallien, W. J., 'The Birth of Glacial Geology', *Nature, Lond.*, CXLVII, (1941), pp. 316–318.

McIntyre, D. B., 'James Hutton and the Philosophy of Geology', pp. 1–11 in Albritton, Claude C., *The Fabric of Geology* (Stanford, California 1963).

Mather, Kirtley F. and Mason, Shirley L., *A Source Book in Geology* (New York 1939).

Moore, Ruth, *The Earth We Live On* (London 1957).

Newbigin, Marion I., and Flett, Sir John S., *James Geikie: The Man and the Geologist* (Edinburgh 1917).

Nicolson, Marjorie H., *Mountain Gloom and Mountain Glory: The Development of the Aesthetics of the Infinite* (Cornell Univ. Press, Ithaca, New York 1959).

North, Frederick J., *Geological Maps* (National Museum of Wales, Cardiff 1928).

North, Frederick J., 'From Giraldus Cambrensis to the Geological Map', *Trans. Cardiff Nat. Soc.*, LXIV (1931), pp. 20–97.

North, Frederick J., 'From the Geological Map to the Geological Survey', *Trans. Cardiff Nat. Soc.*, LXV (1932), pp. 41–115.

North, Frederick J., ' "The Anatomy of the Earth" – A Seventeenth-Century Cosmogony', *Geol. Mag. Lond.*, LXXI (1934), pp. 541–547.

North, Frederick J., 'Centenary of the Glacial Theory', *Proc. Geol. Ass., Lond.*, LIV (1943), pp. 1–28.

Ogden, H. V. S., 'Thomas Burnet's *Telluris Theoria Sacra* and Mountain Scenery', *ELH*, XIV (1947), pp. 139–150.

Phillips, John, *Memoirs of William Smith, LL.D.* (London 1844).

Ramsay, Sir Andrew C., *Passages in the History of Geology: being an Inaugural Lecture at University College, London* (London 1848).

Ramsay, Sir Andrew C., *Passages in the History of Geology: being an Introduc-*

tory Lecture at University College, London, in continuation of the Inaugural Lecture of 1848 (London 1849).

Ramsay, Sir William, *The Life and Letters of Joseph Black, M.D.* (London 1918).

Raven, Charles E., *John Ray*, second edition (Cambridge 1950).

Rossiter, A. P., 'The First English Geologist', *Durham Univ. Journ.*, XXIX (1935), pp. 172–181.

Rudwick, M. J. S., 'Hutton and Werner Compared: George Greenough's Geological Tour of Scotland in 1805', *Brit. J. Hist. Sci.*, I (2) (1962), pp. 117–135.

Schofield, Robert E., *The Lunar Society of Birmingham* (Oxford 1963).

Schneer, C., 'The Rise of Historical Geology in the Seventeenth Century' *Isis*, XLV (1954), pp. 256–268.

Sollas, W. J., 'The Influence of Oxford on the History of Geology', *Sci. Progr.*, VII (1898), pp. 23–52.

Stebbing, W. P. D., 'Some Early References to Geology from the Sixteenth Century Onwards', *Proc. Geol. Ass., Lond.*, LIV (1943), pp. 49–63.

Stoddart, D. R., 'Colonel George Greenwood the Father of Modern Subaerialism', *Scot. geogr. Mag.*, LXXVI (1960), pp. 108–110.

Taylor, E. G. R., 'The English Worldmakers of the Seventeenth Century and their Influence on the Earth Sciences', *Geogr. Rev.*, XXXVIII (1948), pp. 104–112.

Taylor, E. G. R., 'The Origin of Continents and Oceans: A Seventeenth Century Controversy', *Geogr. J.*, CXVI (1950), pp. 193–198.

Tomkeieff, S. I., 'James Hutton's "Theory of the Earth", 1795', *Proc. Geol. Ass., Lond.*, LVII (1946), pp. 322–328.

Tomkeieff, S. I., 'James Hutton and the Philosophy of Geology', *Trans. Edinb. geol. Soc.*, XIV (1938–1951), pp. 253–276.

Tomkeieff, S. I., 'Geology in Historical Perspective', *Advmt. Sci.*, VII (25) (1950), pp. 63–67.

Tomkeieff, S. I., 'Unconformity – An Historical Study', *Proc. Geol. Ass., Lond.*, LXXIII (1962), pp. 386–416.

Toulmin, Stephen and Goodfield, June, *The Discovery of Time* (London 1965).

Tuveson, E., 'Swift and the World-Makers', *J. Hist. Ideas*, XI (1950), pp. 54–74.

Willey, Basil, *The Eighteenth Century Background* (London 1940).

Woodward, Horace B., *The History of the Geological Society of London* (London 1907).

Woodward, Horace B., *History of Geology* (London 1911).

Zittel, Karl A. von, *History of Geology and Palæontology*, translated by Maria M. Ogilvie-Gordon (London 1901).

III. *Periodicals*

The date given is that of the publication of the first volume in each series.

Annals of Philosophy (1813). Combined with *The Philosophical Magazine* in 1827.

The Edinburgh Journal of Science (Brewster's Journal) (1824). Became *The London and Edinburgh Philosophical Magazine and Journal of Science* in 1832.

The Edinburgh Philosophical Journal (1819). Became *The Edinburgh New Philosophical Journal* in 1826.

The Geological Magazine (1864).

The Geologist (1858).

Journal of the Geological Society of Dublin (1838). Became the *Journal of the Royal Geological Society of Ireland* in 1864.

The Journal of the Royal Geographical Society of London (1831).

Memoirs of the Literary and Philosophical Society of Manchester (1789).

Memoirs of the Wernerian Natural History Society (1811).

The Magazine of Natural History (1829).

Memoirs of the Geological Survey of India (1859).

Nicholson's Journal (*A Journal of Natural Philosophy, Chemistry, and the Arts*) (1802).

The Philosophical Magazine (Tilloch's Magazine) (1798). Became *The London and Edinburgh Philosophical Magazine and Journal of Science* in 1832, and *The London, Edinburgh, and Dublin Philosophical Magazine and Journal of Science* in 1840.

Philosophical Transactions of the Royal Society of London (1665).

Prize-Essays and Transactions of the Highland and Agricultural Society of Scotland (1812).

Proceedings of the Geological and Polytechnic Society of the West Riding of Yorkshire (1838).

Proceedings of the Geological Society of London (1826).

Proceedings of the Geologists' Association (1859).

Proceedings of the Liverpool Geological Society (1861).

Proceedings of the Royal Irish Academy (1841).

Quarterly Journal of the Geological Society of London (1845).

Report of the British Association for the Advancement of Science (1833).

Transactions of the Edinburgh Geological Society (1870).

Transactions of the Geological Society of Glasgow (1863).

Transactions of the Geological Society of London (1811).

Transactions of the Manchester Geological Society (1841).

Transactions of the Royal Geological Society of Cornwall (1818).

Transactions of the Royal Irish Academy (1787).

Transactions of the Royal Society of Edinburgh (1788).

Index